Fault Diagnosis and Prognosis Techniques for Complex Engineering Systems

Fault Diagnosis and Prognosis Techniques for Complex Engineering Systems

Edited by

Hamid Reza Karimi
Department of Mechanical Engineering, Politecnico di Milano, Italy

ACADEMIC PRESS

An imprint of Elsevier

elsevier.com/books-and-journals

Academic Press is an imprint of Elsevier
125 London Wall, London EC2Y 5AS, United Kingdom
525 B Street, Suite 1650, San Diego, CA 92101, United States
50 Hampshire Street, 5th Floor, Cambridge, MA 02139, United States
The Boulevard, Langford Lane, Kidlington, Oxford OX5 1GB, United Kingdom

British Library Cataloguing-in-Publication Data
A catalogue record for this book is available from the British Library

Library of Congress Cataloging-in-Publication Data
A catalog record for this book is available from the Library of Congress

ISBN: 978-0-12-822473-1

For Information on all Academic Press publications visit our
website at https://www.elsevier.com/books-and-journals

Publisher: Mara Conner
Acquisitions Editor: Sonnini R. Yura
Editorial Project Manager: Megan Healy
Production Project Manager: Kamesh Ramajogi
Cover Designer: Greg Harris

Typeset by Aptara, New Delhi, India

Contents

8   Quadrotor actuator fault diagnosis and accommodation
     based on nonlinear adaptive state observer
     Sicheng Zhou, Kexin Guo, Xiang Yu, Lei Guo and Youmin Zhang

9   Defect detection and classification in welding using deep
     learning and digital radiography
     *M-Mahdi Naddaf-Sh, Sadra Naddaf-Sh, Hassan Zargaradeh, Sayyed M.
     Zahiri, Maxim Dalton, Gabriel Elpers and Amir R. Kashani*

10   Real-time fault diagnosis using deep fusion of features
     extracted by PeLSTM and CNN
     Funa Zhou, Zhiqiang Zhang and Danmin Chen

Contributors

Danmin Chen, School of Software, Henan University, China

Zhiwen Chen, School of Automation, Central South University, China

Maxim Dalton, Artificial Intelligence Lab, Stanley Oil, and Gas, Stanley Black, and Decker, United States

Gabriel Elpers, Artificial Intelligence Lab, Stanley Oil, and Gas, Stanley Black, and Decker, United States

Ming-Feng Ge, School of Mechanical Engineering and Electronic Information, China University of Geosciences, China

Kexin Guo, School of Automation Science and Electrical Engineering, Beihang University, China

Lei Guo, School of Automation Science and Electrical Engineering, Beihang University; Beijing Advanced Innovation Center for Big Data-Based Precision Medicine, Beihang University, Beijing, China

Hamid Reza Karimi, Department of Mechanical Engineering, Politecnico di Milano, Italy

Amir R. Kashani, Artificial Intelligence Lab, Stanley Oil, and Gas, Stanley Black, and Decker, United States

Yueyang Li, School of Electrical Engineering, University of Jinan, China

Zhichao Li, School of Logistic Engineering, Shanghai Maritime University, China

Ketian Liang, School of Automation, Central South University, China

Paul P. Lin, Fellow of the American Society of Mechanical Engineers (ASME); Professor Emeritus, Mechanical Engineering Department, Cleveland State University, United States; Visiting Scholar, Kaohsiung University of Science and Technology, Taiwan

M-Mahdi Naddaf-Sh, Electrical Engineering Department, Lamar University, United States

Sadra Naddaf-Sh, Electrical Engineering Department, Lamar University, United States

Elias G. Strangas, Michigan State University, United States

Guang Wang, North China Electric Power University – Baoding Campus, China

Tianzhen Wang, School of Logistic Engineering, Shanghai Maritime University, China

Xiang Yu, School of Automation Science and Electrical Engineering, Beihang University; Beijing Advanced Innovation Center for Big Data-Based Precision Medicine, Beihang University, Beijing, China

Sayyed M. Zahiri, Artificial Intelligence Lab, Stanley Oil, and Gas, Stanley Black, and Decker, United States

Hassan Zargaradeh, Electrical Engineering Department, Lamar University, United States

Youmin Zhang, Department of Mechanical, Industrial, and Aerospace Engineering, Concordia University, Montreal, Quebec, Canada

Zhiqiang Zhang, School of Computer and Information Engineering, Henan University, China

Yilai Zheng, School of Logistic Engineering, Shanghai Maritime University, China

Funa Zhou, School of Logistic Engineering, Shanghai Maritime University, China

Sicheng Zhou, School of Automation Science and Electrical Engineering, Beihang University, China

Preface

With the rapid growth of health monitoring technology in various fields such as process industry, energy systems, vehicles, and some other advanced technologies, the problems of fault diagnosis and failure prognosis are receiving much attention in both academic and industrial engineering areas. They are mainly motivated by the enhancement of reliability and resilience capability against different and complex failure modes from theoretical and practical aspects. To achieve reliability requirements, reliability design and resilient control are critical for the development of engineering systems. With the advances in reliability center maintenance and condition-based maintenance techniques, it is opportunistic to exploit them for the benefit of reliability design, fault diagnosis, and failure prognosis to enhance the remaining useful life of systems components. The main core of this book is on the new techniques in reliability modeling, reliability analysis, reliability design, fault and failure detection, signal processing, and fault tolerant control of engineering systems, including mechanical, electrical, hydraulic, and marine systems, for instance.

This book is targeting as a reference for graduate and postgraduate students and for researchers in all engineering disciplines, including mechanical engineering, electrical engineering, and applied mathematics to explore the state-of-the-art techniques for solving problems of integrated fault diagnosis and failure prognosis of complex systems with collective safety and robustness aspects. Thus, it shall be useful as a guidance for system engineering practitioners and system-theoretic researchers alike, today and in the future.

The book chapters are organized as separate contributions and listed according to the order of the list of contents as follows:

Chapter 1 "Quality-Related Fault Detection and Diagnosis: A Technical Review and Summary," conducts a technical review and summary of the classical achievements for quality-related fault detection and diagnosis, including their principles, implementation algorithms, technical advantages, and defects.

Chapter 2 "Canonical Correlation Analysis–Based Fault Diagnosis Method for Dynamic Processes," focuses on the application of canonical correlation analysis (CCA) technique in dynamic process fault diagnosis. Specifically, two variants of the CCA-based method— the dynamical CCA method and the gated recurrent units–aided

CCA method—are presented to deal with the fault diagnosis of dynamic processes.

Chapter 3 "H_∞ Fault Estimation for Linear Discrete Time-Varying Systems With Random Uncertainties," presents fault estimation problems for linear discrete time-varying systems with random uncertainties such as multiplicative noise and packet loss.

Chapter 4 "Fault Diagnosis and Failure Prognosis of Electrical Drives," addresses the faults in the power electronics, the DC link capacitor, batteries, and electrical machines. Specifically, the faults can be an open or short circuit of switches, and they can be identified from current and voltage measurements. For example, in electrical machines, the faults can be in the windings, with either incipient in precipitous degradation, and can be identified and their severity determined using either model- or signal-based techniques. Moreover, mechanical faults detection in bearings is discussed through the measurement of vibrations, whereas eccentricity can be detected through the change of flux and inductances.

Chapter 5 "Intelligent Fault Diagnosis for Dynamic Systems via Extended State Observer and Soft Computing," addresses the common model-based fault diagnosis difficulties encountered in industrial applications. Specifically, this chapter uses an extended state observer to detect faults without exact knowledge of the plant model and a fuzzy inference system to help fault isolation and fault identification.

Chapter 6 "Fault Diagnosis and Failure Prognosis in Hydraulic Systems," reviews the state of the art in diagnostics and prognostics pertaining to hydraulic machinery systems. Attention is given to detailing the application status of sensor detection technology, cavitation research, intelligent evaluation and diagnosis technology, and prognostics research, among others, used by researchers in the main areas of diagnostics and prognostics.

Chapter 7 "Fault Detection and Fault Identification in Marine Current Turbines," develops a Hilbert transform–based detection method to detect the imbalance faults for a marine current turbine's rotor and blade.

Chapter 8 "Quadrotor Actuator Fault Diagnosis and Accommodation Based on Nonlinear Adaptive State Observer," proposes a nonlinear adaptive state observer–based fault-tolerant tracking control system for a quadrotor unmanned aerial vehicle.

Chapter 9 "Defect Detection and Classification in Welding Using Deep Learning and Digital Radiography," presents two realistic welding quality datasets for training deep learning models based on radiography images collected from various projects and nondestructive test expert-annotated datasets: SBD-1 and SBD-2. Then an optimized convolutional neural network was designed to find defects in the weldment and heat-affected zones and was subsequently trained and evaluated based on prepared datasets.

Chapter 10 "Real-Time Fault Diagnosis Using Deep Fusion of Features Extracted by PeLSTM and CNN," focuses on extracting useful features potentially involved in vibration signals using intelligent techniques for safety analysis and health monitoring of rotary machines.

Finally, I would like to express appreciation to all contributors for their excellent contributions to this book.

Hamid Reza Karimi
Milan, November 20, 2020

Chapter 1

Quality-related fault detection and diagnosis: a technical review and summary

Guang Wang[a] and Hamid Reza Karimi[b]

[a] *North China Electric Power University – Baoding Campus, China.* [b] *Department of Mechanical Engineering, Politecnico di Milano, Italy*

1.1 Introduction

Today, industrial production plays a crucial role in the modern age, as it has important influence on every aspect of society. The industrial process is developing rapidly to be further automated and integrated. Plenty of producing processes contain myriad variables and indices and complex structures. As a result, fault detection and diagnosis theories are significant to this issue, alarming the occurrence of the faults and making the analysis to find the faulty variables [1–3].

The development of sensor, data transmission, and storage technology provides great opportunities in the research of data-based fault detection and diagnosis theories. One of most popular methods is multivariate statistical process monitoring [4]. The amount of algorithms are proposed to construct and decompose the feature space of the object systems, among which the representative methods are principal component analysis (PCA), canonical variable analysis, and partial least squares (PLS), among others.

For the fault detection and diagnosis tasks of the industry system, the key performance index (KPI)-related, or quality-related, fault detection and diagnosis attracts more and more attention in the recent research [5–7]. On one hand, the faults with an impact on KPI would influence the quality of the output and other indicators of the product or pose a threat to the safety and stability of the production, to which more attention should be paid quickly with necessary measures. On the other hand, there still exist plenty of faults happening on the variables that do not have any direct relationship to the product or the whole production process, which can be dealt with in the daily maintenance [1, 8].

To solve the quality-related fault detection problem, one of the crucial parts is the algorithm to obtain the relationship between process variables and quality

Fault Diagnosis and Prognosis Techniques for Complex Engineering Systems. DOI: 10.1016/B978-0-12-822473-1.00010-0
1

variables. Multiple linear regression, PLS, canonical variable analysis, and many other methods are proposed to solve this problem. As an effective method to alarm faults, PCA does not consider the correlation with the quality variables when decomposing the feature space, so it cannot distinguish whether the faults relate to the quality variables [9]. At the same time, it is still an important classical theory of dimensionality reduction by extracting the principal components [10], which is widely used in other quality-related algorithms.

The PLS algorithm is a classical theory to acquire the projection directions, reflecting the changes of process variables X that are related to the quality variables. It calculates the max covariance of the process variables and quality variables Y to obtain the scores and makes the decomposition to the feature space [9, 11]. However, the PLS method is not the perfect solution, as it is not a complete orthogonal decomposition. The projection space does not cover all of the directions related to quality variables, and there is still information existing in the residual space, so the alarm results can be inaccurate [12]. Zhou et al. [12] proposed total PLS (T-PLS) to further decompose the subspaces orthogonally into four subspaces to distinguish the quality-related and quality-unrelated parts in the two original subspaces. However, it also faces some problems that it decomposes the residuals obtained from the PLS algorithm without considering the quality-related information remaining in the subspaces. It uses four subspaces to monitor, which makes the judgment logic more complex. Qin and Zheng [13] modified the PLS algorithm and proposed concurrent PLS (C-PLS) to decompose the subspaces according to the contribution or relevancy to the prediction of the quality variables. On the basis of the prediction by PLS, it decomposes the principal components to distinguish the part that has contribution to the predicted quality variables and the part only related to the inputs themselves. At the same time, the quality variables are also decomposed to find the remaining unpredicted part. The multiblock C-PLS algorithm is also proposed to monitor and diagnose the decentralized process [14]. Ding et al. [15] also proposed a modified PLS (M-PLS) algorithm that applies singular value decomposition (SVD) to decompose the feature space into a principal component subspace containing all of the information to predict quality variables and a residual subspace totally unrelated to the prediction of quality variables. This method improves the effectiveness of the KPI prediction, whereas the residual subspace may contain the factors that cannot predict quality variables but are able to affect them. Based on M-PLS, Peng et al. [16] proposed the efficient PLS (E-PLS) algorithm that also makes use of SVD to decompose the feature space according to the contribution to the quality variable prediction and further decompose the residual subspace by PCA to separate the quality-related part.

Apart from the PLS-based method, Peng et al. [17] proposed principal component regression (PCR) theory, which belongs to the multiple linear regression method. It applies the PCA algorithm to extract the principal component of the process variables and constructs the linear regression coefficient matrix to the quality variables. An orthogonal decomposition is conducted to the regression coefficient matrix, obtaining two subspaces. The quality-related part remaining in the residual component of the first PCA function is also considered in the further decomposition of the residuals of the coefficient matrix. Wang et al. [18] proposed a total PCR (T-PCR) algorithm, which extracts the principal components of the predicted quality variables and projects the process variables again to obtain the subspaces more highly related to the quality variables and the corresponding residual subspace.

The canonical correlation analysis (CCA) algorithm is also applied to fault detection. Chen et al. [19] proposed the CCA-based fault detection method to acquire the principal components by maximizing the correlation between process variables and quality variables. Chen et al. [20] further improved the CCA method to solve the fault detection problem in a detailed industrial fault condition. Zhu et al. [21] proposed a concurrent CCA (CCCA) model with regularization to deal with the defect that CCA does not take the variance of the data, which decomposes the feature space into five subspaces. Then Zhu and Qin [22] proposed a supervised diagnosis scheme, applying the CCCA algorithm to realize the fault alarm.

The preceding algorithms are quite effective in the linear process. However, most of the complex industrial processes have strong nonlinear characteristics that cannot be decomposed by linear methods directly. Nonlinear mapping is applied to map the process variables into high-dimensional feature space so that the linear decomposition can be conducted in this high dimension [1]. This method is feasible in principle, but it also brings the problem that the extremely high dimension makes the calculation hard to conduct. In addition, the kernel function method is introduced to solve this problem by forming the kernel matrix to replace the mapping matrix in the calculation, where the Gaussian kernel function is widely used in the modeling of the fault detection problems [23]. Cho et al. [24] proposed a kernel PCA (KPCA) method and applied the nonlinear extension of PCA successfully to the fault identification experiments. Rosipal and Trejo [25] proposed the kernel PLS (KPLS) method and conducted the linear decomposition through a mapping matrix. The nonlinear extension of T-PLS, T-KPL, is also proposed, which provides satisfying detection and diagnosis results to the industrial system [5, 26]. Recently, other theories have been proposed to conduct the decomposition of the sample space. For example, the kernel direct decomposition (KDD) theory is to perform SVD directly on the regression coefficient matrix of the quality variables [27]. Kernel least squares (KLS) theory decomposes the linear regression matrix containing the full correlation between the mapping matrix and quality variables [6]. The two orthogonal subspaces are formed by projecting the mapping matrix to the decomposition results of the regression matrix. One of the subspaces contains the relationship to the quality, whereas the other is quality unrelated. Based on the T-PCR algorithm, Wang et al. [18] also combined this algorithm with the kernel method, through which T-KPCR is proposed to solve the nonlinear problems.

So far, there is a prerequisite for the application of most of the modeling algorithms and methods introduced earlier, which assumes that the process data follows Gaussian distribution. However, it is usually not fulfilled in the practice processes [28]. Much research has been proposed to solve the problem of non-Gaussian process modeling. Independent component analysis (ICA) is one of the widely used algorithms in this field, as Kano et al. [29] first applied this algorithm to the process monitoring task. Lee et al. [30] proposed an ICA-based fault detection method that conducts the measurement of non-Gaussian process by negative entropy and estimates the independent components and the mixing matrix. In their work, a new statistics index I^2 is designed for monitoring with the corresponding confidence limits determined by kernel density estimation, although kernel density estimation is a widely applied method for the estimation of control region of the normal process data [31]. Another method to deal with this problem is the Gaussian mixture model (GMM), which decomposes the process data into different Gaussian components corresponding to different operating modes, respectively [32]. This method has attracted much attention recently. Choi et al. [33] combined the GMM algorithm with PCA for fault detection and further conducted fault isolation with the combination of GMM and discriminant analysis [33]. Yu [32] combined the GMM method with a Bayesian inference strategy to conduct fault isolation [32]. Jiang et al. [34] further modified the GMM with Bayesian inference to decompose the process data, based on which PCA has been performed on each operating mode to select the optimal principal components. In addition, Liu et al. [35] used the support vector data description (SVDD) algorithm to define the control region of the normal data by a minimal spherical volume. Ge and Song [36] proposed the one-class support vector machine method to separate the data with a hyperplane.

As to the dynamic characteristics of the process, there is also much research that deals with this issue [37, 38]. Ku et al. [39] proposed the dynamic PCA (DPCA) method, performing PCA on the process variables with time lags. Li and Qin [40] improved the DPCA into an indirect DPCA to realize the consistent estimation of process variables and quality variables. Dong and Qin [41] proposed a dynamic inner PCA algorithm that extracts the dynamic components with best prediction results from the history data and remains little dynamic relationship information in the corresponding residual components so that the dynamic relationships and static relationships could be processed respectively. In addition, Dong and Qin [42] proposed the dynamic-inner canonical correlation analysis (DiCCA) algorithm. This algorithm selects the dynamic components by the predictability of each latent variable such that the components extracted are sufficient enough to describe the dynamic relationship. The DPCA algorithm is also combined with a dynamic ICA (DICA) algorithm to deal with the process variables obeying Gaussian and non-Gaussian distributions, respectively [43]. As to the PLS-based algorithm, Helland et al. [44] proposed the recursive PLS regression (RPLS) method to update the prediction model with the new calibration objects. This recursion method is also be applied in recent research.

Hu et al. [7] proposed the recursive C-PLS (RCPLS) algorithm, taking the normal testing samples to update the monitoring model. Qin [45] modified the PLS algorithm and proposed the block RPLS algorithm to adapt for the large number of updating data and the new changes of the PLS model. Dong and Qin [46] proposed a dynamic inner PLS (DiPLS) method, which conducts the PLS method with dynamic models for both process variables and quality variables to construct the dynamic relationships between the input and output.

In accordance with the methods introduced previously, when a fault occurs, it can be detected and alarmed in the corresponding statistics. Then it comes to the fault diagnosis, aiming to seek the faulty variables. There are methods proposed to analyze the contribution of each variable done to the fault occurring so that the abnormal variables can be selected [47].

The contribution plot is a widely used method especially for the linear process. It analyzes the contribution values of each process variable to the fault and selects the variables with high contribution values as the faulty variables, according to the corresponding threshold [48]. The effectiveness of this method has been proved in much research. To improve the accuracy of the diagnosis results, the reconstruction-based contribution (RBC) plot method is proposed [49]. RBC calculates the reconstructed faulty amplitude index on each variable direction to seek the variables with an abnormally large index, or compares the contribution value of each original variable and the reconstructed non-fault contribution value of that. This method achieves great improvement of the diagnosis accuracy. Further, some expansion algorithms based on RBC have been proposed. Li et al. [50, 51] proposed a multidirectional RBC method to find the minimum variables that can satisfy the reconstruction condition. In addition, Yoon and MacGregor [52] proposed angle-based contribution (ABC), which analyzes the angle measures between the observed component vector and the preknown fault vectors to judge whether fault occurs or not, of which the diagnosis effectiveness is similar to RBC. Liu [53] proposed an improved contribution plot by the reduction of combined index (RCI) to avoid the influence of the smearing effect. Later, Liu et al. [54] proposed the faulty variable selection method based on Bayesian decision theory.

However, as to the nonlinear processes, the traditional contribution plot can not be used directly. Because of the loss of a corresponding relationship between the process variables and the kernel matrix, the contribution values can not be accurately calculated from the kernel matrix-based statistics. Aiming at the solution of the faulty variable diagnosis for the nonlinear processes, Zhang et al. [55] proposed the partial derivation method, also considered as a kernel gradient method, which takes the partial derivation on every variable direction to obtain contribution values by the gradient decline of each variable direction. Through this method, the faulty variables diagnosis issue of the nonlinear process can be solved, of which the effectiveness is proved in the research. At the same time, the performance of this method also been influenced by the smearing effect. Recently, to construct the revelation of the corresponding relationship, Wang et al. [56] proposed a kernel sample equivalent replacement (KSER) theory that performs the first-order Taylor series expansion on the Gaussian kernel function and acquires the kernel matrix by the variance-covariance matrix of the process variable matrix directly. This theory can solve the problem that the Gaussian kernel does not correspond to the input process variables, which can conduct the detection and diagnosis processing with process variables $\mathbf{X}$ for the nonlinear system.

In this section, several process monitoring relevant theories have been introduced concisely. To solve the problems in this field, as to the quality-relevant detection and diagnosis and the characteristics of the process system and data, plenty of studies have been proposed with valuable theories. In the following section, some classical theories are explained in detail. Then a detailed description of some of the latest research results and progress are introduced.

1.2 Basic methodology

In this section, a detailed explanation of some classical and basic fault detection theories are provided. The mentioned KPLS, T-KPLS, and C-PLS methods are summarized here. The principles of these algorithms are presented concisely in this section.

To initialize, the process samples are input in the form of $\mathbf{x}_i = \left[x_{i,1}, \cdots, x_{i,m}\right]^T \in \mathbb{R}^{m \times 1}$, where $\mathbf{x}_i$ represents a process variable sample with m variables. The quality variable sample is $\mathbf{y}_i = \left[y_{i,1}, \cdots, y_{i,l}\right]^T \in \mathbb{R}^{l \times 1}$, where l is the number of quality variables. N samples would be collected in $\mathbf{X} = [\mathbf{x}_1, \cdots, \mathbf{x}_N]^T \in \mathbb{R}^{N \times m}$ and $\mathbf{Y} = [\mathbf{y}_1, \cdots, \mathbf{y}_N]^T \in \mathbb{R}^{N \times l}$. The centralized $\mathbf{X}$ can be obtained as $\bar{\mathbf{X}} = \left(\mathbf{I}_N - \frac{1}{N}\mathbf{1}_N\mathbf{1}_N^T\right)\mathbf{X}$ that has zero mean and unit standard deviation, where $\mathbf{I}_N$ is a unit matrix of order N and $\mathbf{1}_N$ is a column vector of N order whose elements are all 1.

Aiming at solving the modeling problem of the nonlinear process, the nonlinear mapping method is widely applied, mapping the origin variable samples to the high-dimensional feature space F:

$$\mathbf{x}_i \in \mathbb{R}^{m \times 1} \rightarrow \phi(\mathbf{x}_i) \in \mathbb{R}^{\Omega \times 1}. \tag{1.1}$$

The variables with nonlinear relationships would be linearly decomposable in the high-dimensional space F, and the corresponding linear decomposition could be realized in that space. The mapping matrix $\mathbf{\Phi} \in R^{N \times \Omega}$ can be formed by $\phi(\mathbf{x}_i)$. It can be centralized as

$$\bar{\mathbf{\Phi}} = \left(\mathbf{I}_N - \frac{1}{N}\mathbf{1}_N\mathbf{1}_N^T\right)\mathbf{\Phi}, \tag{1.2}$$

where $\mathbf{I}_{N \times N} \in \mathbb{R}^{N \times N}$ is the unit matrix and $\mathbf{1}_N \in \mathbb{R}^{N \times 1}$ is the all 1 vector.

However, the number of the dimensions of $\mathbf{\Phi}$ can be extremely high, which is not available to conduct the calculation directly. To solve this problem, the kernel function method is applied in much of the research. The kernel matrix can be formed as $\mathbf{K} = \mathbf{\Phi}\mathbf{\Phi}^T$, which consists of $\mathbf{k}_{i,j}$ that can be defined as

$$\mathbf{k}_{i,j} = \langle \phi(\mathbf{x}_i), \phi(\mathbf{x}_j) \rangle = f_{\ker}(\mathbf{x}_i, \mathbf{x}_j) = \exp\left(-\frac{\|\mathbf{x}_i - \mathbf{x}_j\|^2}{c}\right), \tag{1.3}$$

where the Gaussian kernel function is applied to form the kernel matrix, according to plenty of studies, as it can always satisfy the Mercer theorem and obtain good fault detection effectiveness. Unless otherwise stated, the kernel function applied in this section refers to the Gaussian kernel. The normalized kernel matrix can be calculated as

$$\bar{\mathbf{K}} = \bar{\mathbf{\Phi}}\bar{\mathbf{\Phi}}^T = \left(\mathbf{I}_{N\times N} - \frac{1}{N}\mathbf{1}_N\mathbf{1}_N^T\right)\mathbf{K}\left(\mathbf{I}_{N\times N} - \frac{1}{N}\mathbf{1}_N\mathbf{1}_N^T\right). \tag{1.4}$$

Among the fault detection methods of the complex industrial processes, KPLS is a classic algorithm to decompose the feature space according to the quality variables, which is based on the least squares principle. It is a nonlinear extension of the PLS algorithm. KPLS obtains the score matrix $\mathbf{T}$ and the load matrices $\mathbf{P}$ and $\mathbf{Q}$ by Algorithm 1 so that it can map $\bar{\mathbf{\Phi}}$ to the quality-related directions.

Algorithm 1: The principle of KPLS

Initialize $i = 1$ and $\mathbf{u}_i$ as the first column of $\mathbf{Y}_i$.
$\mathbf{w}_i = \bar{\mathbf{\Phi}}_i^T \mathbf{u}_i$;
$\mathbf{t}_i = \bar{\mathbf{\Phi}}_i \mathbf{w}_i$, $\mathbf{t}_i = \frac{\mathbf{t}_i}{\|\mathbf{t}_i\|}$;
$\mathbf{q}_i = \mathbf{Y}_i^T \mathbf{t}_i$;
$\mathbf{u}_i = \mathbf{Y}_i \mathbf{q}_i$, $\mathbf{u}_i = \frac{\mathbf{u}_i}{\|\mathbf{u}_i\|}$.
Repeat steps 2 through 5 until $\mathbf{t}_i$ converges.
Update $\bar{\mathbf{\Phi}}_i$ and $\mathbf{Y}_i$: $\bar{\mathbf{\Phi}}_{i+1} = \left(\mathbf{I}_N - \mathbf{t}_i\mathbf{t}_i^T\right)\bar{\mathbf{\Phi}}_i$, $\mathbf{Y}_{i+1} = \left(\mathbf{I}_N - \mathbf{t}_i\mathbf{t}_i^T\right)\mathbf{Y}_i$.
Collect $\mathbf{t}_i$, $\mathbf{q}_i$, and $\mathbf{u}_i$ to the matrices $\mathbf{T}$, $\mathbf{Q}$, and $\mathbf{U}$, respectively.
$i = i + 1$. If $i \leq A$, repeat steps 1 through 8.

$\mathbf{T}$ is the score matrix, and the load matrix of the process matrix can be expressed as

$$\mathbf{P} = \bar{\mathbf{\Phi}}^T \mathbf{T}. \tag{1.5}$$

Then $\bar{\mathbf{\Phi}}$ can be decomposed into two subspaces: principal component subspace related to the quality variables and the residual subspace. The quality variables can be predicted by $\mathbf{T}$ and its load matrix $\mathbf{Q}$ as

$$\bar{\mathbf{\Phi}} = \mathbf{T}\mathbf{P}^T + \tilde{\mathbf{\Phi}}, \tag{1.6}$$

$$\mathbf{Y} = \mathbf{T}\mathbf{Q}^T + \mathbf{E}, \tag{1.7}$$

where $\hat{\Phi} = \mathbf{T}\mathbf{P}^T$ is the principal component of $\bar{\Phi}$, which is monitored by the Hotelling's T^2 statistics. In addition, $\tilde{\Phi}$ is the residual subspace, using *SPE* statistics. $\hat{\mathbf{Y}} = \mathbf{T}\mathbf{Q}^T$ is the predicted $\mathbf{Y}$, and $\mathbf{E}$ is the residual part of $\mathbf{Y}$.

However, the KPLS algorithm does not realize the orthogonal decomposition actually. It extracts the principal components without considering the extent of their influence on the quality variables. Thus, there are quality-related components remaining in the residual subspace. The principal component subspace also has the quality-unrelated part. It is a kind of oblique decomposition.

One of the solutions is the T-KPLS algorithm, which is developed from the T-PLS theorem. T-KPLS further decomposed the two subspaces $\hat{\Phi}$ and $\tilde{\Phi}$ of KPLS into four by the PCA theorem.

$$\bar{\Phi} = \hat{\Phi}_y + \hat{\Phi}_o + \hat{\Phi}_r + \tilde{\Phi}_r = \mathbf{T}_y \mathbf{P}_y^T + \mathbf{T}_o \mathbf{P}_o^T + \mathbf{T}_r \mathbf{P}_r^T + \tilde{\Phi}_r \qquad (1.8)$$

$$\mathbf{Y} = \mathbf{T}_y \mathbf{Q}_y^T + \mathbf{E}_y \qquad (1.9)$$

As Eq. (1.8) and Eq. (1.9) show, the principal component $\hat{\Phi}$ is further decomposed to quality-related subspace $\hat{\Phi}_y$ and quality-unrelated subspace $\hat{\Phi}_o$, whereas $\hat{\Phi}_r$ is extracted from $\tilde{\Phi}$ as the principal subspace to monitor the components with large variation in $\tilde{\Phi}$. $\tilde{\Phi}_r$ consists of the residual components with small variation.

According to Algorithm 2, $\hat{\Phi}$ and $\tilde{\Phi}$ are decomposed into $\hat{\Phi}_y$, $\hat{\Phi}_o$, $\hat{\Phi}_r$, and $\tilde{\Phi}_r$. $\hat{\Phi}_y$, $\hat{\Phi}_o$, and $\hat{\Phi}_r$ are monitored by T^2 statistics to observe the variation in

---

Algorithm 2: The principle of T-KPLS

Perform PCA on $\hat{\mathbf{Y}} = \mathbf{T}\mathbf{Q}^T$, $\mathbf{Y} = \mathbf{T}_y \mathbf{Q}_y^T + \mathbf{E}_y$. The number of the principal components $A_y = rank(\mathbf{Q})$.
$\mathbf{P}_y^T = \left(\mathbf{T}_y^T \mathbf{T}_y\right)^{-1} \mathbf{T}_y^T \hat{\Phi}$, and it has $\hat{\Phi}_y = \mathbf{T}_y \mathbf{P}_y^T$.
$\hat{\Phi}_o = \hat{\Phi} - \hat{\Phi}_y$, and perform PCA on $\hat{\Phi}_o$, attracting $\left(A_o = A - A_y\right)$ components;
$\hat{\Phi}_o = \mathbf{T}_o \mathbf{P}_o^T$.
Perform PCA on $\tilde{\Phi}$, whose number of components A_r is settled by the PCA algorithm. $\hat{\Phi}_r = \mathbf{T}_r \mathbf{P}_r^T$ is obtained.
$\tilde{\Phi}_r = \tilde{\Phi} - \hat{\Phi}_r$.

---

these subspaces. At the same time, the *SPE* statistic is designed for the residual subspace $\tilde{\Phi}_r$. Although the T-KPLS method realizes the further decomposition of the subspaces, it only focuses on the predicted $\hat{\mathbf{Y}}$ from the process variables and decomposes the process variables into four subspaces, whereas the unpredicted part of $\mathbf{Y}$ is not analyzed.

Another improvement of PLS is the method of C-PLS that is proposed by Qin and Zheng [13] on the basis of PLS to obtain the principal component and the residual components of the unpredicted part of $\mathbf{Y}$, and it also attracts the components of $\bar{\mathbf{X}}$ that have no contribution to the prediction of $\mathbf{Y}$ and the residual components $\tilde{\mathbf{X}}$ with some potential relation to the quality variable $\mathbf{Y}$. The corresponding nonlinear method is proposed by Zhang et al. [57]. The decomposition is realized as

$$\bar{\mathbf{\Phi}} = \hat{\mathbf{\Phi}} + \hat{\mathbf{\Phi}}_x + \tilde{\mathbf{\Phi}}, \tag{1.10}$$

$$\mathbf{Y} = \hat{\mathbf{Y}} + \hat{\mathbf{Y}}_y + \tilde{\mathbf{Y}}. \tag{1.11}$$

The detailed algorithm is shown in Algorithm 3.

Algorithm 3: The principle of C-KPLS

Perform KPLS and obtain matrices $\mathbf{T}$, $\mathbf{U}$, and $\mathbf{Q}$.
Conduct SVD on $\hat{\mathbf{Y}} = \mathbf{TQ}^T$ so that $\hat{\mathbf{Y}} = \mathbf{TDV} = \mathbf{TQ}^T$ and $\mathbf{P} = \mathbf{PQ}^T\mathbf{VD}^{-1}$.
Conduct PCA on $\tilde{\mathbf{Y}}_c = \mathbf{Y} - \hat{\mathbf{Y}}$, so $\tilde{\mathbf{Y}}_c = \mathbf{T}_y\mathbf{P}_y^T + \tilde{\mathbf{Y}}$ with A_y principal components.
It has $\mathbf{t}_y = \mathbf{P}_y^T\hat{\mathbf{y}}$ and $\mathbf{e}_y = (\mathbf{I} - \mathbf{P}_y\mathbf{P}_y^T)\hat{\mathbf{y}}$.
$\hat{\mathbf{\Phi}}_c = \bar{\mathbf{\Phi}} - \mathbf{TP}^T$ and conduct PCA on $\hat{\mathbf{\Phi}}_c$. It holds that $\hat{\mathbf{\Phi}}_c = \mathbf{T}_x\mathbf{P}_x^T + \tilde{\mathbf{\Phi}}$, whose number of principal components is A_x. In addition, $\mathbf{e}_x = (\mathbf{I} - \mathbf{P}_x\mathbf{P}_x^T)\hat{\phi}_c$.

Through Algorithm 3, Eq. (1.10) and Eq. (1.11) can be rewritten as

$$\bar{\mathbf{\Phi}} = \hat{\mathbf{\Phi}}_c + \hat{\mathbf{\Phi}}_x + \tilde{\mathbf{\Phi}} = \mathbf{TP}^T + \mathbf{T}_x\mathbf{P}_x^T + \tilde{\mathbf{\Phi}}, \tag{1.12}$$

$$\mathbf{Y} = \hat{\mathbf{Y}} + \hat{\mathbf{Y}}_y + \tilde{\mathbf{Y}} = \mathbf{TQ}^T + \mathbf{T}_y\mathbf{P}_y^T + \tilde{\mathbf{Y}}. \tag{1.13}$$

Here, $\mathbf{T}_x$ describes the variation in process variables unrelated to $\mathbf{Y}$ and is monitored by T_x^2 statistics for the quality-unrelated faults. $\tilde{\mathbf{\Phi}}$ is the residual part of process variables that contains the variation potentially related to $\mathbf{Y}$. SPE_x statistics is applied to $\tilde{\mathbf{\Phi}}$, monitoring the fault potentially related to quality variables. There also remains variation in $\mathbf{Y}$ unpredicted by process variables, which are taken into account by $\mathbf{T}_y$ and $\tilde{\mathbf{Y}}$. T_y^2 and SPE_y statistics are designed for the principal components of unpredicted quality variables $\mathbf{T}_y$ and the residual part of unpredicted quality variables $\tilde{\mathbf{Y}}$, respectively.

1.3   Recent research

1.3.1   The KDD algorithm

The KDD algorithm is a simple and effective method to solve the problem of nonlinear quality-related fault detection. The main idea of this algorithm is to decompose the feature matrix into two orthogonal parts directly, according to the full correlation between the feature matrix and the output. This method does not need to construct any regression model, so it is much simpler than other conventional nonlinear methods. In addition, the detection performance is more stable. In this part, the principle of the KDD algorithm is explained in detail.

The KDD algorithm processes the cross-covariance matrix of $(\bar{\Phi}, \mathbf{Y})$, which can be estimated as follows:

$$\Upsilon = \frac{\mathbf{Y}^T \bar{\Phi}}{N - 1}. \tag{1.14}$$

Obviously, Υ contains the full correction between Φ and $\mathbf{Y}$.

Perform SVD on Υ, and it gives

$$\Upsilon = \mathbf{V}_1 \Sigma_1 \mathbf{U}_1^T = \mathbf{V}_1 \begin{bmatrix} \Lambda & 0 \end{bmatrix} \begin{bmatrix} \mathbf{U}_{11}^T \\ \mathbf{U}_{12}^T \end{bmatrix} = \mathbf{V} \Lambda \mathbf{U}_{11}^T, \tag{1.15}$$

where $\mathbf{U}_{11} \in \mathbb{R}^{M \times s}$, $\mathbf{U}_{12} \in \mathbb{R}^{M \times (M-s)}$, $\lambda = \begin{bmatrix} \lambda_1 & & \\ & \ddots & \\ & & \lambda_s \end{bmatrix} \in \mathbb{R}^{s \times s}$, where

s is determined by the number of the eigenvalues that could cover most of the characteristics in this cross-covariance matrix. If $\frac{\lambda_n}{\lambda_{n+1}} \geqslant 10$, s is equal to n.

According to the principle of SVD, it can be obtained that

$$\mathbf{V}_1^T \mathbf{V}_1 = \mathbf{I}_s, \ \mathbf{U}_{11}^T \mathbf{U}_{11} = \mathbf{I}_s, \ \mathbf{U}_{11}^T \mathbf{U}_{12} = 0, \tag{1.16}$$

$$\mathbf{U}_{11} \mathbf{U}_{11}^T + \mathbf{U}_{12} \mathbf{U}_{12}^T = \mathbf{I}_M. \tag{1.17}$$

In addition, $\bar{\Phi}$ can be projected by $\mathbf{U}_{11} \mathbf{U}_{11}^T$ and $\mathbf{U}_{12} \mathbf{U}_{12}^T$ to two orthogonal parts, as shown in the following formula:

$$\hat{\Phi} = \bar{\Phi} \mathbf{U}_{11} \mathbf{U}_{11}^T, \ \tilde{\Phi} = \bar{\Phi} \mathbf{U}_{12} \mathbf{U}_{12}^T. \tag{1.18}$$

Therefore, the remaining task is to calculate $\mathbf{U}_{11} \mathbf{U}_{11}^T$ and $\mathbf{U}_{12} \mathbf{U}_{12}^T$. From Eq. (1.14), we have

$$\Upsilon^T \Upsilon = \frac{\bar{\Phi}^T \mathbf{Y} \mathbf{Y}^T \bar{\Phi}}{(N-1)^2} = \frac{\bar{\Phi}^T \Pi \bar{\Phi}}{(N-1)^2}, \tag{1.19}$$

where $\Pi = \mathbf{Y} \mathbf{Y}^T = \begin{bmatrix} \pi_{11} & \cdots & \pi_{1N} \\ \vdots & \ddots & \vdots \\ \pi_{N1} & \cdots & \pi_{NN} \end{bmatrix}$.

Then $\Upsilon^T \Upsilon$ is decomposed by SVD as

$$\begin{aligned} \Upsilon^T \Upsilon &= \mathbf{V}_2 \Sigma_2 \mathbf{U}_2^T \\ &= \begin{bmatrix} \mathbf{V}_{21} & \star \end{bmatrix} \begin{bmatrix} \Lambda^2 & 0 \\ 0 & 0 \end{bmatrix} \begin{bmatrix} \mathbf{V}_{21}^T \\ \star \end{bmatrix}, \\ &= \mathbf{V}_{21} \Lambda^2 \mathbf{V}_{21}^T \end{aligned} \tag{1.20}$$

where $\Lambda^2 = \begin{bmatrix} \lambda_1^2 & & \\ & \ddots & \\ & & \lambda_s^2 \end{bmatrix} \in \mathbb{R}^{s \times s}$. It is clear that

$$col(\mathbf{V}_{21}) = col(\mathbf{U}_{11}). \tag{1.21}$$

Consequently, it holds that

$$\mathbf{U}_{11}\,\Xi\,\mathbf{U}_{11}^{T} = \mathbf{V}_{21}\,\Xi\,\mathbf{V}_{21}^{T}, \, \mathbf{U}_{11}^{T}\,\Xi\,\mathbf{U}_{11} = \mathbf{V}_{21}^{T}\,\Xi\,\mathbf{V}_{21}, \tag{1.22}$$

where Ξ is an arbitrary real symmetric matrix with proper rows and columns. Here, $\mathbf{V}_{21}$ can be expressed as

$$\mathbf{V}_{21} = [\mathbf{v}_1, \cdots, \mathbf{v}_s]. \tag{1.23}$$

At the same time, it has

$$\mathbf{V}_{21}\mathbf{V}_{21}^{T} = \mathbf{I}_s. \tag{1.24}$$

Combined with Eq. (1.20), we have

$$\lambda_i \mathbf{v}_i = \mathbf{\Upsilon}^{T} \mathbf{\Upsilon} \mathbf{v}_i, \tag{1.25}$$

where $\mathbf{v}_i \in \mathbf{V}_{21}, \lambda_i \in \left\{\lambda_1^2, \cdots, \lambda_s^2\right\}, i = 1, \cdots, s.$
According to Eq. (1.19), it holds that

$$\mathbf{\Upsilon}^{T}\mathbf{\Upsilon} = \frac{\bar{\mathbf{\Phi}}^{T}\mathbf{\Pi}\bar{\mathbf{\Phi}}}{(N-1)^2} = \frac{1}{(N-1)^2} \sum_{i=1}^{N} \sum_{j=1}^{N} \pi_{ji}\bar{\phi}(\mathbf{x}_j)\bar{\mathbf{\Phi}}^{T}(\mathbf{x}_i), \tag{1.26}$$

so Eq. (1.25) can be rewritten as the following expression:

$$\lambda\mathbf{v} = \mathbf{\Upsilon}^{T}\mathbf{\Upsilon}\mathbf{v} = \frac{1}{(N-1)^2} \sum_{i=1}^{N} \sum_{j=1}^{N} \pi_{ji}\bar{\phi}(\mathbf{x}_j)\beta_i, \tag{1.27}$$

where $\beta_i = \bar{\phi}^{T}(\mathbf{x}_i)\mathbf{v} \in \mathbb{R}$ is a scalar. Thus,

$$
\begin{aligned}
\mathbf{v} &= \frac{1}{(N-1)^2} \sum_{i=1}^{N} \sum_{j=1}^{N} \pi_{ji}\bar{\phi}(\mathbf{x}_j)\frac{\beta_i}{\lambda}\\
&= \frac{1}{(N-1)^2} \sum_{i=1}^{N} \mathbf{u}_i \sum_{j=1}^{N} \pi_{ji}\bar{\phi}(\mathbf{x}_j)\\
&= \frac{1}{(N-1)^2} \bar{\mathbf{\Phi}}^{T}\mathbf{\Pi}\mathbf{u},
\end{aligned}
\tag{1.28}
$$

where $\mathbf{u}_i = \frac{\beta_i}{\lambda} \in R$ is a scalar and $\mathbf{u} = [\mathbf{u}_1, \cdots, \mathbf{u}_N]^{T} \in R^{N}$ is a column vector. Then we have

$$
\begin{aligned}
\lambda\bar{\phi}^{T}(\mathbf{x}_m)\mathbf{v} &= \frac{1}{(N-1)^2}\lambda\bar{\phi}^{T}(\mathbf{x}_m)\bar{\mathbf{\Phi}}^{T}\mathbf{\Pi}\mathbf{u}\\
&= \frac{1}{(N-1)^2}\lambda\bar{\mathbf{k}}_m^{T}\mathbf{\Pi}\mathbf{u},
\end{aligned}
\tag{1.29}
$$

$$
\begin{aligned}
\bar{\phi}^{T}(\mathbf{x}_m)\mathbf{\Upsilon}^{T}\mathbf{\Upsilon}\mathbf{v} &= \frac{1}{(N-1)^2}\bar{\phi}^{T}(\mathbf{x}_m)\bar{\mathbf{\Phi}}^{T}\mathbf{\Pi}\bar{\mathbf{\Phi}}\mathbf{v}\\
&= \frac{1}{(N-1)^4}\bar{\mathbf{k}}_m^{T}\mathbf{\Pi}\bar{\mathbf{K}}\mathbf{\Pi}\mathbf{u},
\end{aligned}
\tag{1.30}
$$

where $m = 1, \cdots, N$. In accordance with Eqs. (1.27), (1.29), and (1.30), it is clear that

$$\lambda \bar{\mathbf{K}} \mathbf{\Pi} \mathbf{u} = \frac{1}{(N-1)^2} \bar{\mathbf{K}} \mathbf{\Pi} \bar{\mathbf{K}} \mathbf{\Pi} \mathbf{u}. \tag{1.31}$$

Thus,

$$\lambda \mathbf{u} = \frac{\bar{\mathbf{K}} \mathbf{\Pi}}{(N-1)^2} \mathbf{u}. \tag{1.32}$$

After solving the eigenvalue-eigenvector problem of Eq. (1.32), the first s largest eigenvalues $(\lambda_1 \geqslant \lambda_2 \geqslant \cdots \geqslant \lambda_s)$ are selected with their corresponding eigenvectors $(\mathbf{u}^1, \mathbf{u}^2, \cdots, \mathbf{u}^s)$ to calculate $\mathbf{V}_{21}$. According to Eq. (1.23) and Eq. (1.60), $\mathbf{V}_{21}$ is obtained as follows:

$$\mathbf{V}_{21} = [\mathbf{v}_1, \cdots, \mathbf{v}_s] = \frac{1}{(N-1)^2} \bar{\mathbf{\Phi}}^T \mathbf{\Pi} \hat{\mathbf{U}}, \tag{1.33}$$

where $\hat{\mathbf{U}} = [\mathbf{u}^1, \cdots, \mathbf{u}^s]$. Here, $\mathbf{V}_{21}$ can be calculated. On the basis of Eq. (1.17) and Eq. (1.22), it can be noted that $\mathbf{U}_{11} \mathbf{U}_{11}{}^T$ and $\mathbf{U}_{12} \mathbf{U}_{12}^T$ can be indirectly computed from $\mathbf{V}_{21}$.

When it comes to the online testing, the online sample $\mathbf{x}_{new} \in \mathbb{R}^{m \times 1}$ would be mapped into feature space F to get $\phi(\mathbf{x}_{new}) \in \mathbb{R}^{\Omega \times 1}$ first. In addition, the zero mean of $\phi(\mathbf{x}_{new})$ is $\bar{\phi}(\mathbf{x}_{new}) = \phi(\mathbf{x}_{new}) - \bar{\phi}$. As well, the online centralized kernel sample can be calculated as $\bar{\mathbf{k}}_{new}^T = \bar{\phi}^T(\mathbf{x}_{new}) \bar{\mathbf{\Phi}}^T$.

According to Eq. (1.18), the online space can be decomposed into two subpaces:

$$\hat{\phi}^T(\mathbf{x}_{new}) = \bar{\phi}^T(\mathbf{x}_{new}) \mathbf{U}_{11} \mathbf{U}_{11}^T, \tag{1.34}$$

$$\tilde{\phi}^T(\mathbf{x}_{new}) = \bar{\phi}^T(\mathbf{x}_{new}) \mathbf{U}_{12} \mathbf{U}_{12}^T. \tag{1.35}$$

Then the T^2 statistics in the subspace corresponding to $\hat{\mathbf{\Phi}}$ is defined as

$$T_{kdd}^2 = \bar{\phi}^T(\mathbf{x}_{new}) \mathbf{U}_{11} \left(\frac{\mathbf{U}_{11}^T \bar{\mathbf{\Phi}}^T \bar{\mathbf{\Phi}} \mathbf{U}_{11}}{N-1} \right)^{-1} \mathbf{U}_{11}^T \bar{\phi}(\mathbf{x}_{new}). \tag{1.36}$$

According to Eq. (1.22) and Eq. (1.33), T_{kdd}^2 can be calculated as

$$T_{kdd}^2 = \bar{\phi}^T(\mathbf{x}_{new}) \mathbf{V}_{21} \left(\frac{\mathbf{V}_{21}^T \bar{\mathbf{\Phi}}^T \bar{\mathbf{\Phi}} \mathbf{V}_{21}}{N-1} \right)^{-1} \mathbf{V}_{21}^T \bar{\phi}(\mathbf{x}_{new})$$

$$= \bar{\mathbf{k}}_{new}^T \mathbf{\Pi} \hat{\mathbf{U}} \left(\frac{\hat{\mathbf{U}}^T \mathbf{\Pi}^T \bar{\mathbf{K}} \bar{\mathbf{K}}^T \mathbf{\Pi} \hat{\mathbf{U}}}{N-1} \right)^{-1} \hat{\mathbf{U}}^T \mathbf{\Pi}^T \bar{\mathbf{k}}_{new}, \tag{1.37}$$

whose corresponding threshold is

$$J_{th,T^2} = \frac{s(N^2 - 1)}{N(N-s)} F_\alpha(s, N-s). \tag{1.38}$$

Similarly, the *SPE* statistics corresponding to $\tilde{\Phi}$ are calculated as follows:

$$
\begin{aligned}
SPE_{kdd} &= \tilde{\phi}^T(\mathbf{x}_{new})\tilde{\phi}(\mathbf{x}_{new}) \\
&= \bar{\phi}^T(\mathbf{x}_{new})\bar{\phi}(\mathbf{x}_{new}) - 2\bar{\phi}^T(\mathbf{x}_{new})\mathbf{U}_{11}\mathbf{U}_{11}^T\bar{\phi}(\mathbf{x}_{new}) \\
&\quad +\bar{\phi}^T(\mathbf{x}_{new})\mathbf{U}_{11}\mathbf{U}_{11}^T\mathbf{U}_{11}\mathbf{U}_{11}^T\bar{\phi}(\mathbf{x}_{new}) \\
&= C_0 - C_1 + C_2,
\end{aligned}
\tag{1.39}
$$

where

$$
\begin{aligned}
C_0 &= \bar{\phi}^T(\mathbf{x}_{new})\bar{\phi}(\mathbf{x}_{new}) \\
&= \left(\phi(\mathbf{x}_{new}) - \bar{\phi}\right)^T\left(\phi(\mathbf{x}_{new}) - \bar{\phi}\right) \\
&= \phi^T(\mathbf{x}_{new})\phi(\mathbf{x}_{new}) - \frac{2}{N}\bar{\mathbf{k}}_{new}^T\mathbf{1}_N + \frac{1}{N^2}\mathbf{1}_N^T\mathbf{K}\mathbf{1}_N,
\end{aligned}
\tag{1.40}
$$

$$
\begin{aligned}
C_1 &= 2\bar{\phi}^T(\mathbf{x}_{new})\mathbf{U}_{11}\mathbf{U}_{11}^T\bar{\phi}(\mathbf{x}_{new}) \\
&= \frac{2}{(N-1)^4}\bar{\mathbf{k}}_{new}^T\mathbf{\Pi}\hat{\mathbf{U}}\hat{\mathbf{U}}^T\mathbf{\Pi}^T\bar{\mathbf{k}}_{new},
\end{aligned}
\tag{1.41}
$$

$$
\begin{aligned}
C_2 &= \bar{\phi}^T(\mathbf{x}_{new})\mathbf{U}_{11}\mathbf{U}_{11}^T\mathbf{U}_{11}\mathbf{U}_{11}^T\bar{\phi}(\mathbf{x}_{new}) \\
&= \frac{2}{(N-1)^8}\bar{\mathbf{k}}_{new}^T\mathbf{\Pi}\hat{\mathbf{U}}\hat{\mathbf{U}}^T\mathbf{\Pi}^T\bar{\mathbf{K}}\mathbf{\Pi}\hat{\mathbf{U}}\hat{\mathbf{U}}^T\mathbf{\Pi}^T\bar{\mathbf{k}}_{new}.
\end{aligned}
\tag{1.42}
$$

The threshold of *SPE* is

$$
J_{th,SPE_{kdd}} = \chi_\alpha^2(h),
\tag{1.43}
$$

where

$$
g = \frac{\overline{SPE_{kdd}^2} - \overline{SPE}_{kdd}^2}{2\overline{SPE}_{kdd}}, h = \frac{2\overline{SPE}_{kdd}^2}{\overline{SPE_{kdd}^2} - \overline{SPE}_{kdd}^2},
\tag{1.44}
$$

$$
\overline{SPE_{kdd}} = \frac{1}{N}\sum_{i=1}^{N}\left(\bar{C}_o + \bar{C}_1 + \bar{C}_2\right),
\tag{1.45}
$$

$$
\overline{SPE_{kdd}^2} = \frac{1}{N}\sum_{i=1}^{N}\left(\bar{C}_0 + \bar{C}_1 + \bar{C}_2\right)^2.
\tag{1.46}
$$

When calculating $\bar{C}_n(n = 0, 1, 2)$, it uses the similar form with C_n, except the testing samples $\bar{\mathbf{k}}_{new}$ and $\mathbf{k}_{new}$ are replaced by the training samples $\bar{\mathbf{k}}_i^T$ and $\mathbf{k}_i^T$, respectively.

Based on the statistics, it can be assumed that the fault exists when the statistic exceeds its threshold. If the T_{kdd}^2 statistic exceeds J_{th,T_{kdd}^2}, it indicates that a quality-related fault occurs at least. Or if only the SPE_{kdd} statistics exceeds $J_{th,SPE_{kdd}}$ and the T_{kdd}^2 has no alarm, it indicates that there is one quality-unrelated fault occurring at least. The detailed algorithm is presented in Algorithm 4.

Algorithm 4: The KDD-based fault detection method

The training process:

1) Perform the nonlinear mapping to the training samples, and use the kernel function to process the mapping matrix. Normalize the training samples $\mathbf{K}$ and obtain $\bar{\mathbf{K}}$.

2) Obtain $\mathbf{\Pi}$ according to Eq. (1.19), and solve the eigenvalue-eigenvector problem of Eq. (1.32).

3) Calculate $\mathbf{V}_{21}$ and form the matrix $\hat{\mathbf{U}}$, according to Eq. (1.33).

4) Set the confidence limit α, and calculate the thresholds with Eq. (1.38) and Eq. (1.43).

The testing process:

1) Input the testing samples, and build the kernel matrix. Normalize the testing samples to get $\bar{\mathbf{K}}_{new}$.

2) Calculate the statistics $T_{kdd,n}^2$ and $SPE_{kdd,n}$ of each process sample, according to Eq. (1.37) and Eq. (1.39), respectively.

3) Compare the statistics with their corresponding thresholds for each process sample, and make the judgment by the following logic:

if $T_{kdd,n}^2 \leq J_{th,T_{kdd}^2}$ and $SPE_{kdd,n} \leq J_{th,SPE_{kdd}} \Rightarrow$ fault free;

if $T_{kdd,n}^2 > J_{th,T_{kdd}^2} \Rightarrow$ quality-related fault occurs;

if $SPE_{kls,n} > J_{th,SPE_{kdd}} \Rightarrow$ quality-unrelated fault occurs.

1.3.2 The KLS-based approach

This section introduces a KLS-based nonlinear method. Similar to the KPLS-based approaches, this method also deals with the nonlinear relationships among process variables by mapping these original variables into high-dimensional feature space. Then a KLS model is modeled to extract the full correlation between the process matrix and quality matrix to ensure the accuracy of the designed method. In the process, KPLS theory is not applied and it is unnecessary to set the number of latent variables, which indicates the simplicity of engineering implementation. Afterward, the feature matrix is decomposed into two orthogonal parts by SVD according to its full correlation with the quality matrix. Finally, test statistics are appropriately designed in the subspaces corresponding to the decomposed two parts for the purpose of quality-related fault detection. Detailed descriptions about the KLS method are presented in the following.

First, the original training samples $\mathbf{X}$ are mapped into the high-dimensional feature space F to obtain $\mathbf{\Phi}$. The proposed method aims to extract the full correlation between $\bar{\mathbf{\Phi}}$ and $\mathbf{Y}$ by least squares regression, and then utilizes the full correlation to decompose $\bar{\mathbf{\Phi}}$ and $\mathbf{Y}$ into the following forms:

$$\bar{\mathbf{\Phi}} = \hat{\mathbf{\Phi}} + \tilde{\mathbf{\Phi}}, \tag{1.47}$$

$$\mathbf{Y} = \hat{\mathbf{Y}} + \mathbf{E}_y, \tag{1.48}$$

where $\hat{\mathbf{Y}}$ is the full part of $\mathbf{Y}$ that correlated with $\bar{\mathbf{\Phi}}$, whereas $\mathbf{E}_y$ is the noise or disturbance that completely uncorrelated with $\bar{\mathbf{\Phi}}$. $\hat{\mathbf{\Phi}}$ and $\tilde{\mathbf{\Phi}}$ are orthogonal, and $\hat{\mathbf{\Phi}}$ is fully responsible for predicting $\hat{\mathbf{Y}}$, whereas $\tilde{\mathbf{\Phi}}$ has no relationship with $\hat{\mathbf{Y}}$. As such, we realize the decomposition on $\bar{\mathbf{\Phi}}$ according to $\mathbf{Y}$ without losing any correlation between them.

$\hat{\mathbf{Y}}$ can be expressed as

$$\hat{\mathbf{Y}} = \bar{\mathbf{\Phi}}\mathbf{M}_{kls}, \tag{1.49}$$

where $\mathbf{M}_{kls}$ is the regression coefficient matrix. It can be noted that $\mathbf{E}_y$ is completely unrelated to $\bar{\mathbf{\Phi}}$, so it has

$$\text{cov}\left(\mathbf{e}_y, \bar{\phi}(\mathbf{x})\right) = \varepsilon\left\{\mathbf{e}_y \bar{\phi}^T(\mathbf{x})\right\} = 0, \tag{1.50}$$

where $\mathbf{e}_y^T$ and $\bar{\phi}^T(\mathbf{x})$ are the row vectors of $\mathbf{E}_y$ and $\bar{\mathbf{\Phi}}$, respectively.

In this model, $N \geq m > l \geq 1$. According to Eq. (1.48) and Eq. (1.49), it holds that

$$\frac{1}{N}\mathbf{Y}^T\bar{\mathbf{\Phi}} = \frac{1}{N}\mathbf{M}_{kls}^T\bar{\mathbf{\Phi}}^T\bar{\mathbf{\Phi}} + \frac{1}{N}\mathbf{E}_y^T\bar{\mathbf{\Phi}} \approx \mathbf{M}_{kls}^T\frac{\bar{\mathbf{\Phi}}^T\bar{\mathbf{\Phi}}}{N}, \tag{1.51}$$

based on which $\mathbf{M}_{kls}$ can be obtained with the pseudo-inverse function $(\cdot)^{\dagger}$:

$$\mathbf{M}_{kls} = \left(\bar{\mathbf{\Phi}}^T\bar{\mathbf{\Phi}}\right)^{\dagger}\bar{\mathbf{\Phi}}^T\mathbf{Y}. \tag{1.52}$$

On the basis of the definition of the pseudo-inverse, $\left(\bar{\mathbf{\Phi}}^T\bar{\mathbf{\Phi}}\right)^{\dagger}$ could be obtained by a matrix $\mathbf{Z}$ with the same type as $\bar{\mathbf{\Phi}}^T\bar{\mathbf{\Phi}}$, which can satisfy that $\left(\bar{\mathbf{\Phi}}^T\bar{\mathbf{\Phi}}\right)\mathbf{Z}\left(\bar{\mathbf{\Phi}}^T\bar{\mathbf{\Phi}}\right) = \left(\bar{\mathbf{\Phi}}^T\bar{\mathbf{\Phi}}\right)$ and $\mathbf{Z}\left(\bar{\mathbf{\Phi}}^T\bar{\mathbf{\Phi}}\right)\mathbf{Z} = \mathbf{Z}$.

To realize the pseudo-inverse calculation, the following process should be conducted. First, SVD is performed on $\bar{\mathbf{\Phi}}^T\bar{\mathbf{\Phi}}$:

$$\bar{\mathbf{\Phi}}^T\bar{\mathbf{\Phi}} = \begin{bmatrix} \mathbf{W}_1 & \mathbf{W}_2 \end{bmatrix}\begin{bmatrix} \mathbf{\Lambda} & 0 \\ 0 & 0 \end{bmatrix}\begin{bmatrix} \mathbf{W}_1^T \end{bmatrix} = \mathbf{W}_1\mathbf{\Lambda}\mathbf{W}_1^T, \tag{1.53}$$

where $\mathbf{\Lambda}$ is a diagonal matrix with the A nonzero singular values, whereas $\mathbf{W}_1$ contains the corresponding A vectors. The remaining singular values are collected in $\mathbf{W}_2$. In accordance with the principle of SVD, $\mathbf{W}_1$ satisfies that

$$\mathbf{W}_1^T\mathbf{W}_1 = \mathbf{I}_A. \tag{1.54}$$

Here, define that

$$\mathbf{O} = \bar{\mathbf{\Phi}}^T\bar{\mathbf{\Phi}} \tag{1.55}$$

and

$$\mathbf{Z} = \mathbf{W}_1\mathbf{\Lambda}^{-1}\mathbf{W}_1^T. \tag{1.56}$$

Considering Eq. (1.53) and Eq. (1.54), it holds that

$$\mathbf{O}\mathbf{Z}\mathbf{O} = \left(\mathbf{W}_1\mathbf{\Lambda}\mathbf{W}_1^T\right)\left(\mathbf{W}_1\mathbf{\Lambda}^{-1}\mathbf{W}_1^T\right)\left(\mathbf{W}_1\mathbf{\Lambda}\mathbf{W}_1^T\right) = \mathbf{W}_1\mathbf{\Lambda}\mathbf{W}_1^T = \mathbf{O} \tag{1.57}$$

and

$$\mathbf{ZOZ} = \left(\mathbf{W}_1 \boldsymbol{\Lambda}^{-1} \mathbf{W}_1^T\right)\left(\mathbf{W}_1 \boldsymbol{\Lambda} \mathbf{W}_1^T\right)\left(\mathbf{W}_1 \boldsymbol{\Lambda}^{-1} \mathbf{W}_1^T\right) = \mathbf{W}_1 \boldsymbol{\Lambda}^{-1} \mathbf{W}_1^T = \mathbf{Z}. \quad (1.58)$$

It is obvious that $\mathbf{Z}$ is the pseudo-inverse of $\mathbf{O}$. Therefore, it has

$$\left(\bar{\boldsymbol{\Phi}}^T \bar{\boldsymbol{\Phi}}\right)^{\dagger} = \mathbf{W}_1 \boldsymbol{\Lambda}^{-1} \mathbf{W}_1^T. \quad (1.59)$$

Then, according to the SVD of $\bar{\boldsymbol{\Phi}}^T \bar{\boldsymbol{\Phi}}$, as shown in Eq. (1.53), the following theorem can be acquired.

Theorem 1. Define $\mathbf{w} \in \mathbf{W}_1$ and $\lambda \in diag\{\boldsymbol{\Lambda}\}$. It holds that

$$\lambda \mathbf{w} = \bar{\boldsymbol{\Phi}}^T \bar{\boldsymbol{\Phi}} \mathbf{w}. \quad (1.60)$$

Proof.. Based on Eq. (1.53) and Eq. (1.54), we have

$$\bar{\boldsymbol{\Phi}}^T \bar{\boldsymbol{\Phi}} \mathbf{W}_1 = \mathbf{W}_1 \boldsymbol{\Lambda} \mathbf{W}_1^T \mathbf{W}_1 = \mathbf{W}_1 \boldsymbol{\Lambda}, \quad (1.61)$$

which can be rewritten as

$$\left[\bar{\boldsymbol{\Phi}}^T \bar{\boldsymbol{\Phi}} \mathbf{w}_1, \cdots, \bar{\boldsymbol{\Phi}}^T \bar{\boldsymbol{\Phi}} \mathbf{w}_A\right] = \left[\mathbf{w}_1, \cdots, \mathbf{w}_A\right] \begin{bmatrix} \lambda_1 & 0 & 0 \\ 0 & \ddots & 0 \\ 0 & 0 & \lambda_A \end{bmatrix}$$

$$= [\lambda_1 \mathbf{w}_1, \cdots, \lambda_A \mathbf{w}_A]. \quad (1.62)$$

Therefore, Eq. (1.60) holds.

On the basis of this theorem, it has

$$\lambda \mathbf{w} = \bar{\boldsymbol{\Phi}}^T \bar{\boldsymbol{\Phi}} \mathbf{w} = \sum_i^N \bar{\phi}(\mathbf{x}_i) \bar{\phi}^T(\mathbf{x}_i) \mathbf{w} = \sum_j^N \bar{\phi}(\mathbf{x}_i) \beta_i, \quad (1.63)$$

where $\beta_i = \bar{\phi}^T(\mathbf{x}_i)\mathbf{w}$ is a scalar. It can be further transformed as

$$\mathbf{w} = \sum_i^N \frac{\beta_i}{\lambda_i} \bar{\phi}(\mathbf{x}_i) = \sum_i^N u_i \bar{\phi}(\mathbf{x}_i) = \bar{\boldsymbol{\Phi}}^T \mathbf{u}, \quad (1.64)$$

where $u_i = \frac{\beta_i}{\lambda_i}$ is also a scalar, and $\mathbf{u} = [u_1, \cdots, u_N]^T \in \mathbb{R}^{N \times 1}$.

As with other nonlinear algorithms, the Gaussian kernel method is introduced here to realize the calculation. According to Eq. (1.64), for $n = 1, \cdots, N$, multiply $\bar{\phi}(\mathbf{x}_n)$ on both sides of Eq. (1.60). Thus, the left side turns to

$$\lambda \bar{\phi}^T(\mathbf{x}_n)\mathbf{w} = \lambda \bar{\phi}^T(\mathbf{x}_n)\bar{\boldsymbol{\Phi}}^T \mathbf{u} = \lambda \bar{\mathbf{k}}_n^T \mathbf{u}, \quad (1.65)$$

whereas the right side changes to

$$\bar{\phi}^T(\mathbf{x}_n)\bar{\boldsymbol{\Phi}}^T \bar{\boldsymbol{\Phi}} \mathbf{w} = \bar{\phi}^T(\mathbf{x}_n)\bar{\boldsymbol{\Phi}}^T \bar{\boldsymbol{\Phi}} \bar{\boldsymbol{\Phi}}^T \mathbf{u} = \bar{\mathbf{k}}_n^T \bar{\mathbf{K}} \mathbf{u}. \quad (1.66)$$

Thus, it has

$$\lambda \bar{\mathbf{k}}_n^T \mathbf{u} = \bar{\mathbf{k}}_n^T \bar{\mathbf{K}} \mathbf{u}. \quad (1.67)$$

For all of the N samples, the equation can be expressed as

$$\lambda \bar{\mathbf{K}}\mathbf{u} = \bar{\mathbf{K}}\bar{\mathbf{K}}\mathbf{u}. \tag{1.68}$$

Finally, it has

$$\lambda \mathbf{u} = \bar{\mathbf{K}}\mathbf{u}. \tag{1.69}$$

So far, $\mathbf{\Lambda}$ and $\mathbf{U}$ can be calculated by solving the eigenvalue-eigenvector problem of Eq. (1.69). The largest A singular values are selected as the main elements of the diagonal matrix $\mathbf{\Lambda}$, and the corresponding A singular vectors form the matrix $\mathbf{U}$. With $\mathbf{u}$, $\mathbf{W}_1$ can be obtained:

$$\mathbf{W}_1 = \bar{\mathbf{\Phi}}^T \mathbf{U}^T \tag{1.70}$$

Therefore, the pseudo-inverse calculation of Eq. (1.59) can be transformed as

$$\left(\bar{\mathbf{\Phi}}^T \bar{\mathbf{\Phi}}\right)^{\dagger} = \bar{\mathbf{\Phi}}^T \, \Xi \, \bar{\mathbf{\Phi}}, \tag{1.71}$$

where

$$\Xi = \mathbf{U}^T \mathbf{\Lambda}^{-1} \mathbf{U}. \tag{1.72}$$

In addition, Eq. (1.52) is rewritten as

$$\mathbf{M}_{kls} = \left(\bar{\mathbf{\Phi}}^T \bar{\mathbf{\Phi}}\right)^{\dagger} \bar{\mathbf{\Phi}}^T \mathbf{Y} = \bar{\mathbf{\Phi}}^T \, \Xi \, \bar{\mathbf{\Phi}}\bar{\mathbf{\Phi}}^T \mathbf{Y} = \bar{\mathbf{\Phi}}^T \, \Xi \, \bar{\mathbf{K}}\mathbf{Y}. \tag{1.73}$$

At this point, $\mathbf{Y}$ has been decomposed by $\mathbf{M}$ completely.

In the following process, $\bar{\mathbf{\Phi}}$ is projected onto the space $span\{\mathbf{M}_{kls}\}$ and the corresponding orthogonal complement space $span\{\mathbf{M}_{kls}\}^{\perp}$. It has

$$\hat{\mathbf{\Phi}} \equiv span\{\mathbf{M}_{kls}\}, \tag{1.74}$$

$$\tilde{\mathbf{\Phi}} \equiv span\{\mathbf{M}_{kls}\}^{\perp}, \tag{1.75}$$

where

$$\tilde{\mathbf{\Phi}}\mathbf{M}_{kls} = 0. \tag{1.76}$$

Thus,

$$\hat{\mathbf{Y}} = \bar{\mathbf{\Phi}}\mathbf{M}_{kls} = \left(\hat{\mathbf{\Phi}} + \tilde{\mathbf{\Phi}}\right)\mathbf{M}_{kls} = \hat{\mathbf{\Phi}}\mathbf{M}_{kls}. \tag{1.77}$$

The preceding projection can be conducted by the following steps:

(1) Perform SVD on $\mathbf{M}_{kls}\mathbf{M}_{kls}^T$:

$$\mathbf{M}_{kls}\mathbf{M}_{kls}^T = \begin{bmatrix} \bar{\mathbf{w}}_1 & \bar{\mathbf{w}}_2 \end{bmatrix} \begin{bmatrix} \bar{\mathbf{\Lambda}} & 0 \\ 0 & 0 \end{bmatrix} \begin{bmatrix} \bar{\mathbf{w}}_1^T \end{bmatrix} = \bar{\mathbf{W}}_1 \bar{\mathbf{\Lambda}} \bar{\mathbf{W}}_1^T, \tag{1.78}$$

where $\bar{\mathbf{w}}_1 = \begin{bmatrix} \bar{\mathbf{w}}_1, \cdots, \bar{\mathbf{w}}_{\bar{A}} \end{bmatrix}$ and $\bar{\mathbf{w}}_2 = \begin{bmatrix} \bar{\mathbf{w}}_{\bar{A}+1}, \cdots, \bar{\mathbf{w}}_N \end{bmatrix}$. $\bar{A}$ is the number of nonzero singular values.

(2) Project $\bar{\boldsymbol{\Phi}}$ by $\overline{\mathbf{W}}_1 \overline{\mathbf{W}}_1^T$ and $\overline{\mathbf{W}}_2 \overline{\mathbf{W}}_2^T$, respectively. It has

$$\hat{\boldsymbol{\Phi}} = \bar{\boldsymbol{\Phi}} \, \overline{\mathbf{W}}_1 \overline{\mathbf{W}}_1^T, \tag{1.79}$$

$$\tilde{\boldsymbol{\Phi}} = \bar{\boldsymbol{\Phi}} \, \overline{\mathbf{W}}_2 \overline{\mathbf{W}}_2^T. \tag{1.80}$$

Based on the properties of SVD, it holds that

$$\overline{\mathbf{W}}_1 \overline{\mathbf{W}}_1^T + \overline{\mathbf{W}}_2 \overline{\mathbf{W}}_2^T = \mathbf{I}_\Omega \tag{1.81}$$

and

$$\overline{\mathbf{W}}_1^T \overline{\mathbf{W}}_2 = \overline{\mathbf{W}}_2^T \overline{\mathbf{W}}_1 = 0. \tag{1.82}$$

As a result,

$$\hat{\boldsymbol{\Phi}} + \tilde{\boldsymbol{\Phi}} = \bar{\boldsymbol{\Phi}} \left(\overline{\mathbf{W}}_1 \overline{\mathbf{W}}_1^T + \overline{\mathbf{W}}_2 \overline{\mathbf{W}}_2^T \right) = \bar{\boldsymbol{\Phi}}, \tag{1.83}$$

$$\hat{\boldsymbol{\Phi}} \tilde{\boldsymbol{\Phi}}^T = \bar{\boldsymbol{\Phi}} \, \overline{\mathbf{W}}_1 \overline{\mathbf{W}}_1^T \overline{\mathbf{W}}_2 \overline{\mathbf{W}}_2^T \bar{\boldsymbol{\Phi}}^T = 0. \tag{1.84}$$

(3) $\hat{\mathbf{Y}}$ can be acquired such that

$$\hat{\mathbf{Y}} = \bar{\boldsymbol{\Phi}} \mathbf{M}_{kls} = \bar{\boldsymbol{\Phi}} \left(\overline{\mathbf{W}}_1 \overline{\mathbf{W}}_1^T + \overline{\mathbf{W}}_2 \overline{\mathbf{W}}_2^T \right) \mathbf{M}_{kls} = \hat{\boldsymbol{\Phi}} \mathbf{M}_{kls}. \tag{1.85}$$

According to the preceding three steps, the orthogonal decomposition is realized. What remains to be solved is the calculation of $\overline{\mathbf{w}}_1$ and $\overline{\mathbf{w}}_2$. This solution is provided as follows.

According to Eq. (1.73), $\mathbf{M}_{kls}\mathbf{M}_{kls}^T$ can be expressed as

$$\mathbf{M}_{kls}\mathbf{M}_{kls}^T = \bar{\boldsymbol{\Phi}}^T \, \boldsymbol{\Xi} \bar{\mathbf{K}} \mathbf{Y} \mathbf{Y}^T \bar{\mathbf{K}}^T \boldsymbol{\Xi}^T \bar{\boldsymbol{\Phi}} = \bar{\boldsymbol{\Phi}}^T \boldsymbol{\Pi} \bar{\boldsymbol{\Phi}}, \tag{1.86}$$

where

$$\boldsymbol{\Pi} = \boldsymbol{\Xi} \bar{\mathbf{K}} \mathbf{Y} \mathbf{Y}^T \bar{\mathbf{K}}^T \boldsymbol{\Xi}^T = \begin{bmatrix} \pi_{11} & \cdots & \pi_{1N} \\ \vdots & \ddots & \vdots \\ \pi_{N1} & \cdots & \pi_{NN} \end{bmatrix} \in R^{N \times N}. \tag{1.87}$$

Thus, Eq. (1.86) can be expanded as

$$\mathbf{M}_{kls}\mathbf{M}_{kls}^T = \sum_{i=1}^{N} \sum_{j=1}^{N} \pi_{j,i} \bar{\phi}(\mathbf{x}_j) \bar{\phi}^T(\mathbf{x}_i). \tag{1.88}$$

Let $\overline{\mathbf{w}} \in \overline{\mathbf{w}}_1$, and $\frac{1}{\Lambda}$ is the corresponding singular value that $\frac{1}{\Lambda}$. According to Theorem 1, it holds that

$$\frac{1}{\Lambda}\overline{\mathbf{w}} = \mathbf{M}_{kls}\mathbf{M}_{kls}^T\overline{\mathbf{w}} = \sum_{i=1}^{N}\sum_{j=1}^{N}\pi_{j,i}\bar{\phi}(\mathbf{x}_j)\bar{\phi}^T(\mathbf{x}_i)\overline{\mathbf{w}} = \sum_{i=1}^{N}\sum_{j=1}^{N}\pi_{j,i}\bar{\phi}(\mathbf{x}_j)\frac{1}{\beta_i}.$$
(1.89)

Here, $\frac{1}{\beta} = \bar{\phi}^T(\mathbf{x}_i)\overline{\mathbf{w}}$ is a scalar. Therefore, $\overline{\mathbf{w}}$ can be rewritten as

$$\overline{\mathbf{w}} = \sum_{i=1}^{N}\sum_{j=1}^{N}\pi_{j,i}\bar{\phi}(\mathbf{x}_j)\frac{\bar{\beta}_i}{\bar{\Lambda}_i} = \sum_{i=1}^{N}\frac{1}{u_i}\sum_{j=1}^{N}\pi_{j,i}\bar{\phi}(\mathbf{x}_j) = \bar{\mathbf{\Phi}}^T\mathbf{\Pi}\frac{1}{\mathbf{u}},$$
(1.90)

where $\frac{1}{u_i} = \frac{\bar{\beta}_i}{\bar{\Lambda}_i}$ is also a scalar, and $\frac{1}{\mathbf{u}} = \left[\frac{1}{u_1}, \cdots, \frac{1}{u_N}\right]^T \in \mathbb{R}^{N\times 1}$.
Similar to the principle of Eqs. (1.65) through (1.69), it holds that

$$\frac{1}{\Lambda}\bar{\phi}^T(\mathbf{x}_n)\overline{\mathbf{w}} = \frac{1}{\Lambda}\bar{\phi}^T(\mathbf{x}_n)\bar{\mathbf{\Phi}}^T\mathbf{\Pi}\frac{1}{\mathbf{u}} = \frac{1}{\Lambda}\bar{\mathbf{k}}_n^T\mathbf{\Pi}\frac{1}{\mathbf{u}},$$
(1.91)

$$\bar{\phi}^T(\mathbf{x}_n)\mathbf{M}_{kls}\mathbf{M}_{kls}^T\overline{\mathbf{w}} = \bar{\phi}^T(\mathbf{x}_n)\bar{\mathbf{\Phi}}^T\mathbf{\Pi}\bar{\mathbf{\Phi}}\overline{\mathbf{w}} = \bar{\mathbf{k}}_n^T\mathbf{\Pi}\bar{\mathbf{K}}\mathbf{\Pi}\frac{1}{\mathbf{u}},$$
(1.92)

where $n = 1, \cdots, N$. As to the all N samples, it yields that

$$\frac{1}{\Lambda}\bar{\mathbf{K}}\mathbf{\Pi}\frac{1}{\mathbf{u}} = \bar{\mathbf{K}}\mathbf{\Pi}\bar{\mathbf{K}}\mathbf{\Pi}\frac{1}{\mathbf{u}}.$$
(1.93)

The following eigenvalue-eigenvector problem is obtained:

$$\frac{1}{\Lambda}\frac{1}{\mathbf{u}} = \bar{\mathbf{K}}\mathbf{\Pi}\frac{1}{\mathbf{u}}.$$
(1.94)

Obtaining the results of the preceding eigenvalue-eigenvector problem, the largest A singular values are selected as the main elements of the diagonal matrix $\frac{1}{\Lambda}$, and the corresponding A singular vectors forms the matrix $\frac{1}{\mathbf{u}_1}$. The rest of the singular vectors are collected in $\frac{1}{\mathbf{u}_2}$.
Based on Eq. (1.78) and Eq. (1.90), it is obtained that

$$\overline{\mathbf{w}}_1 = \bar{\mathbf{\Phi}}^T\mathbf{\Pi}\frac{1}{\mathbf{u}_1},$$
(1.95)

$$\overline{\mathbf{w}}_2 = \bar{\mathbf{\Phi}}^T\mathbf{\Pi}\frac{1}{\mathbf{u}_2}.$$
(1.96)

The preceding process describes the detailed derivation of the KLS algorithm, which can decompose the feature space into two orthogonal parts.

When it comes to the online testing tasks, the following process can be applied. The online samples are mapped into the high-dimensional space first, as $\mathbf{x}_{new} \in \mathbb{R}^m \to \phi(\mathbf{x}_{new}) \in \mathbb{R}^{\Omega}$. $\mathbf{\Phi}_{new}$ is formed and centralized by $\bar{\mathbf{\Phi}}_{new} = \mathbf{\Phi}_{new} - \frac{1}{N}\mathbf{1}_N\mathbf{1}_N^T\mathbf{\Phi}_{new}$. The kernel matrix $\mathbf{K}_{new} = \mathbf{\Phi}_{new}\mathbf{\Phi}_{new}^T$ is introduced to realize the calculation, and it can be centralized as $\bar{\mathbf{K}}_{new} = \bar{\mathbf{\Phi}}_{new}\bar{\mathbf{\Phi}}_{new}^T = \left(\mathbf{I}_{N\times N} - \frac{1}{N}\mathbf{1}_N\mathbf{1}_N^T\right)\mathbf{K}_{new}\left(\mathbf{I}_{N\times N} - \frac{1}{N}\mathbf{1}_N\mathbf{1}_N^T\right)$. $\bar{\phi}(\mathbf{x}_{new})$ can be decomposed into two orthogonal subspaces as

$$\hat{\phi}^T(\mathbf{x}_{new}) = \bar{\phi}^T(\mathbf{x}_{new})\overline{\mathbf{W}}_1\overline{\mathbf{W}}_1^T, \qquad (1.97)$$

$$\tilde{\phi}^T(\mathbf{x}_{new}) = \bar{\phi}^T(\mathbf{x}_{new})\overline{\mathbf{W}}_2\overline{\mathbf{W}}_2^T. \qquad (1.98)$$

It can be noticed that $\hat{\mathbf{\Phi}}_{new}$ and $\hat{\mathbf{Y}}$ are mutually correlated. Therefore, monitoring $\hat{\phi}^T(\mathbf{x}_{new})$ will provide the fault information related to $\mathbf{Y}$. On the contrary, monitoring $\tilde{\phi}^T(\mathbf{x}_{new})$ will provide us the fault information unrelated to $\mathbf{Y}$.
Since $\overline{A} = rank(\overline{\mathbf{W}}_1\overline{\mathbf{W}}_1^T\bar{\mathbf{\Phi}}^T) \leq rank(\overline{\mathbf{W}}_1^T\bar{\mathbf{\Phi}}^T) \leq \overline{A}$, it follows that $\overline{\mathbf{W}}_1^T\bar{\mathbf{\Phi}}^T = \overline{A} \cdot \overline{\mathbf{W}}_1\bar{\phi}(\mathbf{x}_{new})$ is a suitable candidate for T^2 statistics.
The T^2 statistic in the subspace corresponding to $\hat{\mathbf{\Phi}}$ is calculated as follows:

$$
\begin{aligned}
T_{kls}^2 &= \bar{\phi}^T(\mathbf{x}_{new})\overline{\mathbf{W}}_1\left(\frac{\overline{\mathbf{W}}_1^T\bar{\mathbf{\Phi}}^T\bar{\mathbf{\Phi}}\,\overline{\mathbf{W}}_1}{N-1}\right)\overline{\mathbf{W}}_1^T\bar{\phi}(\mathbf{x}_{new}) \\
&= \bar{\mathbf{k}}_{new}^T\mathbf{\Pi}_{\overline{\mathbf{u}}_1}\left(\frac{\overline{\mathbf{u}}_1^T\mathbf{\Pi}^T\bar{\mathbf{K}}\bar{\mathbf{K}}\mathbf{\Pi}\overline{\mathbf{u}}_1}{N-1}\right)_{\overline{\mathbf{u}}_1}^T\mathbf{\Pi}^T\bar{\mathbf{k}}_{new},
\end{aligned} \qquad (1.99)
$$

whose threshold can be calculated with training samples by the following formula:

$$J_{T_{kls}^2,th} = \frac{\overline{A}(N^2-1)}{N(N-\overline{A})}F_\alpha\left(\overline{A}, N-\overline{A}\right). \qquad (1.100)$$

The confidence level α is settled in advance.
At the same time, $\overline{\mathbf{W}}_2^T\bar{\phi}(\mathbf{x}_{new})$ can also be a feasible candidate for T^2 statistics. However, $\tilde{\mathbf{\Phi}}_{new}$ usually represents the residual portion of $\bar{\mathbf{\Phi}}_{new}$, and the variance of $\tilde{\mathbf{\Phi}}_{new}$ is usually very small. To avoid numerical problem in the inverse process, it is more suitable to use SPE statistics instead of T^2 statistics in this subspace. The SPE statistic is calculated as follows:

$$SPE_{kls} = \bar{\phi}^T(\mathbf{x}_{new})\overline{\mathbf{W}}_2\overline{\mathbf{W}}_2^T\bar{\phi}(\mathbf{x}_{new}) = \bar{\mathbf{k}}_{new}^T\mathbf{\Pi}_{\overline{\mathbf{u}}_2}{}_{\overline{\mathbf{u}}_2}^T\mathbf{\Pi}^T\bar{\mathbf{k}}_{new}, \qquad (1.101)$$

whose threshold can be calculated by

$$J_{SPE_{kls},th} = g\chi_\alpha^2(h), \qquad (1.102)$$

$$g = \frac{\overline{SPE_{kls}^2} - \overline{SPE_{kls}}^2}{2\overline{SPE_{kls}}}, h = \frac{2\overline{SPE_{kls}}^2}{\overline{SPE_{kls}^2} - \overline{SPE_{kls}}^2}.$$

Algorithm 5 shows the specific KLS-based fault detection process.

Algorithm 5: The KLS-based fault detection method

The training process:

1) Perform the nonlinear mapping to the training samples, and use the kernel function to process the mapping matrix. Normalize the training samples **K** to zero mean and unit variance, and obtain $\bar{\mathbf{K}}$.
2) Solve the eigenvalue-eigenvector problem of Eq. (1.69).
3) Calculate Ξ by Eq. (1.72) and Π by Eq. (1.87).
4) Solve the eigenvalue-eigenvector problem of Eq. (1.94). Form $\bar{\mathbf{u}}_1$ and $\bar{\mathbf{u}}_2$.
5) Set the confidence limit α, and calculate the thresholds with Eq. (1.100) and Eq. (1.102).

The testing process:

1) Map the testing samples to the high-dimensional space, and build the kernel matrix. Normalize the testing samples to get $\bar{\mathbf{K}}_{new}$.
2) Calculate the statistics $T^2_{kls,n}$ and $SPE_{kls,n}$ of each process sample, according to Eq. (1.99) and Eq. (1.101), respectively.
3) Compare the statistics with their corresponding thresholds for each process sample, and make the judgment by the following logic:

if $T^2_{kls,n} \leq J_{T^2_{kls},th}$ and $SPE_{kls,n} \leq J_{SPE_{kls},th} \Rightarrow$ fault free;

if $T^2_{kls,n} > J_{T^2_{kls},th} \Rightarrow$ quality-related fault occurs;

if $SPE_{kls,n} > J_{SPE_{kls},th} \Rightarrow$ quality-unrelated fault occurs.

1.3.3   Reconstruction partial derivative contribution plot

Although the contribution plot approach is popular in the fault diagnosis field as the default method, the diagnosis effectiveness would still be influenced by the fault smearing effect seriously. In particular, a traditional contribution plot cannot be used directly for nonlinear fault diagnosis because the relationship between the original variables and statistical index is cut off. The partial derivative contribution plot method has been proposed to improve the diagnosis results, which applies the kernel gradient as the contribution value. In this section, the partial derivative contribution plot method is advanced by data reconstruction to obtain more accurate faulty variables and suppress the smearing effect.

Here, a column vector is defined as $\mathbf{v} \in \mathbb{R}^{m \times 1}$, where the element $\mathbf{v}_i = 1 (i = 1, 2, \cdots, m)$. Then the partial derivative contribution for the l_{th} variable in the kernel matrix **K** can be obtained as follows:

$$\frac{\partial \mathbf{K}}{\partial \mathbf{v}_l} = \begin{bmatrix} \frac{\partial \mathbf{K}_{1,1}}{\partial \mathbf{v}_l} & \cdots & \frac{\partial \mathbf{K}_{1,n}}{\partial \mathbf{v}_l} \\ \vdots & \vdots & \vdots \\ \frac{\partial \mathbf{K}_{n,1}}{\partial \mathbf{v}_l} & \cdots & \frac{\partial \mathbf{K}_{n,n}}{\partial \mathbf{v}_l} \end{bmatrix}. \tag{1.103}$$

A mathematical algorithm is defined as

$$\mathbf{x}_i \odot \mathbf{v} = \begin{bmatrix} \mathbf{x}_{i,1}\mathbf{v}_1 & \mathbf{x}_{i,2}\mathbf{v}_2 & \cdots & \mathbf{x}_{i,m}\mathbf{v}_m \end{bmatrix}. \tag{1.104}$$

Then the elements in matrix $\frac{\partial \mathbf{K}}{\partial \mathbf{v}_l}$ can be calculated by the following formula:

$$
\begin{aligned}
\frac{\partial \mathbf{K}_{i,j}}{\partial \mathbf{v}_l} &= \frac{\partial \mathbf{k}(\mathbf{x}_i, \mathbf{x}_j)}{\partial \mathbf{v}_l} = \frac{\partial \mathbf{k}(\mathbf{x}_i \odot \mathbf{v}, \mathbf{x}_j \odot \mathbf{v})}{\partial \mathbf{v}_l} \\
&= exp(-\frac{\|\mathbf{x}_i \odot \mathbf{v} - \mathbf{x}_j \odot \mathbf{v}\|^2}{c})(\frac{-\frac{\partial \|\mathbf{x}_i\odot\mathbf{v}-\mathbf{x}_j\odot\mathbf{v}\|^2}{c}}{\partial \mathbf{v}_l}) \\
&= -\frac{1}{c}\mathbf{k}(\mathbf{x}_i, \mathbf{x}_j)\frac{\partial \|\mathbf{x}_i \odot \mathbf{v} - \mathbf{x}_j \odot \mathbf{v}\|^2}{\partial \mathbf{v}_l} \\
&= -\frac{1}{c}\mathbf{k}(\mathbf{x}_i, \mathbf{x}_j)\frac{\partial (\sqrt{(\mathbf{x}_{i,1}\mathbf{v}_1 - \mathbf{x}_{j,1}\mathbf{v}_1)^2 + \cdots + (\mathbf{x}_{i,m}\mathbf{v}_m - \mathbf{x}_{j,m}\mathbf{v}_m)^2})^2}{\partial \mathbf{v}_l} \\
&= -\frac{1}{c}\mathbf{k}(\mathbf{x}_i, \mathbf{x}_j)\frac{\partial ((\mathbf{x}_{i,1}\mathbf{v}_1 - \mathbf{x}_{j,1}\mathbf{v}_1)^2 + \cdots + (\mathbf{x}_{i,m}\mathbf{v}_m - \mathbf{x}_{j,m}\mathbf{v}_m)^2)}{\partial \mathbf{v}_l} \\
&= -\frac{1}{c}\mathbf{k}(\mathbf{x}_i, \mathbf{x}_j)\frac{\partial (\mathbf{x}_{i,l}\mathbf{v}_l - \mathbf{x}_{j,i}\mathbf{v}_l)^2}{\partial \mathbf{v}_l} \\
&= -\frac{2}{c}\mathbf{k}(\mathbf{x}_i, \mathbf{x}_j)(\mathbf{x}_{i,l} - \mathbf{x}_{j,l})^2. \tag{1.105}
\end{aligned}
$$

The centralized $\frac{\partial \mathbf{K}}{\partial \mathbf{v}_l}$ can be obtained as follows:

$$\frac{\partial \bar{\mathbf{K}}}{\partial \mathbf{v}_l} = \frac{\partial \mathbf{K}}{\partial \mathbf{v}_l} - \frac{1}{N}\frac{\partial \mathbf{K}}{\partial \mathbf{v}_l}\mathbf{1}_N\mathbf{1}_N^T - \frac{1}{N}\mathbf{1}_N\mathbf{1}_N^T\frac{\partial \mathbf{K}}{\partial \mathbf{v}_l} + \frac{1}{N^2}\mathbf{1}_N\mathbf{1}_N^T\frac{\partial \mathbf{K}}{\partial \mathbf{v}_l}\mathbf{1}_N\mathbf{1}_N^T. \tag{1.106}$$

The combined index based on the KLS model is used in the diagnosis process: $\Theta_{kls} = \frac{T^2}{J_{th,T^2}} + \frac{SPE}{J_{th,SPE}}$. The contributions of the l_{th} variable for this combined index Θ_{kls} in the i_{th} sample can be computed as

$$cont_{i,l} = |\frac{\partial \bar{\mathbf{K}}}{\partial \mathbf{v}_l}|_i\Theta_{kls}(\bar{\mathbf{K}}|_i)^T + \bar{\mathbf{K}}|_i\Theta_{kls}(\frac{\partial \bar{\mathbf{K}}}{\partial \mathbf{v}_l}|_i)^T|. \tag{1.107}$$

The partial derivative contribution values of the training samples conform to the χ^2 distribution. As a result, the threshold of the l_{th} variable can be calculated by training samples by the following formula:

$$C_{UCL,l} = 9 * m(cont_l), \tag{1.108}$$

where $m(cont_l)$ is the average of the contribution values. When the contribution of a variable exceeds its threshold, the variable can be considered as a fault variable.

At the same time, the partial derivative contribution value can also be affected by the smearing effect. Therefore, the reconstruction of the normal testing samples is a solution to suppress the smearing effect. Aiming to solve this problem, the reconstruction-based method is applied. The first step of this method is to reconstruct the variables. It reconstructs one single variable at a time. For this reconstructed variable, a detection process is conducted, whose data consists of this reconstructed variable and the other original variables. This reconstruction detection process is conducted on every variable, obtaining the detection statistics indices. As a result, the variables can be reordered by the descending order of their statistics indices. Then the sequence $[\eta_1, \ldots, \eta_m]$ can be obtained, where η_1 represents the variable with the maximum statistics value and η_m is the variable with the minimum statistics value. When it comes to the diagnosis process, it calculates the detection statistics q times. Every time the statistic is calculated, one more variable is replaced by the reconstructed variable data. The order of the variable replaced each time is determined by the sequence $[\eta_1, \ldots, \eta_m]$. If the statistic detected drops below the corresponding threshold when the first q variables in the sequence are replaced, these q variables are considered as the faulty variables for this fault.

When the fault amplitude is small, partial reconstruction may occur. At this point, only some of the actual faulty variables replaced can make the statistics value satisfy the termination condition of data reconstruction, resulting in the missing diagnosis of other fault variables. Facing to this problem, the original termination condition $J_{0.99}$, the threshold with a confidence level of 0.99, is changed into $J_{0.99} - 3s_t$, where s_t is the variance of the statistics.

Using the reconstruction matrix $\boldsymbol{\Psi}$ proposed in the work of Yu [32], predictive value of fault variables can be obtained as follows:

$$\xi \boldsymbol{\Psi} \mathbf{x}_{new} = 0, \tag{1.109}$$

where $\xi \in \mathbb{R}^{n_f \times m}$. n_f is the number of fault variables. The corresponding element value of the l_{th} fault variable in row vector ξ_l is 1, and the rest are 0. If the sample $\mathbf{x}_{new}$ is transformed to $\mathbf{x}_{new} = \boldsymbol{\Xi} \mathbf{x}_{new} + (\mathbf{I} - \boldsymbol{\Xi}) \mathbf{x}_{new}$, then Eq. (1.109) can be expressed as follows:

$$\xi \boldsymbol{\Psi} \boldsymbol{\Xi} \mathbf{x}_{new} = -\xi \boldsymbol{\Psi}(\mathbf{I} - \boldsymbol{\Xi}) \mathbf{x}_{new}, \tag{1.110}$$

where $\boldsymbol{\Xi}$ is a diagonal matrix, the main elements corresponding to the faulty variable are 1, and the others are 0. It holds that $\boldsymbol{\Xi} \mathbf{x}_{new} = \xi^T \mathbf{x}_{n_f}$, where $\mathbf{x}_{n_f}$ is a column vector comprised of faulty variables. According to Eq. (1.110), it has that

$$\mathbf{x}_{nf}^* = -(\xi \boldsymbol{\Psi} \xi^T)^{-1} \xi \boldsymbol{\Psi}(\mathbf{I} - \boldsymbol{\Xi}) \mathbf{x}_{new}. \tag{1.111}$$

Next, the resulting $\mathbf{x}_{n_f}^*$ is replaced by the corresponding variable in $\mathbf{x}_{new}$:

$$\hat{\mathbf{x}}^* = \xi^T \mathbf{x}_{n_f}^* + \mathbf{x}_{new}(\mathbf{I} - \boldsymbol{\Xi}), \tag{1.112}$$

where $\hat{\mathbf{x}}^*$ is the fault-free predictive value of the testing sample.

During the diagnosis process, one of the variables is replaced by the fault-free predictive data, which is the reconstructed variable. This new sample contains this reconstructed l_{th} variable as follows:

$$\mathbf{x}^l_{new} = \mathbf{x}_{new} \odot \delta_l + \hat{\mathbf{x}}^* \odot (\mathbf{1}^T_m - \delta_l), \tag{1.113}$$

where $\delta_l \in \mathbb{R}^{1 \times m}$ is a row vector, whose l_{th} element is 1 and the others are 0. The new kernel vector $\mathbf{k}^l_{new}$ can be calculated and centralized as follows:

$$\mathbf{k}^l_{\mathbf{new,j}} = \mathbf{k}(\mathbf{x}^l_{new}, \mathbf{x}_j) = exp(-\frac{\|\mathbf{x}^l_{new} - \mathbf{x}_j\|^2}{c}),$$

$$\bar{\mathbf{k}}^l_{new} = \mathbf{k}^l_{new} - \frac{1}{N}\mathbf{1}^T_N\mathbf{K} - \frac{1}{N}\mathbf{k}^l_{new}\mathbf{1}_N\mathbf{1}^T_N + \frac{1}{N^2}\mathbf{1}^T_N\mathbf{K}\mathbf{1}_N\mathbf{1}^T_N. \tag{1.114}$$

Based on the preceding processing, the contribution of l_{th} variables can be obtained as follows:

$$\frac{\partial \mathbf{k}^l_{new}}{\partial \mathbf{v}_l} = \begin{bmatrix} \frac{\partial \mathbf{k}^l_{new,1}}{\partial \mathbf{v}_l} & \cdots & \frac{\partial \mathbf{k}^l_{new,N}}{\partial \mathbf{v}_l} \end{bmatrix}. \tag{1.115}$$

Then the elements of $\frac{\partial \mathbf{k}^l_{new}}{\partial \mathbf{v}_l}$ can be calculated as

$$\frac{\partial \mathbf{k}^l_{new,j}}{\partial \mathbf{v}_l} = \frac{\partial \mathbf{k}(\mathbf{x}^l_{new}, \mathbf{x}_j)}{\partial \mathbf{v}_l}$$

$$= \frac{\partial \mathbf{k}(\mathbf{x}^l_{new} \odot \mathbf{v}, \mathbf{x}_j \odot \mathbf{v})}{\partial \mathbf{v}_l}$$

$$= -\frac{2}{c}\mathbf{k}(\mathbf{x}^l_{new}, \mathbf{x}_j)(\mathbf{x}^l_{new,l} - \mathbf{x}_{j,l})^2. \tag{1.116}$$

As to the new testing sample, $\mathbf{K}$ is fixed so that it would not influence the contribution. The centralized $\bar{\mathbf{K}}_{new}$ has

$$\frac{\partial \bar{\mathbf{k}}^l_{new}}{\partial \mathbf{v}_l} = \frac{\partial \mathbf{k}^l_{new}}{\partial \mathbf{v}_l} - \frac{1}{N}\frac{\partial \mathbf{k}^l_{new}}{\partial \mathbf{v}_l}\mathbf{1}_N\mathbf{1}^T_N. \tag{1.117}$$

On the basis of Eq. (1.107), the contribution to combined index Θ_{kls} of l_{th} variables can be calculated as

$$cont_{new,l} = |\frac{\partial \bar{\mathbf{k}}^l_{new}}{\partial \mathbf{v}_l}\Theta_{kls}(\bar{\mathbf{k}}^l_{new})^T + \bar{\mathbf{k}}^l_{new}\Theta_{kls}(\frac{\partial \bar{\mathbf{k}}^l_{new}}{\partial \mathbf{v}_l})^T|. \tag{1.118}$$

According to the thresholds of each variable determined by Eq. (1.108), the relative contribution can be acquired by

$$C^l_r = \frac{cont_{new}}{C_{UCL}}. \tag{1.119}$$

If the relative contribution of one variable exceeds 1, indicating the contribution value is larger than the threshold, this variable would be selected as a fault variable.

1.3.4 Kernel sample equivalent replacement

Kernel-based methods are the mainstream methods for solving data modeling and fault detection of nonlinear systems. However, they also suffer from some problems when applying different kernel functions in practice. The most widely used kernel function is the Gaussian kernel function, as all of the algorithms introduced earlier are based on this kind of kernel. The Gaussian kernel is constructed on the basis of Euclidean norm, which leads to the uncertain relationship between the original variables and the kernel vectors. The partial derivative contribution is proposed to solve this problem. Here, the KSER algorithm is introduced to provide another way to solve this issue. The key idea of this algorithm is to construct the relationship between the original variable $\mathbf{X}$ and the Gaussian kernel matrix $\mathbf{K}$. The detailed principle is expressed as follows.

The core of this algorithm is to conduct the first-order Taylor expansion on the Gaussian kernel function. As a result, the following results can be obtained:

$$\mathbf{k_{i,j}} = \mathbf{k}(\mathbf{x}_i, \mathbf{x}_j) = exp(-\frac{\|\mathbf{x}_i - \mathbf{x}_j\|^2}{c})$$
$$= 1 - \frac{\|\mathbf{x}_i - \mathbf{x}_j\|^2}{c} + o(\Delta \mathbf{x}^2). \tag{1.120}$$

Then the Gaussian matrix can be expressed as

$$\mathbf{K} = \mathbf{1}_N \mathbf{1}_N^T - \frac{1}{c}\mathbf{S}, \tag{1.121}$$

where $\mathbf{S} = \{\mathbf{s}_{i,j}\}$, $(i, j = 1, \cdots, N)$ and $\mathbf{s}_{i,j} = \|\mathbf{x}_i - \mathbf{x}_j\|^2 = \mathbf{x}_i^T\mathbf{x}_i + \mathbf{x}_j^T\mathbf{x}_j - 2\mathbf{x}_i^T\mathbf{x}_j$.
The kernel matrix $\mathbf{K}$ would be centralized as

$$\bar{\mathbf{K}} = (\mathbf{I}_N - \frac{1}{N}\mathbf{1}_N\mathbf{1}_N^T)\mathbf{K}(\mathbf{I}_N - \frac{1}{N}\mathbf{1}_N\mathbf{1}_N^T)$$
$$= (\mathbf{I}_N - \frac{1}{N}\mathbf{1}_N\mathbf{1}_N^T)(\mathbf{1}_N\mathbf{1}_N^T - \frac{1}{c}\mathbf{S})(\mathbf{I}_N - \frac{1}{N}\mathbf{1}_N\mathbf{1}_N^T). \tag{1.122}$$

It holds that

$$(\mathbf{I}_N - \frac{1}{N}\mathbf{1}_N\mathbf{1}_N^T)\mathbf{1}_N = \mathbf{1}_N - \frac{1}{N}\mathbf{1}_N N = 0,$$
$$\mathbf{1}_N^T(\mathbf{I}_N - \frac{1}{N}\mathbf{1}_N\mathbf{1}_N^T) = \mathbf{1}_N^T - \frac{1}{N}N\mathbf{1}_N^T = 0. \tag{1.123}$$

As mentioned previously, $\mathbf{H} = (\mathbf{I}_N - \frac{1}{N}\mathbf{1}_N\mathbf{1}_N^T)$, so Eq. (1.122) can be simplified as

$$\bar{\mathbf{K}} = \mathbf{H}(\mathbf{1}_N\mathbf{1}_N^T - \frac{1}{c}\mathbf{S})\mathbf{H}. \tag{1.124}$$

According to Eq. (1.123), $\bar{\mathbf{K}}$ can be rewritten as

$$\bar{\mathbf{K}} = -\frac{1}{c}\mathbf{HSH}$$
$$= -\frac{1}{c}\mathbf{H}\left(diag(\mathbf{XX}^T)\mathbf{1}_N^T + \mathbf{1}_N\left(diag(\mathbf{XX}^T)\right)^T - 2\mathbf{XX}^T\right)\mathbf{H}$$
$$= \frac{2}{c}\bar{\mathbf{X}}\bar{\mathbf{X}}^T. \tag{1.125}$$

As to the testing process, the kernel matrix similar to Equation (1.121) can be expressed as follows:

$$\mathbf{K}_{new} = \mathbf{1}_{N_t}\mathbf{1}_{N_t}^T - \frac{1}{c}\mathbf{S}_{new}, \tag{1.126}$$

where $\mathbf{S}_{new} = \left\{s_{new,i} = \|\mathbf{x}_{new} - \mathbf{x}_i\|^2 | i = 1, \cdots, N_t\right\}$ and can be represented as follows:

$$\mathbf{S}_{new} = \frac{1}{N}\mathbf{1}_{N_t}\mathbf{1}_N^T diag(\mathbf{XX}^T)\mathbf{1}_N^T + \mathbf{1}_N\left(diag(\mathbf{XX}^T)\right)^T - 2\mathbf{X}_{new}\mathbf{X}^T. \tag{1.127}$$

On the basis of Eq. (1.121), Eq. (1.124), Eq. (1.126), and Eq. (1.127), $\mathbf{K}_{new}$ can be transformed into

$$\bar{\mathbf{K}}_{new} = \left(\mathbf{K}_{new} - \frac{1}{N}\mathbf{1}_{N_t}\mathbf{1}_N^T\mathbf{K}\right)\mathbf{H}^T$$
$$= \frac{1}{c}\left(\frac{1}{N}\mathbf{1}_{N_t}\mathbf{1}_N^T\mathbf{S} - \mathbf{S}_{new}\right)\mathbf{H}^T$$
$$= \frac{1}{c}\left(2\mathbf{X}_{new}\mathbf{X}^T - \frac{2}{N}\mathbf{1}_{N_t}\mathbf{1}_N^T\mathbf{XX}^T\right)\mathbf{H}^T$$
$$= \frac{2}{c}\left(\mathbf{X}_{new} - \frac{1}{N}\mathbf{1}_{N_t}\mathbf{1}_N^T\mathbf{X}\right)\left(\mathbf{X} - \frac{1}{N}\mathbf{1}_N\mathbf{1}_N^T\mathbf{X}\right)^T$$
$$= \frac{2}{c}\bar{\mathbf{X}}_{new}\bar{\mathbf{X}}^T. \tag{1.128}$$

So far, the Gaussian kernel can be rewritten by $\mathbf{X}$ as

$$\bar{\mathbf{K}} = \frac{2}{c}\bar{\mathbf{X}}\bar{\mathbf{X}}^T, \bar{\mathbf{K}}_{new} = \frac{2}{c}\bar{\mathbf{X}}_{new}^T\bar{\mathbf{X}}^T, \tag{1.129}$$

Taking the KPLS model as an example, the predicted quality variable can be calculated as

$$\widehat{\mathbf{Y}}_{\text{kpls}} = \bar{\mathbf{\Phi}}\mathbf{M}_k$$
$$= \bar{\mathbf{\Phi}}\bar{\mathbf{\Phi}}^T\mathbf{U}\left(\mathbf{T}^T\bar{\mathbf{K}}\mathbf{U}\right)^{-1}\mathbf{T}^T\mathbf{Y}$$
$$= \bar{\mathbf{K}}\mathbf{U}\left(\mathbf{T}^T\bar{\mathbf{K}}\mathbf{U}\right)^{-1}\mathbf{T}^T\mathbf{Y}$$
$$= \frac{2}{c}\bar{\mathbf{X}}\bar{\mathbf{X}}^T\mathbf{U}\left(\mathbf{T}^T\bar{\mathbf{K}}\mathbf{U}\right)^{-1}\mathbf{T}^T\mathbf{Y}. \tag{1.130}$$

The predicted value of the testing sample quality variable is

$$\widehat{\mathbf{Y}}_{kpls,new} = \bar{\mathbf{\Phi}}\mathbf{M}_k$$
$$= \phi^T(\mathbf{x}_{new})\bar{\mathbf{\Phi}}^T\mathbf{U}(\mathbf{T}^T\bar{\mathbf{K}}\mathbf{U})^{-1}\mathbf{T}^T\mathbf{Y}$$
$$= \mathbf{K}_{new}^T\mathbf{U}(\mathbf{T}^T\bar{\mathbf{K}}\mathbf{U})^{-1}\mathbf{T}^T\mathbf{Y}$$
$$= \frac{2}{c}\bar{\mathbf{x}}_{new}^T\bar{\mathbf{X}}^T\mathbf{U}(\mathbf{T}^T\bar{\mathbf{K}}\mathbf{U})^{-1}\mathbf{T}^T\mathbf{Y}. \tag{1.131}$$

Taking quality-related subspace fault detection as an example, the statistical index is

$$\mathbf{t}_{new}^T = \phi^T(\mathbf{x}_{new})\bar{\mathbf{\Phi}}^T\mathbf{U}(\mathbf{T}^T\bar{\mathbf{K}}\mathbf{U})^{-1}$$
$$= \mathbf{k}_{new}^T\mathbf{U}(\mathbf{T}^T\bar{\mathbf{K}}\mathbf{U})^{-1}$$
$$= \frac{2}{c}\bar{\mathbf{x}}_{new}^T\bar{\mathbf{X}}^T\mathbf{U}(\mathbf{T}^T\bar{\mathbf{K}}\mathbf{U})^{-1}, \tag{1.132}$$

$$T_{kpls,n}^2 = \mathbf{t}_{new}^T\left(\frac{\mathbf{T}^T\mathbf{T}}{N-1}\right)^{-1}\mathbf{t}_{new}$$
$$= \frac{4}{c^2}\bar{\mathbf{x}}_{new}^T\bar{\mathbf{X}}^T\mathbf{U}(\mathbf{T}^T\bar{\mathbf{K}}\mathbf{U})^{-1}\left(\frac{\mathbf{T}^T\mathbf{T}}{N-1}\right)^{-1}(\mathbf{U}^T\bar{\mathbf{K}}^T\mathbf{T})^{-1}\mathbf{U}^T\bar{\mathbf{X}}\bar{\mathbf{x}}_{new}$$
$$= \frac{4}{c^2}\bar{\mathbf{x}}_{new}^T\mathbf{\Lambda}_{kser}^{-1}\bar{\mathbf{x}}_{new}. \tag{1.133}$$

It is obvious that the relationship between Gaussian kernel function and the input variables is settled, so the input variables participate in the calculation directly. The diagnosis or isolation of the faulty variable can be more effective and efficient. The calculation of nonlinear data modeling and fault diagnosis process is greatly reduced after the equivalent replacement.

1.4   Simulation

1.4.1   Introduction of the Tennessee-Eastman process

As a widely used object in the multivariate statistical process monitoring field, the Tennessee-Eastman process is introduced in this section and used to verify the effectiveness of several fault detection and diagnosis algorithms [58, 59, 60]. This is a model extracted from the actual industrial chemical process that is created by the Eastman Chemical Company. The detailed production is shown in Fig. 1.1. It contains five major units: a reactor, condenser, compressor, separator, and stripper. There are eight ingredients related to the production in this process: A, B, C, D, E, F, G, and H. The main reactions are

$$\begin{aligned}
A(g) + C(g) + D(g) &\rightarrow G(liq), \\
A(g) + C(g) + E(g) &\rightarrow H(liq), \\
A(g) + E(g) &\rightarrow F(liq), \\
3D(g) &\rightarrow 2F(liq),
\end{aligned} \tag{1.134}$$

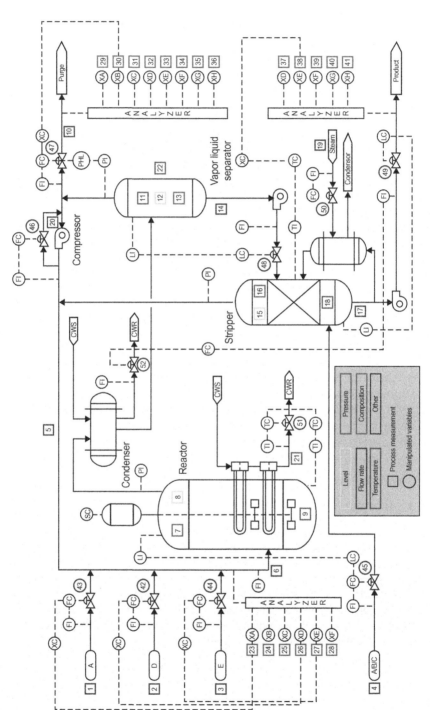

FIGURE 1.1   The production process of Tennessee-Eastman process.

where G and H are two liquid products of this industrial process. The components A, C, D, and E are the gas reactants that are fed into the reactor with inert catalyst B. At the same time, the liquid by-product F is also produced in the reactor. The gas reactants are fed into the reactor with the catalyst to conduct the reaction. The mixture produced and the reactants remaining are processed by the condenser, and then the gas components and liquid components are separated by the separator. The gas components would be recycled after being compressed by the compressor, whereas the liquid components are entered into the stripper for further purification to obtain the product components G and H. The gas components separated by the stripper would also be introduced back to the reactor.

As shown in Fig. 1.1, there are 52 variables in total that can be sorted into two blocks. The XMV block consists of 11 manipulated variables, whereas the XMEAS blocks contain 22 process variables and 11 analysis variables, as shown in Table A.1.1, Table A.1.2, and Table A.1.3. As in most of the previous research, 22 process variables, XMEAS(1--22), and 11 manipulated variables, XMV(1--11), are selected as the input x_1-x_{33} in the simulation, so there are 33 input variables in the simulation dataset. XMEAS(35), the purge gas analysis variable of the product component G, is considered as the output variable y. As to the training samples, 480 normal samples are applied to construct the feature space and the mapping directions. When it comes to the testing process, 960 samples are introduced with the fault occurring at the 161_{th} sample. The faults of this TEP model contain 20 kinds of faults that are described in detail in Table A.1.4.

1.4.2   Fault detection results

In this section, the simulation results of several fault detection methods are provided in detail with the TEP model, particularly the KPLS method, the TKPLS method, the KLS method, and the KDD method. The confidence level is settled as $\alpha = 0.99$, and the parameter in the Gaussian kernel function is $c = 10^4$.

To illustrate the efficiency of the detection methods, two classical faults, IDV(2) and IDV(14), quality related and quality unrelated, respectively, are selected to show the statistics indices curves so that the alarm results can be obtained clearly. At the same time, the fault detection rates (FDRs) and false alarm rates (FARs) of all of the faults are provided and analyzed to compare the efficiency of the different methods. The calculation of FDR and FAR uses the data obtained after the fault occurrence and is conducted through the following formula:

$$\text{Alarm rate} = \frac{\text{Number of alarm samples}}{\text{Total number of faulty samples}} \times 100\%. \qquad (1.135)$$

First, the occurrence of the quality-related faults is discussed. In Fig. 1.2(A-D), the detection results of IDV(2) are noted. This fault would

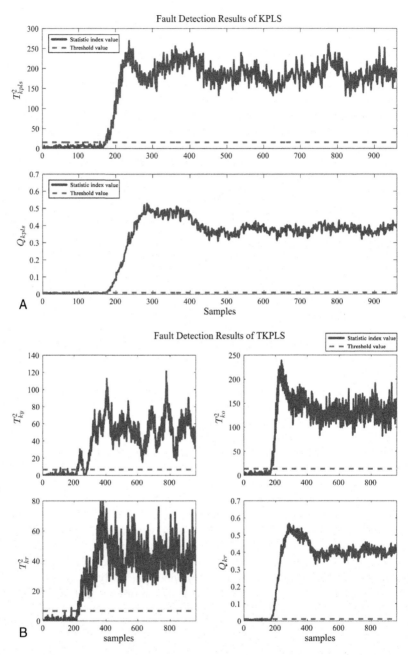

FIGURE 1.2 Fault detection results with IDV(2) occurring. (A) KPLS (B) TKPLS (C) KLS (D) KDD.

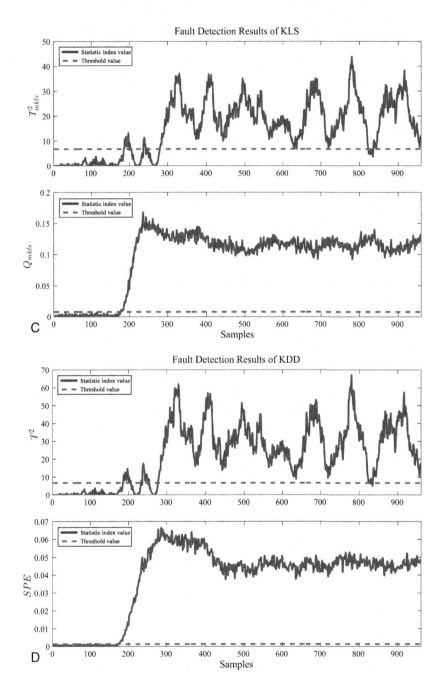

FIGURE 1.2   *(Continued)*

cause alarms both in quality-related subspaces and quality-unrelated spaces. It is clear that the statistics of the four methods alarm obviously after the 161_{th} sample in Fig. 1.2(A-D). According to the FDRs in Table 1.1, the alarm rates are almost up to 100% in the results of four quality-related methods. As to the other quality-related faults, Table 1.1 shows that the KPLS and TKPLS methods have similar detection efficiency, as most of the faults are detected with high alarm rates. Their FDRs for IDV(5), IDV(10), and IDV(20) are consistently lower than the conditions when most other faults occur. At the same time, the alarm rates of the KLS and KDD methods are slightly lower than the previous two classic methods for most of the conditions. When IDV(1), IDV(10), IDV(12), IDV(13), and IDV(20) occur respectively, the FDRs of the T^2 statistics of the KLS and KDD methods are obviously lower than KPLS and TKPLS. This phenomenon partly results from the function of the control system of the process that adjusts the abnormal variables back to the normal level or into a new balance condition, and the statistics may return under the thresholds. In particular, the KLS method shows significantly low detection rates for IDV(6) and IDV(18). In addition, the FDRs of the KPCA method are also shown in Table 1.1, which exhibits low FDRs and illustrates that the performance of KPCA for the quality-related fault detection task is not so reliable.

According to the results and analysis shown previously, it can be known that the KPLS, TKPLS, KLS, and KDD methods can provide efficient quality-related fault detection with satisfactory alarms to make the detection judgment.

Then quality-unrelated faults are taken into consideration, as Fig. 1.3(A-D) and Table 1.2 show. Fig. 1.3(A-D) demonstrates the statistics index curves for IDV(14) detected by the KPLS, TKPLS, KLS, and KDD methods. As IDV(14) is a classical quality-unrelated fault, it can be noted that the KPLS method has obvious alarms both in T^2 and SPE statistics in Fig. 1.3(A), which means that large FARs in T^2 statistics should be normal all the time. As to the TKPLS method, T_{ko}^2 and T_{kr}^2 statistics represent the quality-unrelated subspaces. In Fig. 1.3(B), these two statistics have obvious alarm signals, whereas SPE_{kr} statistics also contain abnormal samples. Although the quality-related T_{ky}^2 statistics have few alarm signals, the SPE_{kr} statistic represents the variation in the residual subspace, whose relationship to the quality variable and alarm conditions are hard to analyze. In Fig. 1.3(C) and (D), the detection results of the KLS and KDD methods show similar characteristics with low FARs in T^2 statistics and obviously FDRs in SPE statistics. The accuracy of the KLS and KDD methods can also be proved by the FDRs and FARs shown in Table 1.2. The quality-related T^2 statistics of KDD achieve low alarm rates for the quality-unrelated faults compared to other methods, whereas the quality-unrelated SPE statistics have high FDRs. The KLS method gives similar detection results to the KDD method, of which the FDRs of SPE statistics are lower than that of the KDD method. Based on the alarm results, the KLS and KDD methods can identify the quality-unrelated faults. In contrast, the FDRs and FARs of KPCA, KPLS, and TKPLS are shown in Table 1.2, within which the T^2 statistics of these

TABLE 1.1 FDRs of the quality related faults (%)

Fault	KPCA		KPLS		TKPLS		KLS		KDD	
	T^2_{kpca}	SPE_{kpca}	T^2_{kpls}	SPE_{kpls}	T^2_{ky} & SPE_{kr}	T^2_{ko} & T^2_{kr}	T^2_{kls}	SPE_{kls}	T^2_{kdd}	SPE_{kdd}
IDV(1)	0.38[a]	0.40	99.63[a]	99.88	99.75[a]	99.5	34.88[a]	99.13	38.38[a]	99.63
IDV(2)	1.88[a]	1.88	98.50[a]	98.63	98.63[a]	98.50	86.88[a]	98.00	89.88[a]	98.50
IDV(5)	76.33[a]	75.60	30.50[a]	25.38	39.38[a]	27.30	23.50[a]	26.88	23.63[a]	37.50
IDV(6)	0[a]	0	99.50[a]	99.88	100.00[a]	99.50	17.75[a]	19.13	98.25[a]	100.00
IDV(8)	0.23[a]	0.23	95.88[a]	97.38	98.13[a]	95.13	79.00[a]	97.00	97.63[a]	99.13
IDV(10)	60.75[a]	33.80	67.25[a]	49.50	61.00[a]	61.88	26.75[a]	47.75	27.00[a]	60.88
IDV(12)	0.10[a]	0.13	98.00[a]	98.88	98.25[a]	97.38	76.63[a]	98.13	77.13[a]	99.63
IDV(13)	0.50[a]	0.55	94.88[a]	95.25	95.50[a]	95.13	79.13[a]	93.88	80.25[a]	90.88
IDV(18)	0.99[a]	10.33	89.13[a]	89.63	91.63[a]	88.75	24.38[a]	22.38	88.63[a]	91.13
IDV(20)	69.33[a]	49.00	44.38[a]	52.88	67.25[a]	39.50	28.38[a]	38.63	28.63[a]	63.63

[a] FDR.

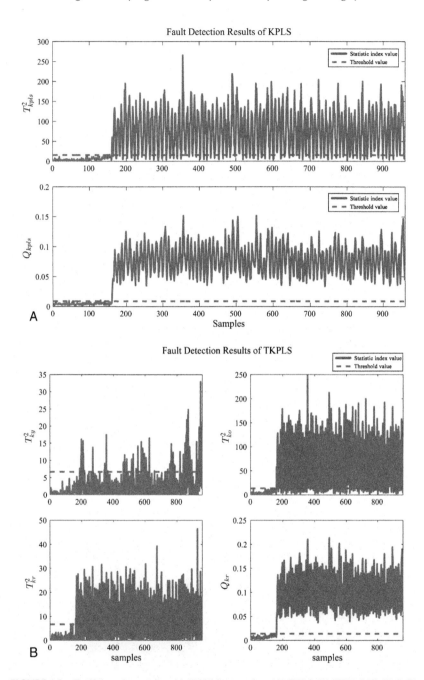

FIGURE 1.3 Fault detection results with IDV(14) occurring. (A) KPLS (B) TKPLS (C) KLS (D) KDD.

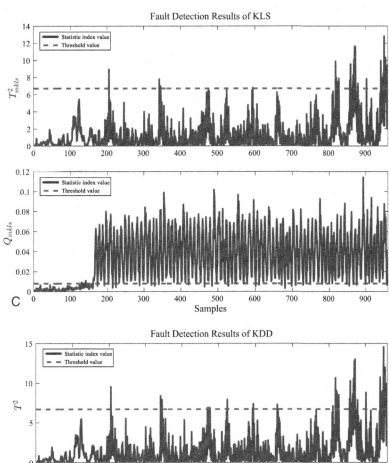

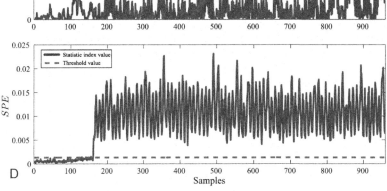

FIGURE 1.3 (*Continued*)

TABLE 1.2 FDRs and FARs of the quality-unrelated faults (%)

Fault	KPCA		KPLS		TKPLS		KLS		KDD	
	T^2_{kpca}	SPE_{kpca}	T^2_{kpls}	SPE_{kpls}	T^2_{ky} & SPE_{kr}	T^2_{ko} & T^2_{kr}	T^2_{kls}	SPE_{kls}	T^2_{kdd}	SPE_{kdd}
IDV(3)	97.13^b	97.88^a	12.63^b	3.25^a	11.38^b	8.88^a	7.25^b	8.88^a	7.63^b	17.38^a
IDV(4)	99.00^b	98.88^a	56.38^b	95.38^a	94.38^b	50.50^a	5.63^b	23.75^a	5.75^b	94.25^a
IDV(7)	59.13^b	59.88^a	95.88^b	99.88^a	100.00^b	88.25^a	32.63^b	68.13^a	33.25^b	100.00^a
IDV(9)	97.50^b	97.00^a	10.88^b	1.00^a	10.75^b	8.00^a	5.63^b	7.00^a	5.88^b	13.38^a
IDV(11)	57.50^b	59.88^a	61.63^b	60.13^a	66.00^b	59.50^a	9.50^b	41.75^a	9.88^b	73.75^a
IDV(14)	0^b	0.13^a	85.13^b	99.88^a	100.00^b	89.13^a	3.38^b	95.75^a	4.38^b	100.00^a
IDV(15)	92.33^b	93.75^a	17.38^b	3.50^a	27.13^b	10.8^a	10.50^b	10.63^a	10.75^b	22.00^a
IDV(16)	70.33^b	68.63^a	49.75^b	18.00^a	27.00^b	41.25^a	20.00^b	37.50^a	20.25^b	53.5^a
IDV(17)	0.40^b	0.64^a	76.00^b	90.50^a	89.38^b	76.75^a	13.25^b	77.88^a	29.38^b	92.25^a
IDV(19)	88.13^b	86.33^a	6.63^b	14.50^a	11.00^b	4.75^a	1.00^b	4.00^a	1.00^b	15.75^a

[a] FDR. [b] FAR. Boldface, highlighting the low FARs of KLS and KDD.

methods are high for several faults occurrence, such as IDV(4), IDV(7), IDV(11), IDV(14), and IDV(17), which can mislead the fault detection results.

In summary, the detection performance of the KPLS, TKPLS, KLS, and KDD methods were compared in this section. It can be concluded that these four methods can realize quality-related fault detection. When quality-related faults occur, they can provide high FDRs to make the judgment. At the same time, it can be noticed that the KPLS and TKPLS methods obtain large FARs in the quality-related subspaces, which would impact the accuracy of the detection results badly. However, the KLS and KDD methods are more reliable by comparison, as they cannot only obtain accurate alarms for quality-related faults but also get low FARs for quality-unrelated faults.

1.4.3   Nonlinear fault detection using KSER

As the preceding section mentioned, plenty of the fault diagnosis methods apply the Gaussian kernel function to solve the nonlinear process detection problem and obtain satisfactory results. However, they also suffer from the unclear relationship between the input and the kernel matrix, when analyzing the contribution of each variable to the fault occurring. Within several solutions, KSER is proposed recently to replace the kernel function in the fault detection and diagnosis process. This section describes the simulation results of the KSER method with the TEP model and compares the detection results of KSER-KPLS with typical kernel-based nonlinear methods to verify the equivalent effect to the original KPLS method and the advantage of the KSER algorithm. In particular, the detection results of KSER-KPLS are compared with those of KPLS, TKPLS, and KPCA. The confidence level and the parameter in the Gaussian kernel function are set as same as in the preceding simulation experiments.

Here, the fault IDV(7) is applied to the simulation. As can be noted in the upper part of Fig. 1.4, both KSER-KPLS and the original KPLS method have achieved satisfactory results in the prediction of the output $\mathbf{Y}$. The effectiveness of the KSER method is even better than that of KPLS, as the deviation between KSER-KPLS (green curve) and the actual output (red curve) is smaller than that between KPLS (blue curve) and the actual output (red curve), which can also be proved in the lower part that shows the deviation of KSER-KPLS is closer to 0.

Fig. 1.5(A) and (B) are the fault diagnosis results of the KSER-KPLS algorithm with the nonlinear diagnosis method and the linear method RCI, respectively. By comparing Fig. 1.5(A) and (B), it can be found that with the KSER algorithm, the linear method can also realize a satisfactory diagnosis effect of the nonlinear process, and the variable x_{26} is diagnosed as the fault variable, which is consistent with the fault diagnosis result of the nonlinear method.

In addition, the application of the KSER algorithm can also improve the calculation efficiency of the fault detection process. Table 1.3 shows the time

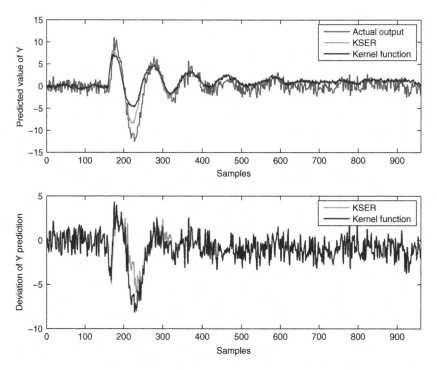

FIGURE 1.4 The prediction effect before and after the application of KSER.

TABLE 1.3 The computational efficiency results for the TE process

Methods	Offline	Online
KPCA/KPLS/TKPLS	$\gg 24hrs$	$> 3hrs$
KPLS-KSER	Unnecessary	0.0245408 s

spent by several conventional methods and the KPLS-KSER method in the online and offline stages of fault detection, respectively. The general kernel-based nonlinear method increases the dimension of the data matrix, so the calculation time increases significantly. However, the KSER method can significantly reduce the time required for fault detection due to the offline part being unnecessary, and the time required for the online process only takes about 0.025 s.

In conclusion, the use of the KSER method can make up for the deficiency of the linear method in dealing with nonlinear problems, and achieves almost the same effect as the nonlinear method. At the same time, it shortens the time cost of fault detection. According to the simulation results, the efficiency of KSER-KPLS is much better than that of the methods without the replacement.

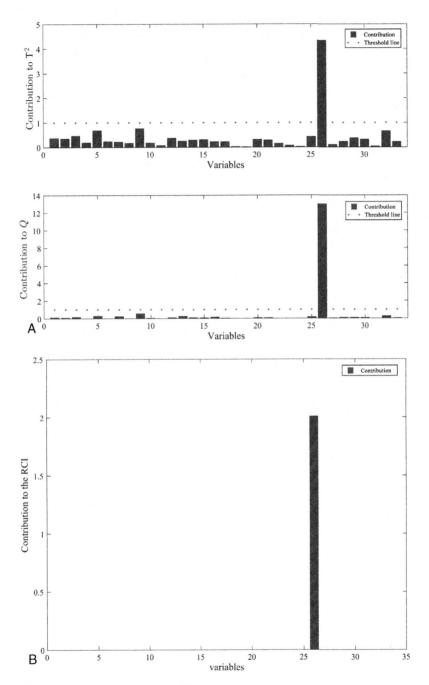

FIGURE 1.5 Fault diagnosis results of KSER-based algorithm with different diagnosis methods. (A) Nonlinear fault diagnosis method (B) Linear fault diagnosis method.

TABLE 1.4 The CDRs of fault diagnosis (%)

Fault ID	Reconstructed partial derivative method	Partial derivative method
IDV(1)	10.13	0
IDV(2)	5.75	0.88
IDV(3)	84.50	25.75
IDV(4)	73.62	18.38
IDV(5)	65.25	20.50
IDV(6)	2.13	0
IDV(7)	50.12	1.87
IDV(8)	4.75	0
IDV(9)	86.50	28.38
IDV(10)	49.63	14.37
IDV(11)	59.50	10.88
IDV(12)	6.88	0.13
IDV(13)	13.13	1.63
IDV(14)	93.42	1.75
IDV(15)	79.87	29.88

1.4.4 Fault diagnosis without smearing effect

The smearing effect is a problem that can lead to false diagnosis results, as the abnormal contribution values may smear over the nonfaulty variables. According to the reconstructed partial derivative contribution plot introduced in Section 1.3.3, the simulation results of this method are provided in this section to prove the effectiveness of avoiding the smearing effect. In the simulation, IDV(14) is selected, which causes the reactor cooling water valve to stick and leads to the change of the water temperature of the reaction tower (XMEAS(21), variable x_{21} in the simulation), the temperature of the reaction tower (XMEAS(9), variable x_9 in the simulation), and affect purge gas analysis D (XMV(10), variable x_{32} in the simulation) eventually. Fig. 1.6 shows the variation of these three faulty variables of this fault. The simulation performs the KLS-based fault detection method to obtain the fault alarm results.

When the partial differential contribution plot is performed, the diagnosis result is shown in Fig. 1.7(A) that other variables also have abnormal samples in red color, except for the three actual faulty variables: x_9, x_{21}, and x_{32}. Fig. 1.7(B) exhibits the diagnosis result of the reconstruction partial derivative contribution plot method. It is obvious that the red bars are concentrated on x_9, x_{21}, and

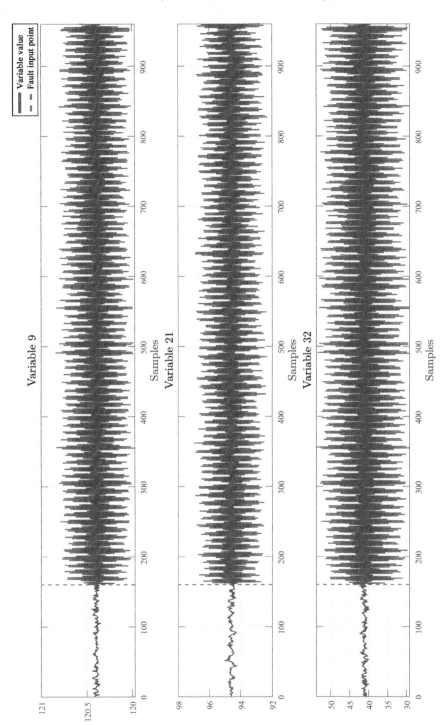

FIGURE 1.6 The changes in the faulty variables of IDV(14).

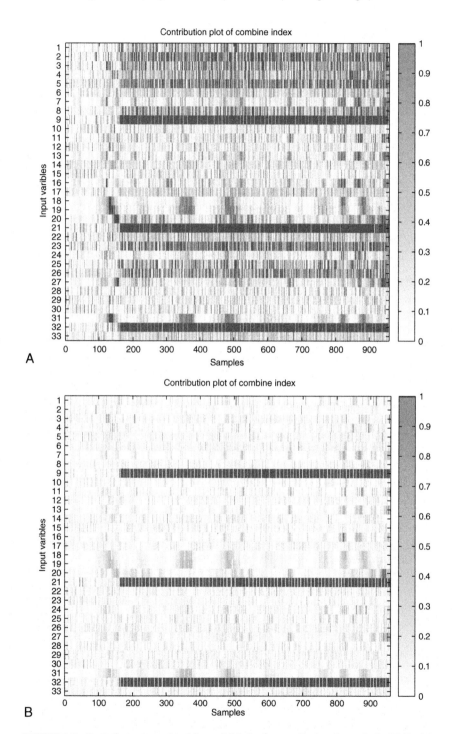

FIGURE 1.7 Fault diagnosis results of the partial derivative contribution plot methods. (A) Partial derivative contribution plot (B) Reconstructed partial derivative contribution plot.

x_{32} and rarely appear in other variables, which is consistent with that shown in Fig. 1.6.

Then the correct diagnosis rate (CDR) index is used to describe the diagnosis effect. This index can be calculated as

$$CDR = \frac{N_c}{N_f}, \qquad (1.136)$$

where N_c is the number of correctly diagnosed faulty samples in total and N_f is the number of all diagnosed faulty samples. Obviously, the larger the CDR, the better the fault diagnosis effect. Table 1.4 provides the CDRs of the partial derivative contribution plot and the reconstructed method. The CDR data of the partial derivative contribution plot is from Li et al. [61]. It is obvious that the CDRs obtained from the reconstructed partial derivative contribution plot in the first column are much larger than that from the original partial derivative method in the second column, which illustrates that the reconstruction can eliminate the smearing effect and improve the accuracy of the diagnosis.

Appendix A
Description of the variables and faults

TABLE A.1.1 Labels and descriptions of the manipulated variables of TEP

Variable no.	Description
XMV(1)	D feed flow
XMV(2)	E feed flow
XMV(3)	A feed flow
XMV(4)	A and C feed flow
XMV(5)	Compressor recycle valve
XMV(6)	Purge valve
XMV(7)	Separator pot liquid flow
XMV(8)	Stripper liquid product flow
XMV(9)	Stripper steam valve
XMV(10)	Reactor cooling water flow
XMV(11)	Condenser cooling water flow

TABLE A.1.2 Labels and descriptions of the process variables of TEP

Variable no.	Description
XMEAS(1)	A feed
XMEAS(2)	D feed
XMEAS(3)	E feed
XMEAS(4)	A and C feed
XMEAS(5)	Recycle flow
XMEAS(6)	Reactor feed rate
XMEAS(7)	Reactor pressure
XMEAS(8)	Reactor level
XMEAS(9)	Reactor temperature
XMEAS(10)	Purge rate
XMEAS(11)	Product separator temperature
XMEAS(12)	Product separator level
XMEAS(13)	Product separator pressure
XMEAS(14)	Product separator underflow
XMEAS(15)	Stripper level
XMEAS(16)	Stripper pressure
XMEAS(17)	Stripper underflow
XMEAS(18)	Stripper temperature
XMEAS(19)	Stripper steam flow
XMEAS(20)	Compressor work
XMEAS(21)	Reactor cooling water outlet temperature
XMEAS(22)	Separator cooling water outlet temperature

TABLE A.1.3 Labels and descriptions of the analysis variables of TEP

Variable no.	Component	Description
XMEAS(23)	A	Reactor feed analysis
XMEAS(24)	B	Reactor feed analysis
XMEAS(25)	C	Reactor feed analysis
XMEAS(26)	D	Reactor feed analysis
XMEAS(27)	E	Reactor feed analysis
XMEAS(28)	F	Reactor feed analysis
XMEAS(29)	A	Purge gas analysis
XMEAS(30)	B	Purge gas analysis
XMEAS(31)	C	Purge gas analysis
XMEAS(32)	D	Purge gas analysis
XMEAS(33)	E	Purge gas analysis
XMEAS(34)	F	Purge gas analysis
XMEAS(35)	G	Purge gas analysis
XMEAS(36)	H	Purge gas analysis
XMEAS(37)	D	Product analysis
XMEAS(38)	E	Product analysis
XMEAS(39)	F	Product analysis
XMEAS(40)	G	Product analysis
XMEAS(41)	H	Product analysis

TABLE A.1.4 Description of the faults

Variable	Description	Type
IDV(1)	A/C feed ratio, B composition constant (stream 4)	Step
IDV(2)	B composition (stream 4)	Step
IDV(3)	D feed temperature (stream 2)	Step
IDV(4)	Reactor cooling water inlet temperature	Step
IDV(5)	Condenser cooling water inlet temperature	Step
IDV(6)	A feed loss	Step
IDV(7)	C header pressure loss-reduced	Step
IDV(8)	A, B, C feed composition (stream 4)	Random variation
IDV(9)	D feed temperature (stream 2)	Random variation
IDV(10)	C feed temperature (stream 4)	Random variation
IDV(11)	Reactor cooling water inlet temperature	Random variation
IDV(12)	Condenser cooling	Random variation
IDV(13)	Reaction kinetics	Slow Drift
IDV(14)	Reactor cooling water valve	Sticking
IDV(15)	Condenser cooling	Sticking
IDV(16)	Unknown	Unknown
IDV(17)	Unknown	Unknown
IDV(18)	Unknown	Unknown
IDV(19)	Unknown	Unknown
IDV(20)	Unknown	Unknown

References

[1] S.J. Qin, Survey on data-driven industrial process monitoring and diagnosis, *Annual Reviews in Control* 36 (2) (2012) 220–234.

[2] Z. Ge, Z. Song, F. Gao, Review of recent research on data-based process monitoring, *Industrial & Engineering Chemistry Research* 52 (10) (2013) 3543–3562.

[3] Z. Ge, Z. Song, S.X. Ding, B. Huang, Data mining and analytics in the process industry: The role of machine learning, *IEEE Access* 5 (2017) 20590–20616.

[4] Z. Ge, Review on data-driven modeling and monitoring for plant-wide industrial processes, *Chemometrics & Intelligent Laboratory Systems* 171 (2017) 16–25.

[5] K. Peng, K. Zhang, G. Li, Quality-related process monitoring based on total kernel PLS model and its industrial application, *Mathematical Problems in Engineering* 2013 (2013) 707953.

[6] G. Wang, J. Jiao, A kernel least squares based approach for nonlinear quality-related fault detection, *IEEE Transactions on Industrial Electronics* 64 (4) (2016) 3195–3204.

[7] C. Hu, Z. Xu, X. Kong, J. Luo, Recursive-CPLS-based quality-relevant and process-relevant fault monitoring with application to the Tennessee Eastman process, *IEEE Access* 7 (2019) 128746–128757.

[8] K. Zhang, H. Hao, Z. Chen, S.X. Ding, K. Peng, A comparison and evaluation of key performance indicator-based multivariate statistics process monitoring approaches, *Journal of Process Control* 33 (2015) 112–126.

[9] S. Yin, S.X. Ding, A. Haghani, H. Hao, P. Zhang, A comparison study of basic data-driven fault diagnosis and process monitoring methods on the benchmark Tennessee Eastman process, *Journal of Process Control* 22 (9) (2012) 1567–1581.

[10] P. Nomikos, J.F. MacGregor, Multivariate SPC charts for monitoring batch processes, *Technometrics* 37 (1) (1995) 41–59.

[11] S. JoeÂ Qin, Statistical process monitoring: Basics and beyond, *Journal ofChemometrics* 17 (8-9) (2003) 480–502.

[12] D. Zhou, G. Li, S.J. Qin, Total projection to latent structures for process monitoring, *AIChE Journal* 56 (1) (2010) 168–178.

[13] S.J. Qin, Y. Zheng, Quality-relevant and process-relevant fault monitoring with concurrent projection to latent structures, *AIChE Journal* 59 (2) (2013) 496–504.

[14] Q. Liu, S.J. Qin, T. Chai, Multiblock concurrent PLS for decentralized monitoring of continuous annealing processes, *IEEE Transactions on Industrial Electronics* 61 (11) (2014) 6429–6437.

[15] S.X. Ding, S. Yin, K. Peng, H. Hao, B. Shen, A novel scheme for key performance indicator prediction and diagnosis with application to an industrial hot strip mill, *IEEE Transactions on Industrial Informatics* 9 (4) (2013) 2239–2247.

[16] K. Peng, K. Zhang, B. You, J. Dong, Quality-relevant fault monitoring based on efficient projection to latent structures with application to hot strip mill process, *IET Control TheoryApplications* 9 (7) (2015) 1135–1145.

[17] K. Peng, K. Zhang, J. Dong, B. You, Quality-relevant fault detection and diagnosis for hot strip mill process with multi-specification and multi-batch measurements, *Journal of the Franklin Institute* 352 (3) (2015) 987–1006.

[18] G. Wang, H. Luo, K. Peng, Quality-related fault detection using linear and nonlinear principal component regression, *Journal of the Franklin Institute* 353 (10) (2016) 2159–2177.

[19] Z. Chen, S.X. Ding, K. Zhang, Z. Li, Z. Hu, Canonical correlation analysis-based fault detection methods with application to alumina evaporation process, *Control Engineering Practice* 46 (2016) 51–58.

[20] Z. Chen, K. Zhang, S.X. Ding, Y.A. Shardt, Z. Hu, Improved canonical correlation analysis-based fault detection methods for industrial processes, *Journal of Process Control* 41 (2016) 26–34.

[21] Q. Zhu, Q. Liu, S.J. Qin, Concurrent quality and process monitoring with canonical correlation analysis, *Journal of Process Control* 60 (2017) 95–103.

[22] Q. Zhu, S.J. Qin, Supervised diagnosis of quality and process faults with canonical correlation analysis, *Industrial & Engineering Chemistry Research* 58 (26) (2019) 11213–11223.

[23] B. Schölkopf, A. Smola, K.-R. Möller, Nonlinear component analysis as a kernel eigenvalue problem, *Neural Computation* 10 (5) (1998) 1299–1319.

[24] J.-H. Cho, J.-M. Lee, S.W. Choi, D. Lee, I.-B. Lee, Fault identification for process monitoring using kernel principal component analysis, *Chemical Engineering Science* 60 (1) (2005) 279–288.

[25] R. Rosipal, L.J. Trejo, Kernel partial least squares regression in reproducing kernel Hilbert space, *Journal of Machine Learning Research* 2 (2001) 97–123.

[26] K. Peng, K. Zhang, G. Li, D. Zhou, Contribution rate plot for nonlinear quality-related fault diagnosis with application to the hot strip mill process, *Control Engineering Practice* 21 (4) (2013) 360–369.

[27] G. Wang, J. Jiao, S. Yin, A kernel direct decomposition-based monitoring approach for nonlinear quality-related fault detection, *IEEE Transactions on Industrial Informatics* 13 (4) (2016) 1565–1574.

[28] G. Li, S.J. Qin, Comparative study on monitoring schemes for non-Gaussian distributed processes, *Journal of Process Control* 67 (2018) 69–82.

[29] M. Kano, S. Tanaka, S. Hasebe, I. Hashimoto, H. Ohno, Monitoring independent components for fault detection, *AIChEJournal* 49 (4) (2003) 1–8.

[30] J.-M. Lee, C. Yoo, I.-B. Lee, Statistical process monitoring with independent component analysis, *Journal of Process Control* 14 (5) (2004) 467–485.

[31] R. Gonzalez, B. Huang, E. Lau, Process monitoring using kernel density estimation and Bayesian networking with an industrial case study, *ISA Transactions* 58 (2015) 330–347.

[32] J. Yu, A new fault diagnosis method of multimode processes using Bayesian inference based Gaussian mixture contribution decomposition, *Engineering Applications of Artificial Intelligence* 26 (1) (2013) 456–466.

[33] S.W. Choi, J.H. Park, I.-B. Lee, Process monitoring using a Gaussian mixture model via principal component analysis and discriminant analysis, *Computers & Chemical Engineering* 28 (8) (2004) 1377–1387.

[34] Q. Jiang, B. Huang, X. Yan, GMM and optimal principal components-based Bayesian method for multimode fault diagnosis, *Computers & Chemical Engineering* 84 (2016) 338–349.

[35] X. Liu, L. Xie, U. Kruger, T. Littler, S. Wang, Statistical-based monitoring of multivariate non-Gaussian systems, *AIChE Journal* 54 (9) (2008) 2379–2391.

[36] Z. Ge, Z. Song, A distribution-free method for process monitoring, *Expert Systems with Applications* 38 (8) (2011) 9821–9829.

[37] S. Ding, Data-driven design of monitoring and diagnosis systems for dynamic processes: A review of subspace technique based schemes and some recent results, *Journal of Process Control* 24 (2) (2014) 431–449.

[38] Y. Dong, S.J. Qin, Dynamic latent variable analytics for process operations and control, *Computers & Chemical Engineering* 114 (2018) 69–80.

[39] W. Ku, R.H. Storer, C. Georgakis, Disturbance detection and isolation by dynamic principal component analysis, *Chemometrics & Intelligent Laboratory Systems* 30 (1) (1995) 179–196.

[40] W. Li, S. Qin, Consistent dynamic PCA based on errors-in-variables subspace identification, *Journal of Process Control* 11 (6) (2001) 661–678.

[41] Y. Dong, S.J. Qin, A novel dynamic PCA algorithm for dynamic data modeling and process monitoring, *Journal of Process Control* 67 (2018) 1–11.

[42] Y. Dong, S.J. Qin, Dynamic-inner canonical correlation and causality analysis for high dimensional time series data, *IFAC-PapersOnLine* 51 (18) (2018) 476–481.

[43] J. Huang, X. Yan, Dynamic process fault detection and diagnosis based on dynamic principal component analysis, dynamic independent component analysis and Bayesian inference, *Chemometrics & Intelligent Laboratory Systems* 148 (2015) 115–127.

[44] K. Helland, H.E. Berntsen, O.S. Borgen, H. Martens, Recursive algorithm for partial least squares regression, *Chemometrics & Intelligent Laboratory Systems* 14 (1) (1992) 129–137.

[45] S.J. Qin, Recursive PLS algorithms for adaptive data modeling, *Computers & Chemical Engineering* 22 (4) (1998) 503–514.

[46] Y. Dong, S.J. Qin, Dynamic-inner partial least squares for dynamic data modeling, *IFAC-PapersOnLine* 48 (8) (2015) 117–122.

[47] C.F. Alcala, S.J. Qin, Analysis and generalization of fault diagnosis methods for process monitoring, *Journal of Process Control* 21 (3) (2011) 322–330.

[48] S.W. Choi, I.-B. Lee, Multiblock PLS-based localized process diagnosis, *Journal of Process Control* 15 (3) (2005) 295–306.

[49] C.F. Alcala, S.J. Qin, Reconstruction-based contribution for process monitoring, *Automatica* 45 (7) (2009) 1593–1600.

[50] G. Li, S.J. Qin, T. Chai. Multi-directional reconstruction based contributions for root-cause diagnosis of dynamic processes. In Proceedings of the 2014 American Control Conference. 3500â¬"3505.

[51] G. Li, T. Yuan, S.J. Qin, T. Chai, Dynamic time warping based causality analysis for root-cause diagnosis of nonstationary fault processes, *IFAC-PapersOnLine* 48 (8) (2015) 1288–1293.

[52] S. Yoon, J.F. MacGregor, Fault diagnosis with multivariate statistical models Part I: Using steady state fault signatures, *Journal of Process Control* 11 (4) (2001) 387–400.

[53] J. Liu, Fault diagnosis using contribution plots without smearing effect on non-faulty variables, *Journal of Process Control* 22 (9) (2012) 1609–1623.

[54] J. Liu, D.S.H. Wong, D.-S. Chen, Bayesian filtering of the smearing effect: Fault isolation in chemical process monitoring, *Journal of Process Control* 24 (3) (2014) 1–21.

[55] Y. Zhang, L. Zhang, R. Lu, Fault identification of nonlinear processes, Industrial & Engineering Chemistry Research 52 (34) (2013) 12072–12081.

[56] G. Wang, J. Jiao, S. Yin, Efficient nonlinear fault diagnosis based on kernel sample equivalent replacement, *IEEE Transactions on Industrial Informatics* 15 (5) (2018) 2682–2690.

[57] Y. Zhang, R. Sun, Y. Fan, Fault diagnosis of nonlinear process based on KCPLS reconstruction, *Chemometrics & Intelligent Laboratory Systems* 140 (2015) 49–60.

[58] J.J. Downs, E.F. Vogel, A plant-wide industrial process control problem, *Computers & Chemical Engineering* 17 (3) (1993) 245–255.

[59] L.H. Chiang, E.L. Russell, R.D. Braatz, Fault diagnosis in chemical processes using Fisher discriminant analysis, discriminant partial least squares, and principal component analysis, *Chemometrics & Intelligent Laboratory Systems* 50 (2) (2000) 243–252.

[60] D. Zhou, G. Li, S.J. Qin, Total projection to latent structures for process monitoring, *AiChE Journal* 56 (1) (2009) 168–178.

[61] G. Li, T. Yuan, S.J. Qin, T. Chai, Dynamic time warping based causality analysis for root-cause diagnosis of nonstationary fault processes, *IFAC-PapersOnLine* 48 (8) (2015) 1288–1293.

Chapter 2

Canonical correlation analysis–based fault diagnosis method for dynamic processes

Zhiwen Chen and Ketian Liang
School of Automation, Central South University, China

2.1 Introduction

With the rapid development of a decentralized control system and the other enabling techniques, such as sensor and computational power, a huge amount of data can be easily obtained in modern industrial processes for control, monitoring, and predictive maintenance tasks, among others [1, 2]. Due to the physical interaction and the role of feedback control, dependence is ubiquitous among the collected variables. In statistics, dependence is any statistical relationship between two random variables. In the broadest sense, correlation is any statistical dependence, although it commonly refers to the degree of which a pair of variables are linearly related [3]. With regard to the fault diagnosis task, correlations are useful because they can indicate a predictive relationship that can be exploited in practice. Over the past decades, multivariate analysis (MVA), which considers the correlation among random variables, has been widely explored in the process monitoring and fault diagnosis community [4]. Parallel with principal component analysis (PCA) and partial least squares, as well as independent component analysis, canonical correlation analysis (CCA) as a typical MVA technique has been widely used for process monitoring and fault diagnosis [5]. The main difference between CCA and the other MVA methods is that CCA is closely related to mutual information [6].

CCA was first introduced by Harold Hotelling in 1936 to study relations between two sets of variables [7], but the motivation is not for industrial data analysis. In an attempt to calculate the amount of information about a random function contained in another random function, CCA has been used by Gelfand and Yaglom [8] with a proper definition of the amount of information. Later, Akaike [9] analyzed the structure of the information interface between the

Fault Diagnosis and Prognosis Techniques for Complex Engineering Systems. DOI: 10.1016/B978-0-12-822473-1.00004-5

future and the past of a discrete-time stochastic processes using CCA. The CCA technique used in Akaike's work is the same as the conventional CCA; however, a study of system dynamics forms a point of departure from the conventional MVA to the dynamic time series analysis. With the aid of the Akaike information criterion for the determination of the state order, Larimore [10] then explored CCA for system identification, filtering, and control of a general linear system represented by state space models. Unlike the CCA method in multivariate statistical analysis, Larimore's method called *canonical variate analysis* (CVA) is more general and involves solving a closely related reduced rank prediction problem, and the solution is given by a CCA of the covariance of past and future. Based on the work of Larimore, the CVA method was explored for process monitoring and fault diagnosis [11–14]. For such a task, the basic idea is to first identify a state space model of the system of interest and then look for abnormal behavior either within the state space or behavior that significantly departs from the state space. For example, Chiang et al. [5] used the CVA-based fault detection method for monitoring a chemical process. In CVA-based process monitoring methods, a state space model should be identified first, and the abnormalities are strictly divided into two types: the one affects the state space, and the other departs from the state space. On one hand, the abnormalities may hardly be strictly divided since the faults are usually unknown and the occurrence of faults may change the process dynamics. On the other hand, for the process monitoring task, a state space model may not need to be identified. Motivated by this discussion, Chen et al. [15] proposed two CCA-based process monitoring methods to detect faults for static and dynamic processes. In the so-called static process, the random variables are assumed to be independent and identical distribution, whereas in the dynamic process, the random variables are autocorrelated—that is, the future variables are correlated with the past and present variables. Instead of identifying a state space model, a residual generator is constructed via the CCA algorithm for obtaining the required parameters.

Except for dynamic characteristic, most processes are nonlinear in nature. Therefore, the conventional process monitoring methods are inappropriate for monitoring such nonlinear dynamic processes. To extend the CCA method to handle the nonlinear characteristic, much information can be found in the literature. For example, in the work of Odiowei and Cao [13], to handle the non-Gaussian issue induced by a nonlinear process, a method based on CVA and the kernel density estimation (KDE) method was proposed. Motivated by the limitation of KDE for determining the threshold, a randomized algorithm was studied to determine an appropriate threshold and was combined with the CCA method to handle the non-Gaussian problem [16]. The other CCA-based advanced fault diagnosis methods can be found in other works [17–20]. Benefiting from nonlinear representation learning ability, the deep neural network (DNN) has been widely explored [21, 22]. Among the methods, CCA combined with DNN has attracted a great deal of attention. In the pioneering work of

Galen Andrew, a deep CCA method was proposed to learn complex nonlinear transformations of two views of data such that the resulting representations are highly linearly correlated [23]. In this deep CCA method, two deep feed-forward neural networks are used. Then a deep canonically correlated autoencoder method was proposed for an extension of dynamical CCA (DCCA) to include the reconstruction objective of multimodal autoencoders with the correlation objective [24]. To improve temporal relationship learning performance, the deep canonically correlated long short-term memory (LSTM) methods were proposed in the work of Mallinar and Rosset [25]. Motivated by the success of the DNN-based CCA method in the multiview learning community, recently a regularized deep correlated representation method that combines two deep belief networks and CCA was proposed for monitoring nonlinear processes [26]. It can be observed that the DNN-based CCA methods are a hot topic and an alternative method for dynamic process fault diagnosis. Therefore, in this work, we introduce two types of CCA-based methods for dynamic processes: one is a DCCA for dynamic processes in steady state, and the other is a DNN-aided CCA method for nonlinear dynamic processes. The performance of the CCA-based fault detection methods will be demonstrated and compared to the conventional CCA method. Two applications will be presented: one using continuously stirred tank reactor (CSTR) data and the other using the traction drive and control system (TDCS) of a high-speed train.

The rest of this chapter is organized as follows. In Section 2.2, the background of CCA-based fault diagnosis is introduced. Two CCA-based fault diagnosis methods for dynamic processes are presented in Section 2.3. In Section 2.4, the introduced methods are validated through two industrial benchmark cases, and furthermore, a comparative study of the presented methods to conventional CCA method is performed. We provide our conclusion in Section 2.5.

2.2 Preliminaries

2.2.1 Basics of conventional CCA

CCA is one of the most well known MVA techniques and is widely used in various disciplines, such as in dimensionality reduction, semantic representation learning, system identification, filtering, and fault detection [27–29]. CCA is a method of correlating linear relationships between two random variables, which could be multidimensional. In mathematics, CCA is used to find basis vectors for two sets of variables such that the correlations between the projections of the variables onto these basis vectors are mutually maximized.

Assume that $u \in \mathbb{R}^l$, $y \in \mathbb{R}^m$ are two set of random vectors with dimensions of l and m, respectively. By collecting N observations of each random vector, we have two data matrices $\mathbf{U} \in \mathbb{R}^{N \times l}$ and $\mathbf{Y} \in \mathbb{R}^{N \times m}$, respectively. The observations are supposed to be jointly sampled from a normal multivariate distribution. Without loss of generality, we assume that the variables are zero mean. In other words,

$$\begin{bmatrix} u \\ y \end{bmatrix} \sim \mathbb{N}\left(\begin{bmatrix} 0 \\ 0 \end{bmatrix}, \begin{bmatrix} \Sigma_u & \Sigma_{uy} \\ \Sigma_{yu} & \Sigma_y \end{bmatrix} \right), \quad \Sigma_{yu} = \Sigma_{uy}^T. \tag{2.1}$$

Let the variable vectors of the n observations be the column vectors $\boldsymbol{u}_i \in \mathbb{R}^n$ for $i = 1, 2, \ldots, l$ and $\boldsymbol{y}_i \in \mathbb{R}^n$ for $i = 1, 2, \ldots, m$, respectively. $\langle u, y \rangle$ or $u^T y$ denotes the inner produce between two vectors. In CCA, the purpose is to find the linear relations between the variables of $\mathbf{U}$ and $\mathbf{Y}$. We consider the following linear transformations,

$$\mathbf{U}\boldsymbol{w}_u = \boldsymbol{z}_u \text{ and } \mathbf{Y}\boldsymbol{w}_y = \boldsymbol{z}_y \tag{2.2}$$

where $\boldsymbol{w}_u \in \mathbb{R}^l$, $\boldsymbol{z}_u \in \mathbb{R}^N$, $\boldsymbol{w}_y \in \mathbb{R}^m$, and $\boldsymbol{z}_y \in \mathbb{R}^N$. Following the work of Uurtio et al. [28], the matrices $\mathbf{U}$ and $\mathbf{Y}$ represent linear transformations of the positions $\boldsymbol{w}_u$ and $\boldsymbol{w}_y$ onto the images $\boldsymbol{z}_u$ and $\boldsymbol{z}_y$ in the space $\mathbb{R}^n$. According to Golub and Zha [30], the solution of Eq. (2.2) when it comes to CCA should satisfy that the position vectors $\boldsymbol{w}_u$ and $\boldsymbol{w}_y$ are unit norm vectors and that the enclosing angle, $\theta \in \left[0, \frac{\pi}{2}\right]$, between $\boldsymbol{z}_u$ and $\boldsymbol{z}_y$ is minimized. In this case, the cosine of the angle, also referred to as the canonical correlations, between the images $\boldsymbol{z}_u$ and $\boldsymbol{z}_y$ is given by the formula

$$\cos\left(z_u, z_y\right) = \langle z_u, z_y \rangle / \|z_u\| \|z_y\| \tag{2.3}$$

and due to the unit norm constraint $\cos(z_u, z_y) = \langle z_u, z_y \rangle$.

Therefore, the basic idea of CCA is to find two positions $\boldsymbol{w}_u \in \mathbb{R}^l$ and $\boldsymbol{w}_y \in \mathbb{R}^m$ that after the linear transformations $\mathbf{U} \in \mathbb{R}^{N \times l}$ and $\mathbf{Y} \in \mathbb{R}^{N \times m}$ are mapped onto an N-dimensional unit ball and located in such a way that the cosine of the angle between the position vectors of their images $z_u \in \mathbb{R}^N$ and $z_y \in \mathbb{R}^N$ is maximized, which results in the smallest angle, θ_1, determine the first canonical correlation that equals $\cos\theta_1$ [31].

We have that

$$\cos\theta_1 = \max_{z_u, z_y \in \mathbb{R}^N} \langle z_u, z_y \rangle$$
$$\|z_u\|_2 = 1 \|z_y\|_2 = 1. \tag{2.4}$$

Let the maximum be obtained by z_u^1 and z_y^1. The second smallest enclosing angle θ_2 is obtained by the pair of images z_u^2 and z_y^2, which can be obtained in the orthogonal complements of z_u^1 and z_y^1. The procedure to find the remaining pairs of images is stopped until no more pairs can be found. Thus, κ angles $\theta_\kappa \in \left[0, \frac{\pi}{2}\right]$

for $\kappa = 1, 2,\ldots,\min\,(l,\,m)$ can be obtained recursively as given by

$$
\begin{aligned}
&\cos\theta_\kappa = \underset{z_u,z_y\in\mathbb{R}^N}{\max}\left\langle z_u^\kappa, z_y^\kappa \right\rangle \\
&\left\| z_u^\kappa \right\|_2 = 1 \left\| z_y^\kappa \right\|_2 = 1 \\
&\left\langle z_u^\kappa, z_u^j \right\rangle = 0 \left\langle z_y^\kappa, z_y^j \right\rangle = 0 \\
&\forall j \neq \kappa : j,\kappa = 1, 2, \cdots, \min\,(l, m).
\end{aligned}
\tag{2.5}
$$

The dimensionality of CCA equals the number of canonical correlations κ. In summary, the principle behind CCA is to respectively find two positions in the two data spaces that have images on a unit ball such that the angle between them is minimized and consequently the canonical correlation is maximized. The number of relevant positions can be determined by analyzing the values of the canonical correlations or by applying statistically significance tests [32].

2.2.2 Obtaining the positions and images in CCA

It is known that to obtain the position vectors w_u and w_y can use techniques from functional analysis, in which the eigenvalue-based methods are widely used depend on various demands. There are three well-accepted eigenvalue-based methods: the methods originally proposed by Hotelling [7], a generalized eigenvalue problem [29], and singular value decomposition (SVD) as presented in the work of Healy [33]. Among them, the SVD-based solution is computationally more tractable for very large datasets. Regarding the efficiency of computation, in this chapter the SVD-based technique is used, which will be briefly introduced in the sequel.

As proposed by Hotelling, both the positions w_u and w_y and images z_u and z_y are obtained by solving a standard eigenvalue problem. The sample covariance matrices Σ_u, Σ_y, and Σ_{uy} in Eq. (2.1) can be obtained as

$$
\begin{aligned}
\Sigma_u &\approx \tfrac{1}{N-1}\mathbf{U}^T\mathbf{U} \\
\Sigma_y &\approx \tfrac{1}{N-1}\mathbf{Y}^T\mathbf{Y} \\
\Sigma_{uy} &\approx \tfrac{1}{N-1}\mathbf{U}^T\mathbf{Y}.
\end{aligned}
\tag{2.6}
$$

The joint covariance matrix is the one given in Eq. (2.1):

$$
\begin{pmatrix} \Sigma_u & \Sigma_{uy} \\ \Sigma_{yu} & \Sigma_y \end{pmatrix}.
\tag{2.7}
$$

The first and greatest canonical correlation that corresponds to the smallest angle is between the first pair of images $z_u = \mathbf{U}w_u$ and $z_y = \mathbf{Y}w_y$. Since the correlation between images z_u and z_y is scale invariant, we can constraint w_u and w_y to be such that z_u and z_y have unit variance. In other words,

$$
z_u^T z_u = w_u^T \mathbf{U}^T \mathbf{U} w_u = w_u^T \Sigma_u w_u = 1,
\tag{2.8}
$$

$$z_y^T z_y = w_y^T \mathbf{Y}^T \mathbf{Y} w_y = w_y^T \boldsymbol{\Sigma}_y w_y = 1. \tag{2.9}$$

Because of the normality assumption, the variables of $\mathbf{U}$ and $\mathbf{Y}$ should be zero means. In such a case, the covariance between two images is given by

$$z_u^T z_y = w_u^T \mathbf{U}^T \mathbf{Y} w_y = w_u^T \boldsymbol{\Sigma}_{uy} w_y. \tag{2.10}$$

Substituting Eq. (2.8), Eq. (2.9), and Eq. (2.10) into the algebraic problem in Eq. (2.4), we have

$$\cos\theta = \max_{z_u, z_y \in \mathbb{R}^N} \langle z_u, z_y \rangle = \max_{w_u \in \mathbb{R}^l, w_y \in \mathbb{R}^m} \langle w_u^T \boldsymbol{\Sigma}_{uy} w_y \rangle$$

$$\|z_u\|_2 = \sqrt{w_u^T \boldsymbol{\Sigma}_u w_u} = 1 \, \|z_y\|_2 = \sqrt{w_y^T \boldsymbol{\Sigma}_y w_y} = 1. \tag{2.11}$$

In the work of Healy [33], the technique of applying SVD to solve the CCA problem was first presented, and later the detailed solutions were described by Ewerbring and Luk [34], which are given as follows. First, the covariance matrices $\boldsymbol{\Sigma}_u$ and $\boldsymbol{\Sigma}_y$ are transformed into identity forms. Based on the symmetric positive definite property, the square roots of the matrices can be found using a Cholesky or eigenvalue decomposition:

$$\boldsymbol{\Sigma}_u = \boldsymbol{\Sigma}_u^{1/2} \boldsymbol{\Sigma}_u^{1/2} \text{ and } \boldsymbol{\Sigma}_y = \boldsymbol{\Sigma}_y^{1/2} \boldsymbol{\Sigma}_y^{1/2}. \tag{2.12}$$

Multiplying the inverses of the square root factors in both sides of the joint covariance matrix (Eq. 2.7), we have

$$\begin{pmatrix} \boldsymbol{\Sigma}_u^{-1/2} & 0 \\ 0 & \boldsymbol{\Sigma}_y^{-1/2} \end{pmatrix} \begin{pmatrix} \boldsymbol{\Sigma}_u & \boldsymbol{\Sigma}_{uy} \\ \boldsymbol{\Sigma}_{yu} & \boldsymbol{\Sigma}_y \end{pmatrix} \begin{pmatrix} \boldsymbol{\Sigma}_u^{-1/2} & 0 \\ 0 & \boldsymbol{\Sigma}_y^{-1/2} \end{pmatrix}$$

$$= \begin{pmatrix} \mathbf{I}_l & \boldsymbol{\Sigma}_u^{-1/2} \boldsymbol{\Sigma}_{uy} \boldsymbol{\Sigma}_y^{-1/2} \\ \boldsymbol{\Sigma}_y^{-1/2} \boldsymbol{\Sigma}_{yu} \boldsymbol{\Sigma}_u^{-1/2} & \mathbf{I}_m \end{pmatrix}. \tag{2.13}$$

Doing an SVD on the following matrix,

$$\boldsymbol{\Sigma}_u^{-1/2} \boldsymbol{\Sigma}_{uy} \boldsymbol{\Sigma}_y^{-1/2} = \mathbf{Q}^T \mathbf{S} \mathbf{V}, \tag{2.14}$$

where the columns of the matrices $\mathbf{Q}$ and $\mathbf{V}$ correspond to the sets of orthonormal left and right singular vectors, respectively. The canonical correlations are the singular values of matrix $\mathbf{S}$. The position vectors w_u and w_y can be obtained as

$$w_u = \boldsymbol{\Sigma}_u^{-1/2} \mathbf{Q} \text{ and } w_y = \boldsymbol{\Sigma}_y^{-1/2} \mathbf{V}. \tag{2.15}$$

It should be noted that there are also some alternatives to solve the CCA problem. In general, the main motivation to improve the eigenvalue-based technique contributes to the computational complexity. The time complexities of the standard and generalized eigenvalue methods are scale with the cube of the input matrix dimension—that is, $O(N^3)$ for a matrix of size $N \times N$. The matrix $\boldsymbol{\Sigma}_u^{-1/2} \boldsymbol{\Sigma}_{uy} \boldsymbol{\Sigma}_y^{-1/2}$ in the SVD-based solution is rectangular, whose time complexity is $O(MN^2)$, for a matrix of size $M \times N$.

2.2.3 Details of the SVD-based technique

After doing SVD of matrix $\boldsymbol{\Sigma}_u^{-1/2}\boldsymbol{\Sigma}_{uy}\boldsymbol{\Sigma}_y^{-1/2}$, we obtain three matrices: $\mathbf{Q}, \mathbf{S}$, and $\mathbf{V}$. They can be expressed as $\mathbf{Q} = (\boldsymbol{q}_1,\ldots, \boldsymbol{q}_l)$, $\mathbf{V} = (\boldsymbol{v}_1,\ldots, \boldsymbol{v}_m)$, $\mathbf{S} = \begin{bmatrix} \mathbf{S}_\kappa & 0 \\ 0 & 0 \end{bmatrix}$, $\mathbf{QQ}^T = \mathbf{I}_l$, and $\mathbf{VV}^T = \mathbf{I}_m$, where $\kappa = rank(\boldsymbol{\Sigma}_{uy})$ denotes the number of nonzero eigenvalues,

$$\mathbf{S}_\kappa = diag(\delta_1, \delta_2, \cdots, \delta_\kappa), \ 1 \geq \delta_1 \geq \delta_2 \geq \cdots \geq \delta_\kappa \geq 0.$$

Note that $\boldsymbol{w}_u = \boldsymbol{\Sigma}_u^{-1/2}\mathbf{Q} \in \mathbb{R}^{l \times l}$, $\boldsymbol{w}_u = \begin{bmatrix} \boldsymbol{w}_u^1, \cdots, \boldsymbol{w}_u^l \end{bmatrix}$, $\boldsymbol{w}_y = \boldsymbol{\Sigma}_y^{-1/2}\mathbf{V} \in \mathbb{R}^{m \times m}$, and $\boldsymbol{w}_y = \begin{bmatrix} \boldsymbol{w}_y^1, \cdots, \boldsymbol{w}_y^m \end{bmatrix}$. It is known that

$$\boldsymbol{w}_u^T\boldsymbol{\Sigma}_u\boldsymbol{w}_u = \mathbf{I}_l, \ \boldsymbol{w}_y^T\boldsymbol{\Sigma}_y\boldsymbol{w}_y = \mathbf{I}_m, \tag{2.16}$$

$$\boldsymbol{w}_u^T\boldsymbol{\Sigma}_{uy}\boldsymbol{w}_y = \mathbf{S} = \begin{bmatrix} diag(\delta_1, \cdots, \delta_\kappa) & 0 \\ 0 & 0 \end{bmatrix}. \tag{2.17}$$

Definition 1. *Given random vectors $\boldsymbol{u} \in \mathbb{R}^l, \boldsymbol{y} \in \mathbb{R}^m$ satisfying Eq. (2.1) and $\boldsymbol{w}_u, \boldsymbol{w}_y$ being defined in Eq. (2.15). Then*

$$\boldsymbol{w}_u^i = \boldsymbol{\Sigma}_u^{-1/2}\boldsymbol{q}_i, \boldsymbol{w}_y^i = \boldsymbol{\Sigma}_y^{-1/2}\boldsymbol{v}_i, i = 1, \cdots, \kappa \tag{2.18}$$

are called canonical correlation vectors,

$$z_u^i = \mathbf{U}\boldsymbol{w}_u^i, z_y^i = \mathbf{Y}\boldsymbol{w}_y^i, i = 1, \cdots, \kappa \tag{2.19}$$

are called canonical correlation variables, and $\delta_1, \delta_2, \ldots, \delta_\kappa$ are called canonical correlation coefficients.

Moreover, it holds for the canonical correlation vectors

$$\begin{aligned} \bar{\boldsymbol{w}}_u &= \begin{bmatrix} \boldsymbol{w}_u^1, \cdots, \boldsymbol{w}_u^\kappa \end{bmatrix}, \bar{\boldsymbol{w}}_y = \begin{bmatrix} \boldsymbol{w}_y^1, \cdots, \boldsymbol{w}_y^\kappa \end{bmatrix} \\ \bar{\boldsymbol{w}}_u^T\boldsymbol{\Sigma}_u\bar{\boldsymbol{w}}_u &= \mathbf{I}_\kappa, \bar{\boldsymbol{w}}_y^T\boldsymbol{\Sigma}_y\bar{\boldsymbol{w}}_y = \mathbf{I}_\kappa \\ \bar{\boldsymbol{w}}_u^T\boldsymbol{\Sigma}_{uy}\bar{\boldsymbol{w}}_y &= diag(\delta_1, \cdots, \delta_\kappa) := \mathbf{S}_\kappa. \end{aligned} \tag{2.20}$$

2.2.4 The CCA-based fault diagnosis method

Here, we introduce the basics of the CCA-based fault diagnosis method based on our previous work; most of details are referred to in the work of Chen et al. [35].

For fault diagnosis, we first define two random vectors:

$$\begin{aligned} \bar{r}_1 &= \bar{\boldsymbol{w}}_u^T\boldsymbol{u} - \mathbf{S}_\kappa\bar{\boldsymbol{w}}_y^T\boldsymbol{y} \\ \bar{r}_2 &= \bar{\boldsymbol{w}}_y^T\boldsymbol{y} - \mathbf{S}_\kappa\bar{\boldsymbol{w}}_u^T\boldsymbol{u}. \end{aligned} \tag{2.21}$$

Let $\varepsilon(\,\bullet\,)$ be the expectation operator; it turns out that

$$
\begin{aligned}
\varepsilon\left(\bar{r}_1\bar{r}_1^T\right) &= \bar{\boldsymbol{w}}_u^T \boldsymbol{\Sigma}_u \bar{\boldsymbol{w}}_u + \mathbf{S}_\kappa \bar{\boldsymbol{w}}_y^T \boldsymbol{\Sigma}_y \bar{\boldsymbol{w}}_y \mathbf{S}_\kappa - \bar{\boldsymbol{w}}_u^T \boldsymbol{\Sigma}_{uy} \bar{\boldsymbol{w}}_y \mathbf{S}_\kappa - \mathbf{S}_\kappa \bar{\boldsymbol{w}}_y^T \boldsymbol{\Sigma}_{yu} \bar{\boldsymbol{w}}_u \\
&= \mathbf{I}_\kappa - \mathbf{S}_\kappa \mathbf{S}_\kappa = diag\left(1 - \delta_1^2, \cdots, 1 - \delta_\kappa^2\right),
\end{aligned}
$$
$$(2.22)$$

$$
\begin{aligned}
\varepsilon\left(\bar{r}_2\bar{r}_2^T\right) &= \bar{\boldsymbol{w}}_y^T \boldsymbol{\Sigma}_y \bar{\boldsymbol{w}}_y + \mathbf{S}_\kappa \bar{\boldsymbol{w}}_u^T \boldsymbol{\Sigma}_u \bar{\boldsymbol{w}}_u \mathbf{S}_\kappa - \bar{\boldsymbol{w}}_y^T \boldsymbol{\Sigma}_{yu} \bar{\boldsymbol{w}}_u \mathbf{S}_\kappa - \mathbf{S}_\kappa \bar{\boldsymbol{w}}_u^T \boldsymbol{\Sigma}_{uy} \bar{\boldsymbol{w}}_y \\
&= \mathbf{I}_\kappa - \mathbf{S}_\kappa \mathbf{S}_\kappa = diag\left(1 - \delta_1^2, \cdots, 1 - \delta_\kappa^2\right).
\end{aligned}
$$
$$(2.23)$$

In general, when we use all columns of the canonical weight matrix, the random vectors can be defined as

$$
\begin{aligned}
\boldsymbol{r}_1 &= \boldsymbol{w}_u^T \boldsymbol{u} - \mathbf{S} \boldsymbol{w}_y^T \boldsymbol{y} \\
\boldsymbol{r}_2 &= \boldsymbol{w}_y^T \boldsymbol{y} - \mathbf{S}^T \boldsymbol{w}_u^T \boldsymbol{u}.
\end{aligned}
$$
$$(2.24)$$

The covariance matrices of both vectors satisfy

$$
\begin{aligned}
\varepsilon\left(\boldsymbol{r}_1\boldsymbol{r}_1^T\right) &= \boldsymbol{w}_u^T \boldsymbol{\Sigma}_u \boldsymbol{w}_u + \mathbf{S} \boldsymbol{w}_y^T \boldsymbol{\Sigma}_y \boldsymbol{w}_y \mathbf{S}^T - \boldsymbol{w}_u^T \boldsymbol{\Sigma}_{uy} \boldsymbol{w}_y \mathbf{S}^T - \mathbf{S} \boldsymbol{w}_y^T \boldsymbol{\Sigma}_{yu} \boldsymbol{w}_u \\
&= \mathbf{I}_l - \mathbf{S}\mathbf{S}^T = diag\left(1 - \delta_1^2, \cdots, 1 - \delta_\kappa^2, \underbrace{1, \cdots, 1}_{l-\kappa}\right),
\end{aligned}
$$
$$(2.25)$$

$$
\begin{aligned}
\varepsilon\left(\boldsymbol{r}_2\boldsymbol{r}_2^T\right) &= \boldsymbol{w}_y^T \boldsymbol{\Sigma}_y \boldsymbol{w}_y + \mathbf{S}^T \boldsymbol{w}_u^T \boldsymbol{\Sigma}_u \boldsymbol{w}_u \mathbf{S} - \boldsymbol{w}_y^T \boldsymbol{\Sigma}_{yu} \boldsymbol{w}_u \mathbf{S} - \mathbf{S}^T \boldsymbol{w}_u^T \boldsymbol{\Sigma}_{uy} \boldsymbol{w}_y \\
&= \mathbf{I}_m - \mathbf{S}^T \mathbf{S} = diag\left(1 - \delta_1^2, \cdots, 1 - \delta_\kappa^2, \underbrace{1, \cdots, 1}_{m-\kappa}\right),
\end{aligned}
$$
$$(2.26)$$

A comparison with

$$
\varepsilon\left(\boldsymbol{w}_u^T \mathbf{U}\mathbf{U}^T \boldsymbol{w}_u\right) = \boldsymbol{w}_u^T \boldsymbol{\Sigma}_u \boldsymbol{w}_u = \mathbf{I}_l, \quad \varepsilon\left(\boldsymbol{w}_y^T \mathbf{Y}\mathbf{Y}^T \boldsymbol{w}_y\right) = \boldsymbol{w}_y^T \boldsymbol{\Sigma}_y \boldsymbol{w}_y = \mathbf{I}_m
$$

which are the normalized covariance matrices of $\boldsymbol{u}, \boldsymbol{y}$, respectively, makes it clear that the covariance matrix of the measurement under consideration becomes smaller when the correlated measurements are taken into account. In fact, $\boldsymbol{r}_1$ and $\boldsymbol{r}_2$ can be rewritten as

$$
\boldsymbol{r}_1 = \boldsymbol{w}_u^T\left(\boldsymbol{u} - \boldsymbol{\Sigma}_{uy} \boldsymbol{w}_y \boldsymbol{w}_y^T \boldsymbol{y}\right) = \mathbf{Q}^T \boldsymbol{\Sigma}_u^{-1/2}\left(\boldsymbol{u} - \boldsymbol{\Sigma}_{uy} \boldsymbol{\Sigma}_y^{-1} \boldsymbol{y}\right),
$$
$$(2.27)$$

$$
\boldsymbol{r}_2 = \boldsymbol{w}_y^T\left(\boldsymbol{y} - \boldsymbol{\Sigma}_{yu} \boldsymbol{w}_u \boldsymbol{w}_u^T \boldsymbol{u}\right) = \mathbf{V}^T \boldsymbol{\Sigma}_y^{-1/2}\left(\boldsymbol{y} - \boldsymbol{\Sigma}_{yu} \boldsymbol{\Sigma}_u^{-1} \boldsymbol{u}\right).
$$
$$(2.28)$$

Note that

$$
\hat{\boldsymbol{u}} = \boldsymbol{\Sigma}_{uy} \boldsymbol{\Sigma}_y^{-1} \boldsymbol{y}, \quad \hat{\boldsymbol{y}} = \boldsymbol{\Sigma}_{yu} \boldsymbol{\Sigma}_u^{-1} \boldsymbol{u}
$$

are least squares estimations for $\boldsymbol{u}$, $\boldsymbol{y}$, and thus the estimation errors $\boldsymbol{u} - \hat{\boldsymbol{u}}$ and $\boldsymbol{y} - \hat{\boldsymbol{y}}$ have the minimum variances. This motivates us to use signals $\boldsymbol{r}_1$ and $\boldsymbol{r}_2$ for fault detection and estimation.

Let

$$u = f_u + \omega_u, \omega_u \sim \mathbb{N}(\mathbf{0}, \mathbf{\Sigma}_u), \tag{2.29}$$

$$y = f_y + \omega_y, \omega_y \sim \mathbb{N}(\mathbf{0}, \mathbf{\Sigma}_y), \tag{2.30}$$

be the process models for (sub)systems y, u, where f_u and f_y represent fault vectors in process measurements u and y, respectively. We assume that f_u and f_y are not present in the process simultaneously. Suppose that ω_u and ω_y are correlated with

$$\varepsilon\left(\omega_u\omega_y^T\right) = \mathbf{\Sigma}_{uy}$$

Then after determining w_u, w_y, S according to Eq. (2.18), the hypothesis testing technique can be used for the fault detection decision [36, 37].

To be concise, fault detection can be formulated as a binary hypothesis testing problem because the main objective in fault detection is to make a yes/no decision about the presence or absence of a fault. It is known that the solution to this hypothesis problem should perform a compromise between two incorrect decisions: a false alarm (i.e., false rejection of the null hypothesis, H_0) and no detection alarm (i.e., missed acceptance of the alternative hypothesis, H_1). Then we should develop a tractable statistical test to aid in making the hypothesis testing decision. The quality of a test can be characterized by two measures: the probability of a false alarm (referred as false alarm rate [FAR]) and the power function, which is the probability of deciding H_1 when H_1 is true (referred as fault detection rate [FDR]). Of course, a good fault detection can be defined as the value of FAR should be as small as possible, and the value of FDR should be as large as possible for each fault [38].

Hence, after getting the residual signals r_1 and r_2, we can first develop a statistical test and then compare the statistical test with a corresponding threshold to make a fault detection decision. The dedicated fault detection solution is as follows:

- Develop the test statistics:

$$J_u = r_1^T \mathbf{\Sigma}_{r1}^{-1} r_1, \tag{2.31}$$

where $\mathbf{\Sigma}_{r1} = \mathbf{I}_l - \mathbf{S}\mathbf{S}^T = \left(diag\left(1 - \delta_1^2, \cdots, 1 - \delta_\kappa^2, \underbrace{1, \cdots, 1}_{l-\kappa}\right)\right)$, and

$$J_y = r_2^T \mathbf{\Sigma}_{r2}^{-1} r_2, \tag{2.32}$$

where $\mathbf{\Sigma}_{r2} = \mathbf{I}_m - \mathbf{S}^T\mathbf{S} = \left(diag\left(1 - \delta_1^2, \cdots, 1 - \delta_\kappa^2, \underbrace{1, \cdots, 1}_{m-\kappa}\right)\right)$.

- Determine the thresholds: for a given acceptable FAR α:

$$J_{th,u} = \chi_\alpha^2(l), \text{prob}\{J_u > \chi_\alpha^2(l)|\text{fault-free}\} = \alpha \qquad (2.33)$$

$$J_{th,y} = \chi_\alpha^2(m), \text{prob}\{J_y > \chi_\alpha^2(m)|\text{fault-free}\} = \alpha \qquad (2.34)$$

where $\chi^2(l)$ stands for stands for the chi-square distribution with l degrees of freedom, and $\text{prob}\{J_y > \chi_{1-\alpha}^2(l)|\text{fault-free}\} = \alpha$ for the probability of $J_y > \chi_{1-\alpha}^2(l)$ equals to α(significance level) given that there is a fault-free case.

- Make the detection logic:

$$\begin{cases} J_u > J_{th,u} \Rightarrow \text{absence of fault, otherwise precense of fault} \\ J_y > J_{th,y} \Rightarrow \text{absence of fault, otherwise precense of fault.} \end{cases} \qquad (2.35)$$

It should be noted that the preceding fault detection solutions only allow a successful fault detection but do not guarantee a perfect fault isolation. This fact can be clearly seen from the following relations:

$$r_1 = w_u^T u - S w_y^T y = w_u^T f_u - S w_y^T f_y + w_u^T \omega_u - S w_y^T \omega_y. \qquad (2.36)$$

$$r_2 = w_y^T y - S^T w_u^T u = w_y^T f_y + S^T w_u^T f_u + w_y^T \omega_y - S^T w_u^T \omega_u. \qquad (2.37)$$

which means that r_1 and r_2 will be influenced by both f_u and f_y. However, it holds that

$$\begin{cases} \varepsilon(J_u) = f_u^T \Sigma_{u,1} f_u + l \\ \varepsilon(J_y) = f_u^T \Sigma_{u,2} f_u + m \end{cases} \text{for } f_u \neq 0, \ f_y = 0$$

$$\begin{cases} \varepsilon(J_y) = f_y^T \Sigma_{y,1} f_y + m \\ \varepsilon(J_u) = f_y^T \Sigma_{y,2} f_y + l \end{cases} \text{for } f_y \neq 0, \ f_u = 0$$

$$\Sigma_{u,1} = \Sigma_u^{-1/2} Q \left(diag(1 - \delta_1^2, \cdots, 1 - \delta_\kappa^2, 1, \cdots, 1)\right)^{-1} Q^T \Sigma_u^{-1/2},$$

$$\Sigma_{u,2} = \Sigma_u^{-1/2} Q \left(diag\left(\frac{\delta_1^2}{1 - \delta_1^2}, \cdots, \frac{\delta_\kappa^2}{1 - \delta_\kappa^2}, 0, \cdots, 0\right)\right)^{-1} Q^T \Sigma_u^{-1/2},$$

$$\Sigma_{y,1} = \Sigma_y^{-1/2} V \left(diag(1 - \delta_1^2, \cdots, 1 - \delta_\kappa^2, 1, \cdots, 1)\right)^{-1} V^T \Sigma_y^{-1/2},$$

$$\Sigma_{y,2} = \Sigma_y^{-1/2} V \left(diag\left(\frac{\delta_1^2}{1 - \delta_1^2}, \cdots, \frac{\delta_\kappa^2}{1 - \delta_\kappa^2}, 0, \cdots, 0\right)\right)^{-1} V^T \Sigma_y^{-1/2},$$

where l and m are the mean value of vectors that follow chi-squared distribution with the degree of freedom of l and m, respectively.

It turns out that on the assumption $\delta_1 < 1$,

$$\Sigma_{u,1} > \Sigma_{u,2}, \Sigma_{y,1} > \Sigma_{y,2} \Rightarrow \varepsilon(J_u) > \varepsilon(J_y) - m + n$$

$$\text{for } f_u \neq 0, \ f_y = 0, \qquad (2.38)$$

$$\varepsilon(J_u) < \varepsilon(J_y) - m + n \text{ for } f_y \neq 0, \, f_u = 0. \tag{2.39}$$

Inequalities (Eq. 2.38) and (Eq. 2.39) can be applied as a decision logic for fault isolation. Note that if J_u and J_y are used, instead of their mean values, false isolation decisions can be made for this purpose. The rate of the false isolation decision depends on f_u and f_y, which are in general unknown. To reduce false isolation decisions, we can collect data and estimate $\varepsilon(J_u)$ and $\varepsilon(J_y)$.

Following the discussion in the work of Chen et al. [39], it is evident that the preceding solutions, $\{J_u, J_{th,u}\}$ and $\{J_y, J_{th,y}\}$, are the optimal solutions for detecting faults f_u and f_y, thanks to the fact that $u - \hat{u}$ and $y - \hat{y}$ have the minimum variances. However, attention should be paid to the assumption that f_u and f_y are not present in the process measurement simultaneously. If this is not the case, then the overall model

$$\begin{bmatrix} u \\ y \end{bmatrix} = \begin{bmatrix} f_u \\ f_y \end{bmatrix} + \begin{bmatrix} \omega_u \\ \omega_y \end{bmatrix}, \begin{bmatrix} \omega_u \\ \omega_y \end{bmatrix} \sim \mathbb{N}\left(\begin{bmatrix} 0 \\ 0 \end{bmatrix}, \begin{bmatrix} \Sigma_u & \Sigma_{uy} \\ \Sigma_{yu} & \Sigma_y \end{bmatrix} \right) \tag{2.40}$$

should be used for the detection purpose.

2.2.5 Main steps of the CCA-based fault diagnosis method

From the preceding discussion, it can be seen that the CCA-based fault diagnosis method usually consists of two main steps: the generation of residual signals based on the use of a process model and then the evaluation of these residuals using a statistical test. To this end, the flowchart of the method is scratched in Fig. 2.1.

It can be seen from Fig. 2.1 that we reasonably partition the CCA-based fault diagnosis method into four steps:

(1) *Data collection.* This is a step of capturing and storing various measurements from different sensors installed on the equipment in the process of interest [40]. It is the first step of CCA-based fault diagnosis, which provides basic information for the following steps. Usually, a data collection system consists of sensors, data transmission devices, and data storage devices. Various sensors are used to capture different types of measurement data, which represent the status of the process and are also able to reflect the information of fault, and if not, this type of fault cannot be detected anyway. In practice, the commonly used sensors include current sensors, voltage sensors, temperature transducers, accelerometers, and flow sensors, among others. Through a data transmission device, the collected data are transmitted into a PC or portable devices and stored into a memory location for further analysis.

(2) *Data preprocessing.* There are two stages in this step. In the first stage, due to the ubiquitous noise in process, errors in measurement devices, and loss and disturbances in data transmission and data storage, the collected

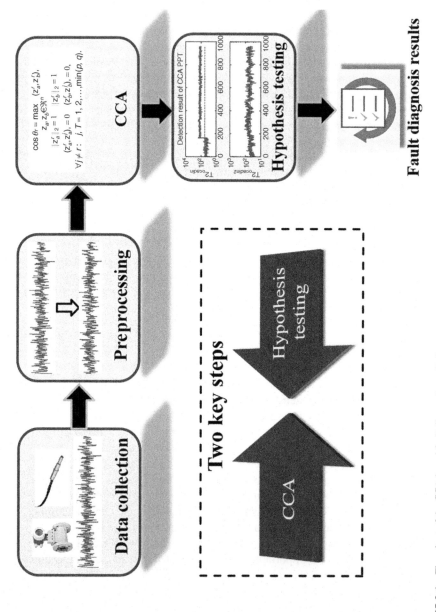

FIGURE 2.1   Flowchart of the CCA-based fault diagnosis method.

data usually is missing data, outliers, and other abnormal data, which will definitely affect the quality of the following steps. In the literature, numerous techniques have been proposed for the removal of outliers and the imputation of missing entries [41]. In the second stage, to handle the case that the data fed into the following steps are too large and redundant, or to find more informative variables than the original measurements, feature variables are extracted from the original measurements to be used in the following steps [42]. For example, the PCA method is a second-order method, which only considers the mean and variance-covariance of the measurement. To provide higher-order representations for non-Gaussian data, some higher-order statistics (kinds of features) can be first extracted and use PCA to analyze the extracted higher-order statistics [43]. Since the removal of outliers and missing data is standard, in the sequel we will not explicitly mention this step. In the second stage, extracting features depends on demands. In this work, we use the original measurements.

The remaining two steps—CCA modeling and hypothesis testing—are the key steps in the CCA-based fault diagnosis method, which we explain in detail in Section 2.3 and 2.4.

The preceding introduction to the conventional CCA method is reliable to the process, in which variables are independent. However, in practice, autocorrelation is present in data due to the dynamics of process of interest. In other words, when the data contains dynamic information, applying CCA on the data will not reveal the exact relations between the measurement vectors but rather a linear static approximation. Like the successful application of DPCA and DPLS methods to detect faults from a dynamic process [44–46], however, the statistical basis is violated because the data break the assumption of time independence. Therefore, to deal with the data, which will be autocorrelated and possibly cross correlated, two variants of the CCA method—the DCCA method and the gated recurrent units (GRU)-aided CCA method—for dealing with the fault diagnosis of the dynamic process are presented in Sections 2.3.1 and 2.3.2. From the viewpoint of the four steps, the DCCA method makes a change in the CCA modeling step, and the GRU-aided CCA method can be viewed as a variant of CCA changed in the preprocessing and CCA modeling steps. In the preprocessing step, the DNN GRU is used as a feature extraction tool, but the GRU's training involves the conventional CCA optimization.

2.3 CCA-based fault diagnosis method for dynamic processes

2.3.1 DCCA-based fault detection

Suppose that the dynamic processes under consideration are linear time invariant and with Gaussian distributed process noise and measurement noise. A standard model form of a dynamic process is the state space representation given by

$$x(k+1) = \mathbf{A}x(k) + \mathbf{B}u(k) + w(k), \tag{2.41}$$

$$y(k) = \mathbf{C}x(k) + \mathbf{D}u(k) + v(k), \tag{2.42}$$

where $x \in \mathbb{R}^n$ is the state vector, $u \in \mathbb{R}^l$ and $y \in \mathbb{R}^m$ are input and output vectors, and $w \in \mathbb{R}^n$ and $v \in \mathbb{R}^m$ are process noise and measurement noise, respectively. Matrices $\mathbf{A}$, $\mathbf{B}$, $\mathbf{C}$, and $\mathbf{D}$ are unknown constant matrices with appropriate dimensions. In this study, we further assume that the process is in the steady state—that is, $\lim_{k \to \infty} \varepsilon(x) = $ constant and $\lim_{k \to \infty} \mathbf{\Sigma}_x = $ constant. As a result, the cross covariance of input and output vectors is constant. In the following, we present DCCA, as an extension of the CCA-based method, to detect faults in such dynamic systems in the steady state.

2.3.1.1 Modeling of input and output datasets

Based on the stochastic system model (Eqs. 2.41 and 2.42), the dependence of future outputs y_f on past inputs- outputs z_p and future inputs u_f is investigated in this section. To this end, we first define the data structures and sets. Suppose that s and s_f are the time lags. Let the lagged variables and corresponding data matrices be defined as

$$z_p(k) = \begin{bmatrix} y(k-s) \\ \cdots \\ y(k-1) \\ u(k-s) \\ \cdots \\ u(k-1) \end{bmatrix}, y_f(k) = \begin{bmatrix} y(k) \\ \cdots \\ y(k+s_f) \end{bmatrix}, u_f(k) = \begin{bmatrix} u(k) \\ \cdots \\ u(k+s_f) \end{bmatrix}$$

$$\begin{aligned} z_p(k) &= \begin{bmatrix} z_p(1), \cdots, z_p(N) \end{bmatrix} \in \mathbb{R}^{(s(m+l)) \times N} \\ y_f(k) &= \begin{bmatrix} y_1(1), \cdots, y_f(N) \end{bmatrix} \in \mathbb{R}^{(s_f+1)m \times N} \\ u_f(k) &= \begin{bmatrix} u_f(1), \cdots, u_f(N) \end{bmatrix} \in \mathbb{R}^{(s_f+1)l \times N}. \end{aligned} \tag{2.43}$$

It is shown in the work of Lehmann [47] that the representation of Eqs. (2.41 and 2.42) can be rewritten as

$$x(k+1) = \mathbf{A}_K x(k) + \mathbf{B}_K u(k) + \mathbf{K}y(k), \tag{2.44}$$

$$y(k) = \mathbf{C}x(k) + \mathbf{D}u(k) + e(k), \tag{2.45}$$

where $\mathbf{A}_K = \mathbf{A} - \mathbf{KC}$, $\mathbf{B}_K = \mathbf{B} - \mathbf{KD}$, with $\mathbf{K}$ as Kalman filter gain matrix to ensure that the eigenvalues of $\mathbf{A}_K$ are all located in the unit circle to make the system stable. $e(k)$ is the innovation sequence. It is straightforward from

Eq. (2.44) that the following equation holds:

$$x(k+1) = \mathbf{A}_K^s x(k-s) + \sum_{i=1}^{s} \mathbf{A}_K^{i-1}\left[\mathbf{KB}_K\right]\begin{bmatrix} y(k-i) \\ u(k-i) \end{bmatrix}. \tag{2.46}$$

Recall that $\mathbf{A}_K$ is stable and a large s leads to $\mathbf{A}_K^s \approx 0$, then

$$x(k) \approx \mathbf{P}^T z_p(k), \tag{2.47}$$

where $\mathbf{P}^T = \begin{bmatrix} \mathbf{P}_y \mathbf{P}_u \end{bmatrix}$, $\mathbf{P}_y = \begin{bmatrix} \mathbf{A}_K^{s-1}\mathbf{K} \cdots \mathbf{A}_K \mathbf{KK} \end{bmatrix}$, $\mathbf{P}_u = \begin{bmatrix} \mathbf{A}_K^{s-1}\mathbf{B}_K \cdots \mathbf{A}_K \mathbf{B}_K \mathbf{B}_K \end{bmatrix}$. The "past" process measurements $z_p(k)$ include the process input and output data in the time period $[k-s, \ k-1]$ as shown in Eq. (2.43). However, from Eqs. (2.44) and (2.45), the following equations also hold:

$$y_f(k) = \mathbf{\Gamma}_{K,s_f}x(k) + \mathbf{H}_{K,u,s_f}u_f(k) + \mathbf{H}_{K,y,s_f}y_f(k) + e_f(k), \tag{2.48}$$

where

$$\mathbf{\Gamma}_{K,s_f} = \begin{bmatrix} \mathbf{C} \\ \mathbf{CA}_K \\ \vdots \\ \mathbf{CA}_K^{s_f} \end{bmatrix}^T, \quad \mathbf{H}_{K,u,s_f} = \begin{bmatrix} \mathbf{D} & 0 & \cdots & 0 \\ \mathbf{CB}_K & \mathbf{D} & \ddots & \vdots \\ \vdots & \ddots & \ddots & 0 \\ \mathbf{CA}_K^{s_f-1}\mathbf{B}_K & \cdots & \mathbf{CB}_K & \mathbf{D} \end{bmatrix}$$

$$\mathbf{H}_{K,y,s_f} = \begin{bmatrix} 0 & 0 & \cdots & 0 \\ \mathbf{CK} & 0 & \ddots & \vdots \\ \vdots & \ddots & \ddots & 0 \\ \mathbf{CA}_K^{s_f-1}\mathbf{K} & \cdots & \mathbf{CK} & 0 \end{bmatrix}, \quad e_f(k) = \begin{bmatrix} e(k) \\ e(k+1) \\ \vdots \\ e(k+s_f) \end{bmatrix}.$$

Based on Eq. (2.47), we obtain

$$\begin{aligned} \left(\mathbf{I} - e_{K,y,s_f}\right)y_f(k) &\approx \mathbf{\Gamma}_{K,s_f}\mathbf{P}^T z_p(k) + \mathbf{H}_{K,u,s_f}u_f(k) + e_f(k) \\ \begin{bmatrix} \mathbf{\Gamma}_{K,s_f}\mathbf{P}^T \mathbf{H}_{K,u,s_f} \end{bmatrix}&\begin{bmatrix} z_p(k) \\ u_f(k) \end{bmatrix} + e_f(k). \end{aligned} \tag{2.49}$$

Eq. (2.49) is further rewritten as

$$\mathbf{L}^T y_f(k) = \mathbf{M}^T \begin{bmatrix} z_p(k) \\ u_f(k) \end{bmatrix} + e_f(k), \tag{2.50}$$

where $\mathbf{L} = \left(\mathbf{I} - \mathbf{H}_{K,y,s_f}\right)^T$, $\mathbf{M} = \begin{bmatrix} \mathbf{\Gamma}_{K,s_f}\mathbf{P}^T \mathbf{H}_{K,u,s_f} \cdot \end{bmatrix}^T$

2.3.1.2   The DCCA-based fault detection method

This section addresses fault detection in dynamic processes by applying the CCA technique for residual generation. The process input and output data are constructed in a time interval, denoted as $\mathbf{Y}_f$ and $\begin{bmatrix} \mathbf{Z}_p \\ \mathbf{U}_f \end{bmatrix}$. Let $\mathbf{Y}_f$ and $\begin{bmatrix} \mathbf{Z}_p \\ \mathbf{U}_f \end{bmatrix}$ be mean centered, then

$$
\begin{bmatrix} \boldsymbol{\Sigma}_z & \boldsymbol{\Sigma}_{z,y_f} \\ \boldsymbol{\Sigma}_{y_f,z} & \boldsymbol{\Sigma}_{y_f} \end{bmatrix} \approx \frac{1}{N-1} \left(\begin{bmatrix} \mathbf{Z}_p \\ \mathbf{U}_f \end{bmatrix} \begin{bmatrix} \mathbf{Z}_p \\ \mathbf{U}_f \end{bmatrix}^T \quad \begin{bmatrix} \mathbf{Z}_p \\ \mathbf{U}_f \end{bmatrix} \mathbf{Y}_f^T \\ \mathbf{Y}_f \begin{bmatrix} \mathbf{Z}_p \\ \mathbf{U}_f \end{bmatrix}^T \quad \mathbf{Y}_f \mathbf{Y}_f^T \right).
$$

By using CCA, the weighting matrices $\boldsymbol{w}_{uf}$ and $\boldsymbol{w}_{yf}$ can be obtained from

$$
\boldsymbol{w}_{uf} = \boldsymbol{\Sigma}_z^{-1/2} \boldsymbol{\Gamma}(:, 1:n), \quad \boldsymbol{w}_{yf} = \boldsymbol{\Sigma}_{y_f}^{-1/2} \boldsymbol{\Delta}(:, 1:n),
$$

$$
\boldsymbol{\Sigma}_z^{-1/2} \boldsymbol{\Sigma}_{z,y_f} \boldsymbol{\Sigma}_{y_f}^{-1/2} = \boldsymbol{\Gamma} \boldsymbol{\Lambda} \boldsymbol{\Delta}, \quad \boldsymbol{\Lambda} = \begin{bmatrix} \boldsymbol{\Lambda}_n & 0 \\ 0 & 0 \end{bmatrix}, \tag{2.51}
$$

where $\boldsymbol{\Lambda}_n = diag(\lambda_1,\ldots, \lambda_n)$. The cumulative percentage value or the Akaike information criterion method [15] can be used to determine n, which is called the *order* of the system. Note that

$$
\boldsymbol{w}_{uf}^T \begin{bmatrix} \mathbf{Z}_p \\ \mathbf{U}_f \end{bmatrix} \begin{bmatrix} \mathbf{Z}_p \\ \mathbf{U}_f \end{bmatrix}^T \boldsymbol{w}_{uf} = \text{Im}, \quad \boldsymbol{w}_{yf}^T \mathbf{Y}_f \mathbf{Y}_f^T \boldsymbol{w}_{yf} = \mathbf{I}
$$

the following equation can be obtained from Eq. (2.51):

$$
\boldsymbol{w}_{uf}^T \begin{bmatrix} \mathbf{Z}_p \\ \mathbf{U}_f \end{bmatrix} \mathbf{Y}_f^T \boldsymbol{w}_{yf} = \boldsymbol{\Lambda}_n. \tag{2.52}
$$

It is reasonable to define a residual vector according to Eq. (2.52),

$$
\boldsymbol{r}(k) = \boldsymbol{w}_{yf}^T \boldsymbol{y}_f(k) - \boldsymbol{\Lambda}_n \boldsymbol{w}_{uf}^T \begin{bmatrix} z_p(k) \\ u_f(k) \end{bmatrix}. \tag{2.53}
$$

Furthermore, the covariance matrix of $\boldsymbol{r}(k)$ can be estimated as

$$
\left(\boldsymbol{w}_{yf}^T \mathbf{Y}_f(k) - \boldsymbol{\Lambda}_n \boldsymbol{w}_{uf}^T \begin{bmatrix} \mathbf{Z}_p \\ \mathbf{U}_f \end{bmatrix} \right) \left(\boldsymbol{w}_{yf}^T \mathbf{Y}_f - \boldsymbol{\Lambda}_n \boldsymbol{w}_{uf}^T \begin{bmatrix} \mathbf{Z}_p \\ \mathbf{U}_f \end{bmatrix} \right)^T
$$

$$
= \boldsymbol{w}_{yf}^T \mathbf{Y}_f \mathbf{Y}_f^T \boldsymbol{w}_{yf} + \boldsymbol{\Lambda}_n^2 \boldsymbol{w}_{uf}^T \begin{bmatrix} \mathbf{Z}_p \\ \mathbf{U}_f \end{bmatrix} \begin{bmatrix} \mathbf{Z}_p \\ \mathbf{U}_f \end{bmatrix}^T \boldsymbol{w}_{uf} - 2\boldsymbol{\Lambda}_n \boldsymbol{w}_{uf}^T \begin{bmatrix} \mathbf{Z}_p \\ \mathbf{U}_f \end{bmatrix} \mathbf{Y}_f^T \boldsymbol{w}_{yf}
$$

$$
= \mathbf{I} - \boldsymbol{\Lambda}_n^2. \tag{2.54}
$$

The residual follows multivariate normal distribution with zero mean and covariance matrix given by Eq. (2.54). It is thus reasonable to apply the following test statistic for the fault detection decision:

$$
T_r^2(k) = (N-1)\boldsymbol{r}^T(k)\left(\mathbf{I} - \boldsymbol{\Lambda}_n^2\right)^{-1}\boldsymbol{r}(k). \tag{2.55}
$$

The corresponding threshold $J_{th,T}$ can be determined by

$$
J_{th,T} = \frac{n(N^2 - n)}{N(N-n)} F_{1-\alpha}(n, N-n), \tag{2.56}
$$

where $F_{1-\alpha}(n, N-n)$ stands for the F-distribution with n and $N-n$ degrees of freedom with the given significance level α.

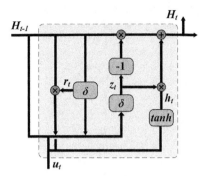

FIGURE 2.2   Illustrations of the GRU model [48].

2.3.2   The GRU-aided CCA fault detection method

In this section, the proposed method is described in detail. Section 2.3.2.1 introduces the basic structure of the GRU with only one layer. In Section 2.3.2.2, the GRU is combined with the CCA technique to form a new method for fault diagnosis of dynamic processes.

2.3.2.1   The GRU

The recurrent neural network (RNN) is a commonly used DNN to deal with time series data analysis. The GRU is a gating mechanism in the RNN, which can overcome the problem of gradient disappearance [48]. The same gate control mechanism is used as LSTM [49]. In the GRU, it has only two gates: the update gate and the reset gate. The pre-memory can be correlated when identifying and predicting the follow-up data. Compared with LSTM, the GRU has fewer parameters, and hence the calculation of the GRU is much less. Its structure is explained in Fig. 2.2.

The update gate vector z_t is used to control the extent to which the state information of the previous moment is brought into the current state. The reset gate vector r_t is adopted to control the degree of ignoring the state information of the previous moment. For the input vector u_t at time t, operating functions in GRU hidden elements are given as follows:

$$z_t = \delta(\mathbf{W}^z u_t + \mathbf{V}^z \mathbf{H}_{t-1} + b^z), \qquad (2.57)$$

$$r_t = \delta(\mathbf{W}^r u_t + \mathbf{V}^r \mathbf{H}_{t-1} + b^r), \qquad (2.58)$$

$$h_t = \tanh(\mathbf{W}^c u_t + \mathbf{V}^c(r_t \circ H_{t-1})), \qquad (2.59)$$

$$\mathbf{H}_t = (1 - z_t) \circ \mathbf{H}_{t-1} + z_t \circ h_t, \qquad (2.60)$$

where $\mathbf{W}^z$, $\mathbf{W}^r$ are weight parameters of z_t and r_t gates, respectively, and $\mathbf{W}^c$ denotes the weight parameter of the output gate. $\mathbf{H}_{t-1}$ is the output states at time

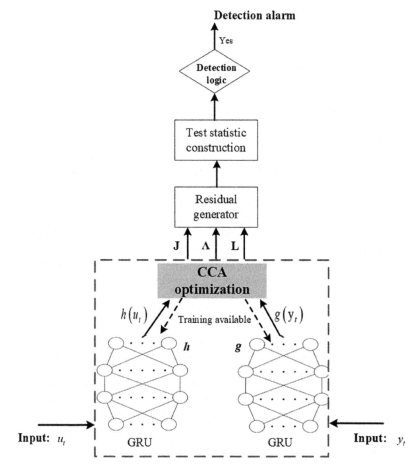

FIGURE 2.3 Structure of GRU-aided CCA.

$t - 1$. "o" represents the Hadamard product, and h_t and H_t are candidate states and output states at time t. $\delta(\bullet)$ and tanh ($\bullet$) denote activation functions, which activate the update gate and reset gate. V^z, V^r, b^z, and b^r are the parameter matrices and vectors, and they are learned through the model training process. For an intuitive illustration, it can be seen from Eqs. (2.57) through (2.60) that when r_t is set to 1 and z_t is set to 0, the GRU degenerates into a simple RNN model.

2.3.2.2 The GRU-aided CCA method

Note that the GRU has been used as fault classifier for fault diagnosis [50, 51]. The structure of the GRU-aided CCA method is illustrated in Fig. 2.3, and the major component of the proposed method includes two deep GRUs and a CCA optimization in the top layer. Before the training, the measured data including input vectors $\boldsymbol{u}_t$ and $\boldsymbol{y}_t$ should be prepared. It is well accepted that the GRU can represent the dynamic in nature; therefore, the input data fed to the GRU has no need to augment with s lagged data as in conventional CVA. Then they are expanded with a batch of N samples, respectively. When the sample interval is short enough in practice, it will also detect the fault in time even if the value of N is large.

Joint representation learning consists of two GRUs and a CCA, the aim of which is to learn the parameters in GRUs so that the correlation between the transformed input data $h(\boldsymbol{u}_t)$ and the transformed output data $g(\boldsymbol{y}_t)$ is as large as possible, where $h(\bullet)$ and $g(\bullet)$ denote the mapping function of the two GRUs. By collecting a batch of samples with length N, the batched input and output data are $\mathbf{U}_N \in \mathbb{R}^{l \times N}$ and $\mathbf{Y}_N \in \mathbb{R}^{m \times N}$, and it is necessary to be initially scaled into the interval between 0 and 1, which is convenient for neural networks to train. For the established process input and output time series dataset, two GRUs with the same structure are constructed to extract the hierarchical representations from the input and output, respectively. Two GRUs' output, $\mathbf{U} = h(\mathbf{U}_N)$ and $\mathbf{Y} = g(\mathbf{Y}_N)$, have the fixed and same dimension. GRU-aided CCA trains h and g based on the following objective, which maximizes the canonical correlation at the output layer between the two data:

$$
\max_{\Theta_h, \Theta_g, \mathbf{J}, \mathbf{L}, \Lambda} \frac{1}{N} trace\left(\mathbf{J}^T \mathbf{N} \mathbf{Y}^T \mathbf{L}\right)
$$

$$
s.t. \mathbf{J}^T \left(\frac{\mathbf{U}\mathbf{U}^T}{N} + r_u \mathbf{I}\right) \mathbf{J} = \mathbf{I}
$$

$$
\mathbf{L}^T \left(\frac{\mathbf{Y}\mathbf{Y}^T}{N} + r_y \mathbf{I}\right) \mathbf{L} = \mathbf{I} \tag{2.61}
$$

$$
r_u > 0, r_y > 0.
$$

where Θ_h and Θ_g are the collection of parameters in both GRUs, and r_u and r_y are small real values and are used as regularization parameters to avoid the possible numerical problem of calculating the covariance matrices of U and Y. Then the conventional CCA algorithm can be used to solve the objective (Eq. 2.61) and train the two GRUs. Ending with the training of two GRUs, the parameters $\hat{\Theta}_h$ and $\hat{\Theta}_g$ are retained, the linear mappings J and L can be obtained, and the diagonal matrix Λ is also obtained.

As shown in Fig. 2.3, the procedures include residual generator building, test statistic construction, and monitoring by detection logic. For implementation, the complete method includes two stages: offline modeling and online monitoring.

(1) *Offline modeling*:

Deep GRU training and construction of the test statistic. Fault-free datasets are used for GRU-aided CCA training in this chapter. To begin with, it is necessary to construct the training set with the form of time series based on the original time series of N samples. At this point, it is particularly important to select the length of each time series in the training set, which may not only affect the amounts of samples in the whole training set but also directly determines the training effect. For the time series training sets constructed by system input and output data, we built two GRU networks with the same structure for joint learning, respectively. The whole neural network consists of multiple GRU layers, dropout layers, and fully connected layers. Then CCA is performed with the output data from the two GRUs. The choice of node number of output layers will directly affect the accuracy and complexity of the subsequent analysis and calculation. In addition, the activation function of the hidden layer in GRUs is chosen as "Sigmoid," whereas the activation function of the output layer is "linear" and the training optimizer uses "Adam."

Construction of the test statistic. For the established process input and output time series dataset, two GRUs with the same structure are constructed to extract the hierarchical representations from the input and output, respectively. Each of the GRUs outputs hierarchical representation vectors u_t^h and y_t^g with the fixed and same dimension.

With the mappings $\mathbf{J}$ and $\mathbf{L}$ calculated, the residual vectors can be obtained as

$$r_{g1} = \mathbf{J}^T u_t^h - \mathbf{\Lambda L}^T y_t^g, \tag{2.62}$$

$$r_{g2} = \mathbf{L}^T y_t^g - \mathbf{\Lambda}^T \mathbf{J}^T u_t^h. \tag{2.63}$$

Then the T^2 test statistics can be constructed as

$$T_{g1}^2 = r_{g1}^T \Sigma_{g1}^{-1} r_{g1}. \tag{2.64}$$

$$T_{g2}^2 = r_{g2}^T \Sigma_{g2}^{-1} r_{g2}, \tag{2.65}$$

where Σ_{g1} and Σ_{g2} are covariance matrices of residual vectors r_{g1} and r_{g2}, respectively. They can be obtained as

$$\Sigma_{g1} = \mathbf{I} - \mathbf{\Lambda\Lambda}^T, \tag{2.66}$$

$$\Sigma_{g2} = \mathbf{I} - \mathbf{\Lambda}^T \mathbf{\Lambda}. \tag{2.67}$$

Threshold determination. Since the process is nonlinear, the Gaussian distribution does not hold. Therefore, KDE can be used to calculate the threshold of statistics [13]:

$$P(x < b) = \int_{-\infty}^{b} \frac{1}{Md} \sum_{k=1}^{M} K\left(\frac{x - x_k}{d}\right) dx, \tag{2.68}$$

where x_k, $k = 1, 2, \ldots, M$ is the sample of the test statistic T^2, function $K(\bullet)$ is a kernel function, and d is the bandwidth of it. Here, a radial basis kernel function is used:

$$K(\beta) = \frac{1}{\sqrt{2\pi}} \exp\left(-\frac{\beta^2}{2}\right). \tag{2.69}$$

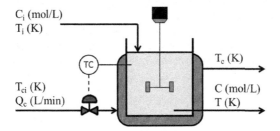

C_i (mol/L)
T_i (K)

TC

T_{ci} (K)
Q_c (L/min)

T_c (K)

C (mol/L)
T (K)

FIGURE 2.4   Schematic of closed-loop CSTR.

Given a significance level α, a threshold $J_{th,g}$ can be calculated by

$$P\left(T^2 < J_{th,g}\right) = \alpha. \tag{2.70}$$

Hence, using Eq. (2.70), the corresponding thresholds of statistics T_{g1}^2 and T_{g2}^2 can be obtained and denoted as $J_{th,g1}$ and $J_{th,g2}$, respectively.

(2) *Online monitoring*:

When a new sample is measured, it is initially scaled with the same methods as normal training sets, noted as u_{new} and y_{new}. The T^2 statistics are calculated on the basis of the established GRU-CCA model. The following decision logic is used to decide if the fault occurs or not:

$$\begin{cases} T_{g1}^2 > J_{th,g1} \text{ or } T_{g2}^2 > J_{th,g2} \Rightarrow \text{presence of fault} \\ T_{g1}^2 \le J_{th,g1} \text{ and } T_{g2}^2 \le J_{th,g2} \Rightarrow \text{absence of fault.} \end{cases} \tag{2.71}$$

2.4 Experimental results and analysis

In this section, two industrial benchmark experiment cases will be used to assess the performance of the proposed methods.

2.4.1 The CSTR process

2.4.1.1 Introduction to CSTR

Controlled CSTR is a second-order nonlinear dynamic simulation system in the chemical industry and is widely used for validation of fault diagnosis methods. In this work, a Simulink model of a CSTR under closed-loop control is used, where the process is represented by three ordinary differential equations that are mass and energy balances around the system. CSTR carries out a hypothetical first-order exothermic reaction, where the tank temperature (T) is maintained using a cooling jacket. Process conditions are being perturbed around the nominal operating point by random disturbances on three input variables.

Fig. 2.4 shows the schematic of the closed-loop CSTR, in which the measurement locations and the control strategy are illustrated, and reactor temperature, T, is maintained by manipulating the coolant flow rate, Q_c. In this simulation,

TABLE 2.1 Fault types in the CSTR dataset

Case	Description
F1	Catalyst deactivation
F2	Heat transfer fouling by exponential decay
F3	Feed temperature with disturbance ramp changes
F4	Coolant feed temperature with disturbance ramp changes

the controller ($K_c = 1.0$ and $\tau_I = 0.2$) is set to saturate below 10 L/min and above 200 L/min. Detailed descriptions of CSTR can be found in the work of Pilario and Cao [52], and no details are given here. According to Fig. 2.4, the inputs are material import concentration, C_i, material import temperature, T_i, coolant inlet temperature, T_{ci} and Q_c, the outputs are coolant outlet temperature, T_c, and concentration of material in the reactor, C and T. Therefore, a dataset comprised of seven variables is manipulated to create incipient faults to evaluate the efficiency of the fault diagnosis method. Four typical incipient faults were simulated, including two multiplicative faults and two additive faults, as show in Table 2.1.

Fault-free and faulty data for every kind of fault are generated from the simulation for 20 hours of operations with sampling interval being 1 minute, so the length of time series of every variable is 1200 points. In a faulty dataset, the fault is introduced after 200 minutes of normal operation, so every faulty data includes 200 points of normal data and 1000 points of faulty data behind. In addition, by perturbing the input variables randomly every fixed length of time, the datasets measured become temporally correlated and dynamic. At the same time, they do not fit a Gaussian distribution due to the nonlinearity of process, making the data suitable for the evaluation of the presented methods.

2.4.1.2 DCCA and GRU-aided CCA training

In the training process of CCA and DCCA, iterative optimization is no need, and SVD is applied to solve the optimization objective, as well the fault detection threshold is determined. However, the training process of the GRU is relatively complex because neural networks contain several hyperparameters and a large number of trainable parameters. In the training phase, the main task of GRU-aided CCA is the hyperparameter adjustment of two GRUs, among which the length of input sequence and the number of hidden layers and neurons play the most important role. In this experiment, the final parameters and structure of GRU-aided CCA are as follows. First, only fault-free data are used for training, and seven variables are divided into two groups and input into two GRU models, respectively. The length of the input sequence is 60. Both GRU models have

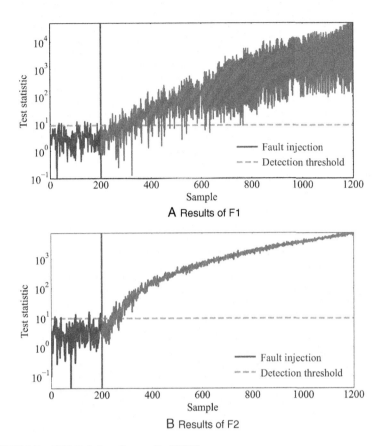

A Results of F1

B Results of F2

FIGURE 2.5   CCA fault detection result of CSTR.

the same structure, which is Input→GRU (64 cells)→Dropout (0.2)→GRU (64 cells)→Dropout (0.2)→fully connected layer (32 nodes)→fully connected layer (20 nodes)→Output. According to the preceding structure, the input sequences will be mapped to a 20-dimensional vector after passing through the GRU. Then the loss function value is calculated by Eq. (2.68), and the Adam gradient descent algorithm is used to optimize the parameters of two neural networks.

2.4.1.3   Results and analysis

The fault detection result of CCA is shown in Fig. 2.5. In the case of F1, the statistics of CCA fluctuate sharply, and a large number of miscalculations occur at the initial stage of the failure. This is because the system variable oscillates violently when F1 occurs, whereas CCA only considers the correlation of the variables of current moment, so the statistics also change with the oscillation of variables. CCA performs well in the case of other faults, indicating that CCA is

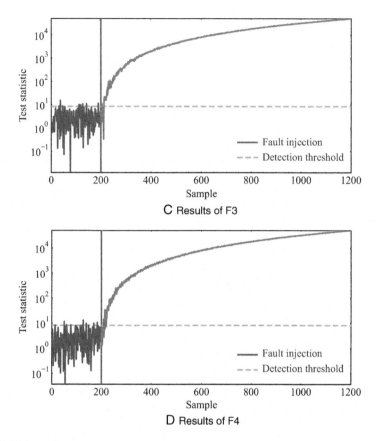

C Results of F3

D Results of F4

FIGURE 2.5   Continued.

not completely useless for nonlinear systems and can still correctly detect faults in some cases.

The fault detection result of DCCA is shown in Fig. 2.6. For F1, the test statistics obtained by DCCA are obviously more stable than that of CCA, with lower detection delay, and there will be no high misjudgment rate in the early stage of failure. For other fault types, DCCA also achieves a good detection effect.

GRU-aided CCA was trained according to Section 2.4.1.2, and the value of loss function during the training process is shown in Fig. 2.7. This means that the neural network is able to perform nonlinear mapping for two sets of input sequences, and the correlation after mapping is close to 1.

After training with fault-free data, GRU-aided CCA was tested by faulty data of cases F1, F2, F3, and F4, and the detection results are shown in Fig. 2.8.

It can be seen from Fig. 2.8 that GRU-aided CCA can effectively detect multiple types of faults, and the T^2 test statistics under normal and fault scenarios

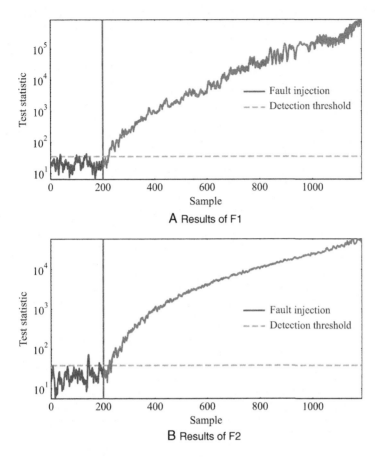

FIGURE 2.6   DCCA fault detection result of CSTR.

are clearly distinguished. The specific evaluations are shown in Table 2.2, in which the three fault detection performance indicators are FAR, FDR, and fault detection delay (FDD), whose formula can be found in the work of Chen et al. [39].

2.4.2   The TDCS process

2.4.2.1   Introduction to TDCS

To verify the feasibility of GRU-aided CCA in more complex nonlinear dynamic systems, another experimental platform of high-speed train TDCS is used for testing. TDCS is jointly developed by Central South University and CRRC Zhuzhou Institute Company, Limited, based on the hardware-in-the-loop (HIL) platform, including a dSPACE real-time simulator, signal conditioner, traction control unit (TCU), and host PC, as shown in Fig. 2.9 [53, 54].

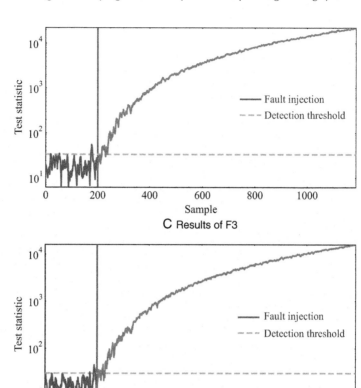

C Results of F3

D Results of F4

FIGURE 2.6   Continued.

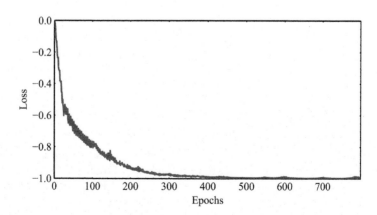

FIGURE 2.7   Loss function value of GRU-aided CCA during training.

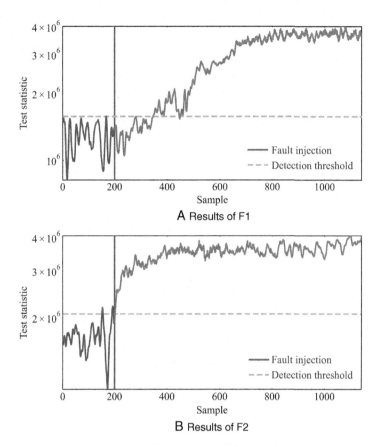

FIGURE 2.8   GRU-aided CCA fault detection result of CSTR.

The platform can simulate a variety of equipment faults; in this case, three motor faults are selected for the experiment: rotor broken bar fault (RBB), interturn short circuit fault (ITSC), and air gap eccentricity fault (AGE). Only the motor stator three-phase current data were used in the experiment, and the severity of the three motor faults was 10%. The experimental data consisted of normal and fault data under fixed working conditions. Due to the complexity of the motor fault characteristic, feature extraction is required before training. In this experiment, variance and kurtosis of stator three-phase current are extracted, so all together there are six variables in the input sequences.

2.4.2.2 DCCA and GRU-aided CCA training

As in the CSTR experiment, GRU-aided CCA requires hyperparameter adjustment and iterative optimization in the training phase. The optimized parameters and structure selected in this case are Input→GRU (64

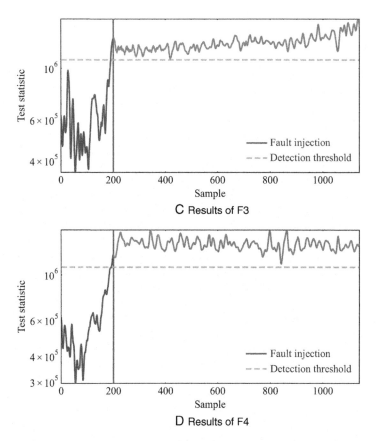

C Results of F3

D Results of F4

FIGURE 2.8 Continued.

cells)→Dropout (0.2)→GRU (64 cells)→Dropout (0.2)→fully connected layer (64 nodes)→fully connected layer (30 nodes)→Output. Since a lot of noise is contained in raw data, the length of input sequences is 256.

2.4.2.3 Results and analysis

The fault detection result of CCA is shown in Fig. 2.10. It can be seen that under normal conditions, the T^2 test statistics fluctuate greatly, but in the RBB and ITSC fault scenario, the statistics change significantly and the fluctuation becomes smaller, in which it can be considered that good detection results have been achieved. However, in the case of AGE fault, the T^2 test statistics is unstable, which may lead to miss detection.

The fault detection results of DCCA are shown in Fig. 2.11. The statistical difference between normal and fault is greater than CCA, but there are also large fluctuations, and the statistics become more unstable under AGE fault.

TABLE 2.2 Evaluations of the fault detection result of CSTR

	CCA			DCCA			GRU-aided CCA		
Fault	FAR	FDR	FDD	FAR	FDR	FDD	FAR	FDR	FDD
F1	5.5%	88.7%	16	4.5%	97.7%	21	0.5%	84.2%	147
F2	5.0%	95.1%	13	5.0%	97.3%	22	5.5%	99.7%	3
F3	5.0%	98.2%	5	5.0%	98.2%	16	5.0%	100%	0
F4	4.5%	98.5%	9	4.0%	98.0%	6	4.0%	100%	0

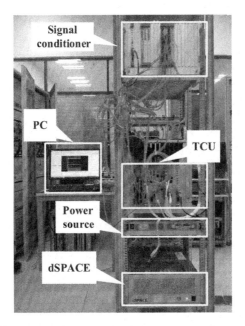

FIGURE 2.9 TDCS experimental setup.

The value of loss function during the training process is shown in Fig. 2.12. Under the influence of environmental noise, the convergence speed of the neural network is slower, and the amplitude fluctuation of the loss function is also larger, but eventually it can converge to around -1.

After the training, the detection results of three faults are shown in Fig. 2.13. As can be seen from this figure, in the normal state, the T^2 test statistics are seriously affected by noise, but in case of motor fault, the statistics will change significantly and the fluctuation caused by noise will decrease, so the distinction between normal and fault states is very obvious.

Table 2.3 shows the evaluation of the preceding three methods.

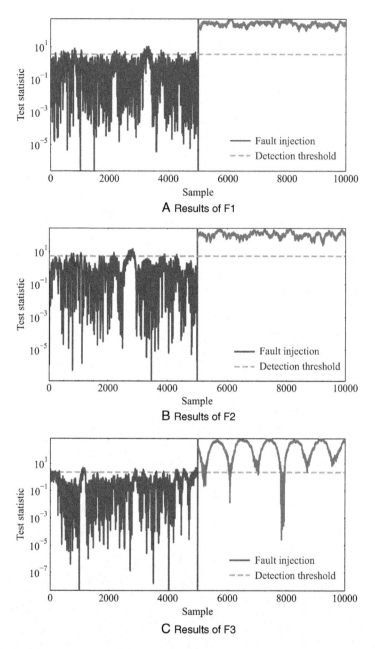

FIGURE 2.10   CCA fault detection result of TDCS.

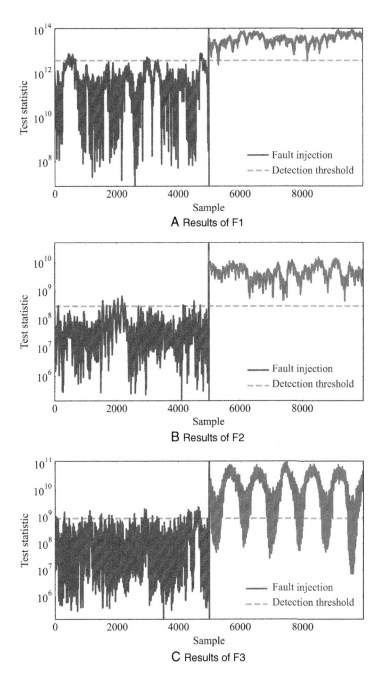

FIGURE 2.11 DCCA fault detection result of TDCS.

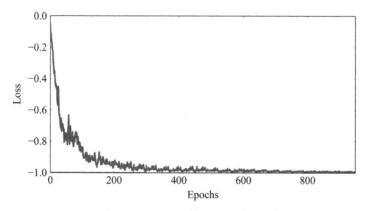

FIGURE 2.12   Loss function value of GRU-aided CCA during training.

According to experiment results, for the nonlinear dynamic system, CCA can still correctly detect faults and achieve good detection performance under some circumstances, but the statistical difference between normal and fault is small, and it is easy to be disturbed by noise. The statistics obtained by DCCA under normal and fault conditions are more diverse, but the influence of noise is still not eliminated. GRU-aided CCA correctly detected faults in the case of a large amount of noise and obtained low FAR and high FDR under the three faults, but there was certain trouble in training, and FDD was relatively large. Therefore, the practical application of the CCA fault detection method can be reasonably selected considering the requirements of system complexity, noise level, and FDD.

2.5   Conclusion

This work investigates the application of the CCA technique for fault diagnosis of dynamic processes. We introduce three CCA methods, namely conventional CCA, DCCA, and GRU-aided CCA, and the main steps of fault diagnosis with these methods. Conventional CCA aims to find basis vectors for two sets of

TABLE 2.3 Evaluations of GRU-aided CCA

	CCA			DCCA			GRU-aided CCA		
Fault	FAR	FDR	FDD	FAR	FDR	FDD	FAR	FDR	FDD
RBB	5.0%	100%	0	4.8%	99.9%	1	4.6%	98.4%	79
ITSC	4.9%	100%	0	2.7%	100%	0	4.9%	99.5%	23
AGE	4.9%	85.5%	1	4.6%	87.3%	1	4.9%	99.9%	1

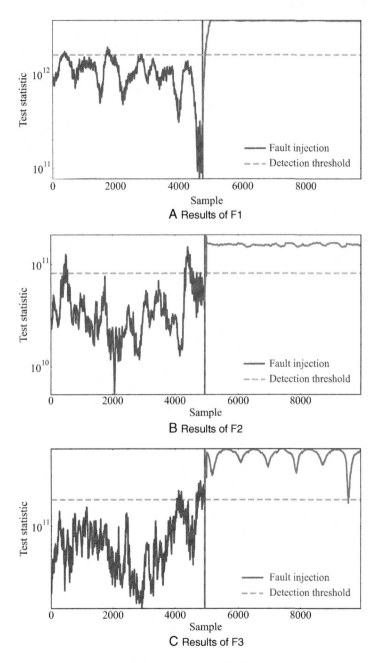

FIGURE 2.13   GRU-aided CCA fault detection result of TDCS.

variables such that the correlations between the projections of the variables onto these basis vectors are mutually maximized. By means of hypothesis testing, a certain test statistic is constructed with normal data and the corresponding threshold is determined. Then the system state can be judged by the value of the test statistic. On this basis, DCCA takes into account the dynamic and autocorrelation of the system and adds past moments of the system into the calculation of CCA. In GRU-aided CCA, the nonlinear dynamic property is considered, and the system variables are nonlinearly mapped by the neural network so that the mapped variables can meet the requirements of conventional CCA and achieve a better fault diagnosis effect. The presented methods are then used in detecting faults in the CSTR and TDCS processes. Experimental results show that for nonlinear dynamic systems, conventional CCA can achieve a good detection effect in some faults, but it is greatly affected by noise and variable fluctuation. The overall diagnostic effect of DCCA is close to that of CCA, and the fluctuation of statistics is smaller, but it is also affected by noise. GRU-aided CCA performs well on all fault data and has a higher tolerance for noise, but it has disadvantages such as difficulty in training and large delay in fault detection. These two benchmark experiments further verify the effectiveness of CCA in complex engineering systems.

Acknowledgments

Financial sponsorship from the project of the National Natural Science Foundation of China (#61803390, #61773407, #61790571, #61621062) is gratefully acknowledged. This work was also partly sponsored by Hunan Provincial Key Laboratory (#2017TP1002), the postdoctoral foundation (#2019T120713), and the Project of State Key Laboratory of High Performance Complex Manufacturing, Central South University (#ZZYJKT2020-14).

References

[1] S.X. Ding, Data-Driven Design of Fault Diagnosis and Fault-Tolerant Control Systems, Springer-Verlag, London, UK, 2014.

[2] S.J. Qin, L.H. Chiang, Advances and opportunities in machine learning for process data analytics, Computers & Chemical Engineering 126 (2019) 465–473.

[3] W. Härdle, L. Simar, Canonical correlation analysis, Applied Multivariate Statistical Analysis, Springer, Berlin, Germany, 2003.

[4] T.W. Anderson, *An Introduction to Multivariate Statistical Analysis* (2nd ed.), John Wiley & Sons, Hoboken, New Jersey, USA, 1984.

[5] L.H. Chiang, E. Russell, R. Braatz, Fault Detection and Diagnosis in Industrial Systems, Advanced Textbooks in Control and Signal Processing, Springer-Verlag, London, UK, 2001.

[6] M. Borga, Canonical Correlation: A Tutorial, 2001.

[7] H. Hotelling, Relation between two sets of variates, Biometrika 28 (1936) 321–377.

[8] I.M. Gelfand, A.M. Yaglom, Calculation of amount of information about a random function contained in another such function, American Mathematical Society Translations: Series 2, 12,

English translation of original in *Uspekhi Matematicheskikh Nauk*, 1975, pp. 3–52. 12:199--246.

[9] H. Akaike, Markovian representation of stochastic processes by canonical variables, SIAM Journal on Control 13 (1) (1975) 162–173.

[10] W.E. Larimore, Canonical variate analysis in identification, filtering, and adaptive control, Proceedings of the 29th IEEE Conference on Decision and Control (1990).

[11] Y. Wang, D.E. Seborg, W.E. Larimore, Process monitoring using canonical variate analysis and principal component analysis, IFAC Proceedings Volumes 30 (9) (1997) 577–582.

[12] B.C. Juricek, D.E. Seborg, W.E. Larimore, Fault detection using canonical variate analysis, Industrial & Engineering Chemistry Research 43 (2004) 458–474.

[13] P.P. Odiowei, Y. Cao, Nonlinear dynamic process monitoring using canonical variate analysis and kernel density estimations, IEEE Transactions on Industrial Informatics 6 (1) (2010) 36–45.

[14] B.B. Jiang, X. Zhu, D.X. Huang, R.D. Braatz, Canonical variate analysis-based monitoring of process correlation structure using causal feature representation, Journal of Process Control 32 (2015) 109–116.

[15] Z.W. Chen, S.X. Ding, K. Zhang, Z.B. Li, Z.K. Hu, Canonical correlation analysis-based fault detection methods with application to alumina evaporation process, Control Engineering Practice 46 (2016) 51–58.

[16] Z.W. Chen, S.X. Ding, T. Peng, C.H. Yang, W.H. Gui, Fault detection for non-Gaussian processes using generalized canonical correlation analysis and randomized algorithms, IEEE Transactions on Industrial Electronics 65 (2) (2018) 6321–6330.

[17] Z.W. Chen, C. Liu, S.X. Ding, T. Peng, C.H. Yang, W.H. Gui, Y. Shardt, A just-in-time-learning aided canonical correlation analysis method for multimode process monitoring and fault detection, IEEE Transactions on Industrial Electronics (2020).

[18] Y.Q. Liu, B. Liu, X.J. Zhao, M. Xie, A mixture of variational canonical correlation analysis for nonlinear and quality-relevant process monitoring, IEEE Transactions on Industrial Electronics 65 (8) (2017) 6478–6486.

[19] X.C. Li, Y.J. Yang, I. Bennett, D. Mba, Condition monitoring of rotating machines under time-varying conditions based on adaptive canonical variate analysis, Mechanical Systems & Signal Processing 131 (2019) 348–363.

[20] Q.C. Jiang, X.F. Yan, Locally weighted canonical correlation analysis for nonlinear process monitoring, Industrial & Engineering Chemistry Research 57 (41) (2018) 13783–13792.

[21] Y. LeCun, Y. Bengio, G. Hinton, Deep learning, Nature 521 (2015) 436–444.

[22] J. Schmidhuber, Deep learning in neural networks: An overview, Neural Networks 61 (2015) 85–117.

[23] G. Andrew, R. Arora, J. Bilmes, K. Livescu, Deep canonical correlation analysis, Proceedings of the 30th International Conference on Machine Learning (2013) 1247–1255.

[24] W.R. Wang, R. Arora, K. Livescu, J. Bilmes, On deep multi-view representation learning, Proceedings of the 32nd International Conference on Machine Learning (2015) 1083–1092.

[25] N. Mallinar, C. Rosset. Deep canonically correlated LSTMs. [Online]. 2018. Available at https://arxiv.org/abs/1801.05407.

[26] Q.C. Jiang, X.F. Yan, Learning deep correlated representations for nonlinear process monitoring, IEEE Transactions on Industrial Informatics 15 (12) (2019) 6200–6209.

[27] K. Zhang, K. Peng, R. Chu, J. Dong, Implementing multivariate statistics-based process monitoring: A comparison of basic data modeling approaches, Neurocomputing 290 (2018) 172–184.

[28] V. Uurtio, J. Monteiro, J. Kandola, J. Shawe-Taylor, D. Fernandez-Reyes, J. Rousu, A tutorial on canonical correlation methods, ACM Computing Surveys 50 (2017) 14–38.

[29] D.R. Hardoon, S. Szedmak, J. Shawe-Taylor, Canonical correlation analysis: An overview with application to learning methods, Neural Computation 16 (12) (2004) 2639–2664.

[30] G.H. Golub, H. Zha, The canonical correlations of matrix pairs and their numerical computation. In *Linear Algebra for Signal Processing*, Springer, 1995.

[31] A. Bjorck, G.H. Golub, Numerical methods for computing angles between linear subspaces, Mathematics of Computation 123 (1973) 579–594.

[32] M.S. Bartlett, The statistical significance of canonical correlations, Biometrika 32 (1) (1941) 29–37.

[33] N.J.R. Healy, A rotation method for computing canonical correlations, Mathematics of Computation 11 (58) (1957) 83–86.

[34] L. M. Ewerbring, F. T. Luk. Canonical correlations and generalized SVD: Applications and new algorithms. In *Proceedings of the 32nd Annual Technical Symposium, 1989*. International Society for Optics and Photonics. 206-222.

[35] Z. W. Chen, S. X. Ding, K. Zhang, C. H. Yang, T. Peng. Generalized CCA with applications for fault detection and estimation. In *Proceedings of the IEEE 7th Data Driven Control and Learning Systems Conference (DDCLS), 2018*. 545--550.

[36] M. Basseville, I. Nikiforov, Detection of Abrupt Changes, PTR Prentice Hall, 1993.

[37] E.L. Lehmann, *Testing Statistical Hypotheses* (2nd ed.), Springer-Verlag, 1986.

[38] Z.W. Chen, K. Zhang, Y.A.W. Shardt, S.X. Ding, X. Yang, C.H. Yang, T. Peng, Comparison of two basic statistics for fault detection and process monitoring, IFAC-PapersOnLine 50 (1) (2017) 14776–14781.

[39] Z.W. Chen, C.H. Yang, T. Peng, H. Dan, C.G. Li, W.H. Gui, A cumulative canonical correlation analysis-based sensor precision degradation detection method, IEEE Transactions on Industrial Electronics 66 (8) (2018) 6321–6330.

[40] Y.G. Lei, N.P. Li, L. Guo, N. Li, T. Yan, J. Lin, Machinery health prognostics: A systematic review from data acquisition to RUL prediction, Mechanical Systems & Signal Processing 104 (2018) 799–834.

[41] J.L. Zhu, Z.Q. Ge, Z.H. Song, F.R. Gao, Review and big data perspectives on robust data mining approaches for industrial process modeling with outliers and missing data, Annual Reviews in Control 46 (2018) 107–133.

[42] Z.W. Chen, R.J. Guo, Z. Lin, T. Peng, X. Peng, A data-driven health monitoring method using multi-objective optimization and stacked autoencoder based health indicator, IEEE Transactions on Industrial Informatics (2020), doi:10.1109/TII.2020.2999323.

[43] J. Wang, Q. He, Multivariate statistical process monitoring based on, statistics pattern analysis. Industrial & Engineering Chemistry Research 49 (2010) 7858–7869.

[44] S. Yin, S.X. Ding, A. Haghani, H.Y. Hao, P. Zhang, A comparison study of basic data-driven fault diagnosis and process monitoring methods on the benchmark Tennessee Eastman process, Journal of Process Control 22 (2012) 1567–1581.

[45] Z.Q. Ge, Z.H. Song, F.R. Gao, Review of recent research on data-based process monitoring, Industrial & Engineering Chemistry Research 52 (10) (2013) 3543–3562.

[46] W.F. Ku, R.H. Storer, C. Georgakis, Disturbance detection and isolation by dynamic principal component analysis, Chemometrics & Intelligent Laboratory Systems 30 (1) (1995) 179–196.

[47] S.J. Qin, An overview of subspace identification, Computers & Chemical Engineering 30 (2006) 1502–1513.

[48] K. Cho, B. van Merrienboer, C. Gulcehre, F. Bougares, H. Schwenk, Y. Bengio, Learning phrase representations using RNN encoder-decoder for statistical machine translation, Proceedings of Empirical Methods in Natural Language Processing (2014) *EMNLP*.

[49] S. Hochreiter, J. Schmidhuber, Long short-term memory, Neural Computation 9 (8) (1997) 1735–1780.

[50] Z.Z. Wang, Y.J. Dong, W. Liu, Z. Ma, A novel fault diagnosis approach for chillers based on 1-D convolutional neural network and gated recurrent unit, Sensors 20 (9) (2020) 2458.

[51] Y. Tao, X. Wang, R. Sánchez, S. Yang, Y. Bai, Spur gear fault diagnosis using a multilayer gated recurrent unit approach with vibration signal, IEEE Access 7 (2019) 56880–56889.

[52] K.E.S. Pilario, Y. Cao, Canonical variate dissimilarity analysis for process incipient fault detection, IEEE Transactions on Industrial Informatics 14 (12) (2018) 5308–5315.

[53] C.H. Yang, C. Yang, T. Peng, X.Y. Yang, W.H. Gui, A fault-injection strategy for traction drive control systems, IEEE Transactions on Industrial Electronics 64 (7) (2017) 5719–5727.

[54] X.Y. Yang, C.H. Yang, T. Peng, Z.W. Chen, B. Liu, W.H. Gui, Hardware-in-the-loop fault injection for traction control system, IEEE Journal of Emerging & Selected Topics in Power Electronics 6 (2) (2018) 696–706.

Chapter 3

$H\infty$ fault estimation for linear discrete time-varying systems with random uncertainties

Yueyang Li
School of Electrical Engineering, University of Jinan, China

3.1 Introduction

Research on observer-based robust fault detection and isolation (FDI) problems for linear time-invariant (LTI) systems has received much attention over the past three decades (see [1–3] and the references therein). Basically, the fault detection (FD) issue concerns designing a fault detection filter (FDF) for generating a residual signal such that the sensitivity of residual to fault is intensified by enhancing the robustness to the disturbance. In reviewing the development of the observer-based FDI for LTI systems with various characteristics such as time-delay, model inaccuracy, time-dependent switching mode, and uncertain observations, H_∞ optimization and H_∞ filtering techniques are two primary approaches that are widely used for LTI systems with l_2-norm bounded unknown inputs and faults [4–6]. Recently, some contributions have been devoted to linear time-varying (LTV) systems since most practical industrial processes can be represented or well approximated by time-varying dynamics (see [7–10] and related works). In the work of Zhong et al. [8], a finite horizon $H_-/H_\infty, H_\infty/H_\infty$ FDI formulation was proposed for linear discrete time-varying (LDTV) systems and an optimal solution was derived by solving a Riccati equation. In another work of Zhong et al. [9], a Krein space–based approach was proposed to H_∞-filtering-based FDI for LDTV systems. In the work of Shen et al. [7], the H_∞-filtering-based fault estimation methods are proposed for LDTV systems in virtue of the Krein space–based reorganized innovation analysis and projection theory in the background of Zhang et al. [10].

On another research front line, multiplicative noise is widely used to represent model uncertainty in state space representation and plays a significant role in many practical engineering fields, such like aerospace, machinery, chemical

Fault Diagnosis and Prognosis Techniques for Complex Engineering Systems. DOI: 10.1016/B978-0-12-822473-1.00001-X reaction, and communication. Naturally, problems of stability analysis, control, and filtering for systems with multiplicative noise have been widely investigated [11–15]. For example, the statistic testing scheme is proposed in the work of Ding et al. [16] and the H_∞-filtering-based FDI is implemented in the work of Ma et al. [17].

With the rapid progress of networked control systems and distributed sensor/actuator systems, the packet dropout caused by sensor gain reductions may happen when transmitting information under unreliable links. The so-called packet dropout refers to the incomplete measurements phenomenon described by Bernoulli random distribution as the multiplicative factor, which is the special case of multiplicative noise. The random uncertainty introduced by packet dropouts evidently deteriorates the performance of the FDF [18]. Many contributions are dedicated to the FD issue for systems with incomplete measurements by employing the LMI-formulated H_∞ fault estimation approach over infinite horizon (refer to [19–21] and references therein). For example, He et al. [22] discuss the problem of fault detection for LTI systems with both random delay and packet loss. Ruan et al. [23] design FDF for LTI systems with multistep packet loss. Wang et al. [24] and He et al. [25] solve the problem of fault detection for a class of LTI systems with packet loss and Markov jump characteristics by using the equivalent space method and observer method, respectively.

In this chapter, we are devoted to solving H_∞ fault estimation problems for LDTV systems with random uncertainties (i.e., multiplicative noise and packet loss). The rest of the content is organized as follows. Section 3.2 deals with the problem of robust FDI for LDTV systems subject to multiplicative noise, and l_2-norm bounded unknown input will be dealt with. The FDF design of the LDTV system with packet loss is studied in Section 3.3. The problem of H_∞ fixed-lag fault estimator design for LDTV systems subject to intermittent observations is dealt with in Section 3.4. Finally, Section 3.5 presents some conclusions.

Notations. Throughout this chapter, vectors in the Krein space are represented by boldface letters, and vectors in the Euclidean space are denoted by normal letters. For a matrix X, X^{T} and X^{-1} stand for the transpose and inverse of X, respectively. $X > 0$ ($X < 0$) denotes that X is positive (negative) definite. R^n means the set of n-dimensional real vectors. I and 0 denote the identity matrix and zero matrix with appropriate dimensions, respectively. $\mathrm{E}\{\vartheta(\mathrm{k})\}$ means the mathematical expectation of $\vartheta(k)$. $\vartheta(k) \in l_2[0, N]$ means $\sum_{k=0}^{N} \vartheta^{\mathrm{T}}(k)\vartheta(k) < \infty$, where N is a positive integer. The symbol $\mathcal{L}\{\{\vartheta(i)\}_{i=j}^{k}\}$ represents the linear space spanned by the sequence $\vartheta(k)$ taking values in the time interval $[j, k]$. $\mathrm{Prob}\{\Upsilon\}$ denotes the occurrence probability of the event "Υ." δ_{ij} represents the Kronecker delta function, which is equal to unity for $i = j$ and zero for $i \neq j$. $diag\{S_1, S_2, \ldots, S_n\}$ means a block diagonal matrix with diagonal blocks $S_1, S_2, \ldots, S_n$.

3.2 Robust H_∞ fault detection for LDTV systems with multiplicative noise

3.2.1 Problem formulation

Consider the following LDTV system:

$$
\begin{cases}
x(k+1) = (A(k) + A_v(k)v(k))x(k) + (B_f(k) + B_{fv}(k)v(k))f(k) \\
\qquad + (B_d(k) + B_{dv}(k)v(k))d(k) \\
y(k) = (C(k) + C_v(k)v(k))x(k) + (D_f(k) + D_{fv}(k)v(k))f(k) \qquad (3.2.1) \\
\qquad + (D_d(k) + D_{dv}(k)v(k))d(k) \\
x(0) = x_0
\end{cases}
$$

where $x(k) \in R^n$, $y(k) \in R^{n_y}$, $d(k) \in R^{n_d}$, and $f(k) \in R^{n_f}$ denote the state, measurement output, unknown input, and fault to be detected, respectively; $d(k) \in l_2[0, N]$, $f(k) \in l_2[0, N]$. $A(k)$, $B_f(k)$, $B_d(k)$, $C(k)$, $D_f(k)$, $D_d(k)$, $A_v(k)$, $B_{fv}(k)$, $B_{dv}(k)$, $C_v(k)$, $D_{fv}(k)$, and $D_{dv}(k)$ are known matrices with appropriate dimensions.

We first introduce the definition of exponential stability in mean square for system (3.2.1).

Definition 3.2.1 [26]. System (3.2.1) with $f(k) = 0$ and $d(k) = 0$ is said to be exponentially stable in mean square if there exist $c \geq 0$ and $q \in (0, 1)$ such that

$$
\mathrm{E}\{\|x(k)\|^2\} \leq cq^k\|x(0)\|^2
$$

Throughout this chapter, it is assumed that $C(k)$ is full row rank for all k, $\{v(k)\}$ is a scalar zero-mean white noise sequence, and $\mathrm{E}\{v(i)v(j)\} = \varepsilon\delta_{ij}$, where ε is a known positive scalar, and δ_{ij} denotes the Kronecker delta function; system (3.2.1) is exponential stable in mean square in finite horizon $[0, N]$.

For the purpose of fault detection, the following observer-based FDF is considered for system (3.2.1):

$$
\begin{cases}
\hat{x}(k+1) = A(k)\hat{x}(k) + L(k)(y(k) - C(k)\hat{x}(k)) \\
r(k) = W(k)(y(k) - C(k)\hat{x}(k)) \qquad\qquad (3.2.2) \\
\hat{x}(0) = \hat{x}_0
\end{cases}
$$

where $\hat{x}(k)$ is an estimate for $x(k)$, $r(k) \in R^r$ is the generated residual, $\hat{x}_0$ is a guess of initial state, and observer gain matrix $L(k)$ and post-filter $W(k)$ are parameters to be determined. Define

$$
e(k) = x(k) - \hat{x}(k), \eta(k) = \begin{bmatrix} x^\mathrm{T}(k) & e^\mathrm{T}(k) \end{bmatrix}^\mathrm{T}
$$

$$
r_e(k) = r(k) - f(k), w(k) = [f^\mathrm{T}(k) \quad d^\mathrm{T}(k)]^\mathrm{T}
$$

It follows from (3.2.1) and (3.2.2) that

$$
\begin{cases}
\eta(k+1) = (A_\eta(k) + A_{\eta v}(k)v(k))\eta(k) + (B_\eta(k) + B_{\eta v}(k)v(k))w(k) \\
r_e(k) = (C_\eta(k) + C_{\eta v}(k)v(k))\eta(k) + (D_\eta(k) + D_{\eta v}(k)v(k))w(k)
\end{cases} \quad (3.2.3)
$$

where

$$A_\eta(k) = \begin{bmatrix} A(k) & 0 \\ 0 & A(k) - L(k)C(k) \end{bmatrix}, \quad A_{\eta v}(k) = \begin{bmatrix} A_v(k) & 0 \\ A_v(k) - L(k)C_v(k) & 0 \end{bmatrix}$$

$$B_\eta(k) = \begin{bmatrix} B_f(k) & B_d(k) \\ B_f(k) - L(k)D_f(k) & B_d(k) - L(k)D_d(k) \end{bmatrix}$$

$$B_{\eta v}(k) = \begin{bmatrix} B_{fv}(k) & B_{dv}(k) \\ B_{fv}(k) - L(k)D_{fv}(k) & B_{dv}(k) - L(k)D_{dv}(k) \end{bmatrix}$$

$$C_\eta(k) = \begin{bmatrix} 0 & W(k)C(k) \end{bmatrix}, \quad C_{\eta v}(k) = \begin{bmatrix} W(k)C_v(k) & 0 \end{bmatrix}$$

$$D_\eta(k) = \begin{bmatrix} W(k)D_f(k) - I & W(k)D_d(k) \end{bmatrix}$$

$$D_{\eta v}(k) = \begin{bmatrix} W(k)D_{fv}(k) & W(k)D_{dv}(k) \end{bmatrix}$$

Now the problem of H_∞-FDF design can be formulated to find $L(k)$ and $W(k)$ such that system (3.2.3) is exponential stable in mean square and satisfies

$$\sup_{\|w(k)\|_2 \neq 0} \frac{\|r_e(k)\|_{2,E}^2}{\eta^T(0)S\eta(0) + \|w(k)\|_2^2} < \gamma^2 \tag{3.2.4}$$

where γ is a positive scalar and S is a given positive definite initial state weighting matrix.

3.2.2 H_∞ performance analysis

To derive the main results, the following lemma that indicates the condition on exponential stability in mean square for system (3.2.3) will first be given.

Lemma 3.2.1. *The stochastic parameter system (3.2.3) is exponentially stable in mean square if there exists a symmetric positive definite matrix $P(\cdot)$ such that the following inequality holds:*

$$A_\eta^T(k)P(k+1)A_\eta(k) + \varepsilon A_{\eta v}^T(k)P(k+1)A_{\eta v}(k) - P(k) < 0 \tag{3.2.5}$$

Proof. Let F_k be the minimal σ-algebra generated by $\{v(k), 0 \leq k \leq N\}$. Suppose (3.2.5) holds, since $P(\cdot) > 0$, then there exist $\kappa_1(\cdot) > 0$ and $\kappa_2(\cdot) > 0$ such that

$$\kappa_1(k)I \leq P(k) \leq \kappa_2(k)I$$

and

$$\kappa_1(k+1)I \leq P(k+1)\kappa_2(k+1)I,$$

then

$$\kappa_1(k)E\{\eta^T(k)\eta(k)\} \leq E\{\eta^T(k)P(k)\eta(k)\} \leq \kappa_2(k)E\{\eta^T(k)\eta(k)\}.$$

In addition, we have

$$E\{\eta^T(k+1)P(k+1)\eta(k+1)|F_k\} = E\{\eta^T(k)(A_\eta^T(k)P(k+1)A_\eta(k)$$
$$+ \varepsilon A_{\eta v}^T(k)P(k+1)A_{\eta v}(k)$$
$$- P(k))\eta(k)\} + E\{\eta^T(k)P(k)\eta(k)\}.$$

If (1.4.4) holds, then there exists $0 < \kappa_3(k) < \kappa_2(k)$ such that

$$E\{\eta^T(k+1)P(k+1)\eta(k+1)\} \le -\kappa_3(k)E\{\eta^T(k)\eta(k)\} + E\{\eta^T(k)P(k)\eta(k)\}$$
$$< \frac{-\kappa_3(k)}{\kappa_2(k)}E\{\eta^T(k)P(k)\eta(k)\} + E\{\eta^T(k)P(k)\eta(k)\}$$
$$= (1 - \frac{\kappa_3(k)}{\kappa_2(k)})E\{\eta^T(k)P(k)\eta(k)\}.$$

Thus, we can find that

$$\kappa_1(k+1)E\{\eta^T(k+1)\eta(k+1)\}leE\{\eta^T(k+1)P(k+1)\eta(k+1)\}$$
$$\le \left(1 - \frac{\kappa_3(k)}{\kappa_2(k)}\right)E\{\eta^T(k)P(k)\eta(k)\}$$
$$\le \left(1 - \frac{\kappa_3(k)}{\kappa_2(k)}\right)\left(1 - \frac{\kappa_3(k-1)}{\kappa_2(k-1)}\right)$$
$$\times E\{\eta^T(k-1)P(k-1)\eta(k-1)\}$$
$$\le \cdots \le \left(1 - \frac{\kappa_3(k)}{\kappa_2(k)}\right)$$
$$\cdots \left(1 - \frac{\kappa_3(0)}{\kappa_2(0)}\right)E\{\eta^T(0)P(0)\eta(0)\}.$$

Let $q_1 = \max\{(1 - \frac{\kappa_3(k)}{\kappa_2(k)}), \cdots, (1 - \frac{\kappa_3(0)}{\kappa_2(0)})\}$, then

$$E\{\eta^T(k+1)P(k+1)\eta(k+1)\} \le q_1^{k+1}E\{\eta^T(0)P(0)\eta(0)\} = q_1^{k+1}\eta^T(0)P(0)\eta(0).$$

Furthermore, we have

$$\kappa_1(k+1)E\{\eta^T(k+1)\eta(k+1)\} \le \kappa_2(0)q_1^{k+1}\|\eta(0)\|^2,$$

which leads to $E\{\|\eta(k)\|^2\} \le cq^k\|\eta(0)\|^2$ with

$$c = \frac{\kappa_2(0)}{\kappa_1(k)} > 0$$
$$q = \max\left\{\left(1 - \frac{\kappa_3(k-1)}{\kappa_2(k-1)}\right), \cdots, \left(1 - \frac{\kappa_3(0)}{\kappa_2(0)}\right)\right\} \in (0, 1).$$

This completes the proof. □

Based on Lemma 3.2.1, the following two theorems that play important roles in deriving the main results will be obtained in terms of Riccati equations.

Theorem 3.2.1. *If there exist a positive scalar ζ and a solution $P(k) > 0$ such that*

$$
\begin{cases}
P(k) = A_\eta^{\mathrm{T}}(k)P(k+1)A_\eta(k) + \varepsilon C_{\eta v}^{\mathrm{T}}(k)C_{\eta v}(k) + \varepsilon A_{\eta v}^{\mathrm{T}}(k)P(k+1)A_{\eta v}(k) \\
\quad + C_\eta^{\mathrm{T}}(k)C_\eta(k) + E^{\mathrm{T}}(k)\Theta^{-1}(k)E(k) + \zeta I \\
P(N+1) = S_{N+1}
\end{cases}
$$

(3.2.6)

where

$$
E(k) = (A_\eta^{\mathrm{T}}(k)P(k+1)B_\eta(k) + \varepsilon C_{\eta v}^{\mathrm{T}}(k)D_{\eta v}(k) + \varepsilon A_{\eta v}^{\mathrm{T}}(k)P(k+1)B_{\eta v}(k)
$$
$$
\quad + C_\eta^{\mathrm{T}}(k)D_\eta(k))^{\mathrm{T}}
$$
$$
\Theta(k) = \gamma^2 I - B_\eta^{\mathrm{T}}(k)P(k+1)B_\eta(k) - D_\eta^{\mathrm{T}}(k)D_\eta(k) - \varepsilon B_{\eta v}^{\mathrm{T}}(k)P(k+1)B_{\eta v}(k)
$$
$$
\quad - \varepsilon D_{\eta v}^{\mathrm{T}}(k)D_{\eta v}(k) > 0,
$$

$S_{N+1} > 0$ *is a terminal state weighting matrix, then system (3.2.3) with $\eta(0) = 0$ is exponentially stable in mean square and, for given $\gamma > 0$, the following H_∞ performance is satisfied:*

$$
\sup_{\|\omega(k)\|_2 \neq 0} \frac{\|\gamma_e(k)\|_{2,E}^2 + E\{\eta^T(N+1)S_{N+1}\eta(N+1)\}}{\|\omega(k)\|_2^2} < \gamma^2.
$$

(3.2.7)

Proof. Define

$$
V(\eta(k), k) = \eta^{\mathrm{T}}(k)P(k)\eta(k),\ P(k) > 0,
$$

then

$$
\Delta V(k) = E\{V(k+1)|F_k\} - V(k)
$$
$$
= \eta^{\mathrm{T}}(k)A_\eta^{\mathrm{T}}(k)P(k+1)A_\eta(k)\eta(k)
$$
$$
+ \eta^{\mathrm{T}}(k)A_\eta^{\mathrm{T}}(k)P(k+1)B_\eta(k)w(k) + \varepsilon\eta^{\mathrm{T}}(k)A_{\eta v}^{\mathrm{T}}(k)(k)P(k+1)A_{\eta v}(k)\eta(k)
$$
$$
+ \varepsilon\eta^{\mathrm{T}}(k)A_{\eta v}^{\mathrm{T}}(k)P(k+1)B_{\eta v}(k)w(k) + w^{\mathrm{T}}(k)B_\eta^{\mathrm{T}}(k)P(k+1)A_\eta(k)\eta(k)
$$
$$
+ w^{\mathrm{T}}(k)B_\eta^{\mathrm{T}}(k)P(k+1)B_\eta(k)w(k) + \varepsilon w^{\mathrm{T}}(k)B_{\eta v}^{\mathrm{T}}(k)P(k+1)A_{\eta v}(k)\eta(k)
$$
$$
+ \varepsilon w^{\mathrm{T}}(k)B_{\eta v}^{\mathrm{T}}(k)P(k+1)B_{\eta v}(k)w(k) - \eta^{\mathrm{T}}(k)P(k)\eta(k),
$$

which leads to the following identical equation with the aid of (3.2.3):

$$
E\{\Delta V\} = E\{\Delta V\} - E\{\gamma^2 w^{\mathrm{T}}(k)w(k)\} + E\{\gamma^2 w^{\mathrm{T}}(k)w(k)\}
$$
$$
- E\{r_e^{\mathrm{T}}(k)r_e(k)\} + E\{r_e^{\mathrm{T}}(k)r_e(k)\}
$$
$$
= \eta^{\mathrm{T}}(k)\Pi_{11}(k)\eta(k) + w^{\mathrm{T}}(k)\Pi_{21}(k)\eta(k) + \eta^{\mathrm{T}}(k)\Pi_{12}(k)w(k)
$$
$$
- w^{\mathrm{T}}(k)(-\Pi_{22}(k))w(k) + E\{\gamma^2 w^{\mathrm{T}}(k)w(k)\} - E\{r_e^{\mathrm{T}}(k)r_e(k)\},
$$

(3.2.8)

where

$$\Pi_{11}(k) = A_\eta^T(k)P(k+1)A_\eta(k) - P(k) + C_\eta^T(k)C_\eta(k) + \varepsilon C_{\eta v}^T(k)C_{\eta v}(k)$$
$$+ \varepsilon A_{\eta v}^T(k)P(k+1)A_{\eta v}(k)$$
$$\Pi_{12}(k) = A_\eta^T(k)P(k+1)B_\eta(k) + C_\eta^T(k)D_\eta(k) + \varepsilon C_{\eta v}^T(k)D_{\eta v}(k)$$
$$+ \varepsilon A_{\eta v}^T(k)P(k+1)B_{\eta v}(k)$$
$$\Pi_{22}(k) = B_\eta^T(k)P(k+1)B_\eta(k) - \gamma^2 I + D_\eta^T(k)D_\eta(k) + \varepsilon D_{\eta v}^T(k)D_{\eta v}(k)$$
$$+ \varepsilon B_{\eta v}^T(k)P(k+1)B_{\eta v}(k)$$
$$\Pi_{21}(k) = \Pi_{12}^T(k).$$

Based on (3.2.8), taking the sum of both sides of $E\{\Delta V\}$ from zero to N by the completing squares method, we have

$$\sum_{k=0}^{N} E\{\Delta V\} = E\{\eta^T(N+1)P(N+1)\eta(N+1)\} - \eta^T(0)P(0)\eta(0)$$

$$= \sum_{k=0}^{N} E\{\eta^T(k)R(P(k))\eta(k)\} + E\sum_{k=0}^{N-1}\{\gamma^2 w^T(k)w(k) - r_e^T(k)r_e(k)\}$$

$$- E\sum_{k=0}^{N}\left\{(w(k)-\mu^*(k))^T\Theta(k)(w(k)-\mu^*(k))\right\} \qquad (3.2.9)$$

where

$$\Theta(k) = -\Pi_{22}(k), \quad \mu(k) = \Theta^{-1}(k)\Pi_{21}(k)\eta(k)$$
$$R(P(k)) = \Pi_{11}(k) + \Pi_{12}(k)\Theta^{-1}(k)\Pi_{21}(k).$$

(i) *Stability analysis.* Let $w(k) = 0$, then from Lemma 3.2.1, we know that system (3.2.3) is exponentially stable in mean square if (3.2.5) holds. It is clear that when (3.2.6) holds with $\Theta(k) > 0$, (3.2.5) is satisfied, which leads to the exponential stability in mean square.

(ii) *H_∞ performance analysis.* When $w(k) \neq 0$, define

$$J_N = E\left\{\sum_{k=0}^{N} r_e^T(k)r_e(k) - \gamma^2\sum_{k=0}^{N} w^T(k)w(k)\right\}$$
$$+ E\{\eta^T(N+1)S(N+1)\eta(N+1)\},$$

then from (3.2.9), under zero initial condition, we have

$$J_N = -E\sum_{k=0}^{N}\left\{(w(k)-\mu^*(k))^T\Theta(k)(w(k)-\mu^*(k))\right\}$$

$$+ E\sum_{k=0}^{N}\{\eta^T(k)R(P(k))\eta(k)\}$$

$$+ \text{E}\{\eta^{\text{T}}(N+1)S(N+1)\eta(N+1)\} - \text{E}\{\eta^{\text{T}}(N+1)P(N+1)\eta(N+1)\}$$

$$\leq -\text{E}\sum_{k=0}^{N}\{(w(k) - \mu^*(k))^{\text{T}}\Theta(k)(w(k) - \mu^*(k))\}$$

$$+ \text{E}\sum_{k=0}^{N}\{\eta^{\text{T}}(k)(R(P(k)) + \zeta I)\eta(k)\} + \text{E}\{\eta^{\text{T}}(N+1)S(N+1)\eta(N+1)\}$$

$$- \text{E}\{\eta^{\text{T}}(N+1)P(N+1)\eta(N+1)\}. \tag{3.2.10}$$

From (3.2.10), we can conclude that if the equation $R(P(k)) + \zeta I = 0$ (i.e., if there exists $P(k) > 0$ such that (3.2.6) holds with the constraint condition that $\Theta(k) > 0$), then system (3.2.3) is exponentially stable in mean square and $J_N < 0$ (i.e., H_∞ performance (3.2.7) holds). This completes the proof. $\square$

Remark 3.2.1. Theorem 3.2.1 establishes a relationship between backward Riccati equation (3.2.6) and the H_∞ performance (3.2.7). Next, a solution to $L(k)$ and $W(k)$ in terms of a forward Riccati equation will be derived in the following Theorem 3.2.2 by applying an adjoint operator [27,28].

Theorem 3.2.2. *If there exist a positive scalar ζ and matrix $Q(k) > 0$ satisfying the following forward Riccati equation,*

$$\begin{cases} Q(k+1) = A_\eta(k)Q(k)A_\eta^{\text{T}}(k) + \varepsilon B_{\eta v}(k)B_{\eta v}^{\text{T}}(k) + \varepsilon A_{\eta v}(k)Q(k)A_{\eta v}^{\text{T}}(k) \\ \qquad\qquad + B_\eta(k)B_\eta^{\text{T}}(k) + M(k)\Xi^{-1}(k)M^{\text{T}}(k) + \zeta I \\ Q(0) = S^{-1} \end{cases}$$

$$\tag{3.2.11}$$

where

$$M(k) = (C_\eta(k)Q(k)A_\eta^{\text{T}}(k) + \varepsilon D_{\eta v}(k)B_{\eta v}^{\text{T}}(k)$$
$$\qquad + \varepsilon C_{\eta v}(k)Q(k)A_{\eta v}^{\text{T}}(k) + D_\eta(k)B_\eta^{\text{T}}(k))^{\text{T}}$$
$$\Xi(k) = \gamma^2 I - C_\eta(k)Q(k)C_\eta^{\text{T}}(k) - D_\eta(k)D_\eta^{\text{T}}(k) - \varepsilon C_{\eta v}(k)Q(k)C_{\eta v}^{\text{T}}(k)$$
$$\qquad - \varepsilon D_{\eta v}(k)D_{\eta v}^{\text{T}}(k) > 0,$$

then system (3.2.3) is exponentially stable in mean square and, for given $\gamma > 0$, H_∞ performance (3.2.4) is satisfied.

Proof. Define G to be the linear operator that maps $(\eta(0), w(k))$ to $r_e(k)$ based on the following definition of inner product:

$$\langle(\eta_1(0), w_1(k)), ((\eta_2(0), w_2(k)))\rangle = \text{E}\{\eta_1^{\text{T}}(0)S\eta_2(0)\} + \langle w_1(k), w_2(k)\rangle$$

$$\langle\omega_1(k), \omega_2(k)\rangle = E\left\{\sum_{k=0}^{N}\omega_1^T(k)\omega_2(k)\right\}$$

Let $G^\sim$ be the adjoint operator of G, and denote $G^\sim r_e(k) = [\eta_a^T(0) \ w_a^T(k)]^T$. The inner product of $G^\sim$ and G has the following property [27]:

$$\langle G(\eta(0), w(k)), r_e(k) \rangle = \langle (\eta(0), w(k)), G^\sim r_e(k) \rangle. \tag{3.2.12}$$

Applying (3.2.12) for $w(k)$ and $r_e(k)$ in $l_2[0, N]$, we have

$$\langle G(\eta(0), w(k)), r_e(k) \rangle = \langle (\eta(0), w(k)), G^\sim r_e(k) \rangle$$

$$= E\{\eta^T(0)S\eta_a(0)\} + E\left\{ \sum_{k=0}^{N} w^T(k)w_a(k) \right\}.$$

In other words,

$$\sum_{k=0}^{N} E\left\{ r_e^T(k) \left[(C_\eta(k) + v(k)C_{\eta v}(k))\Phi(k, 0)\eta(0) + (C_\eta(k) \right. \right.$$

$$+ v(k)C_{\eta v}(k)) \sum_{j=0}^{k-1} \Phi(k, j+1)(B_\eta(j) + v(j)B_{\eta v}(j))w(j)$$

$$\left. \left. + (D_\eta(k) + v(k)D_{\eta\alpha}(k))w(k) \right] \right\}$$

$$= E\{\eta^T(0)S\eta_a(0)\} + E\left\{ \sum_{k=0}^{N} w^T(k)w_a(k) \right\},$$

which implies that

$$\eta_a(0) = S^{-1} \sum_{j=0}^{N-1} \Phi^T(j, 0)(C_\eta(j) + v(j)C_{\eta v}(j))^T r_e(j)$$

$$w_a(k) = (B_\eta(k) + v(k)B_{\eta v}(k))^T \sum_{j=k+1}^{N} \Phi^T(j, k+1)(C_\eta(k) + v(k)C_{\eta v}(k))^T r_e(j)$$

$$+ (D_\eta(k) + v(k)D_{\eta v}(k))^T r_e(k).$$

Let

$$\lambda_a(k) = \sum_{j=k+1}^{N} \Phi^T(j, k+1)(C_\eta(k) + v(k)C_{\eta v}(k))^T r_e(j),$$

and thus the state-space realization of $G^\sim$ can be obtained as

$$\begin{cases} \lambda_a(k-1) = (A_\eta^T(k) + v(k)A_{\eta v}^T(k))\lambda_a(k) + (C_\eta^T(k) + v(k)C_{\eta v}^T(k))r_e(k) \\ w_a(k) = (B_\eta^T(k) + v(k)B_{\eta v}^T(k))\lambda_a(k) + (D_\eta^T(k) + v(k)D_{\eta v}^T(k))r_e(k) \end{cases}$$
$$\tag{3.2.13}$$

Denote $\bar{k} = N - k$, then (3.2.13) can be rewritten as follows:

$$
\begin{cases}
\bar{\lambda}_a(\bar{k} + 1) = (\bar{A}_\eta^{\mathrm{T}}(\bar{k}) + \upsilon(\bar{k})\bar{A}_{\eta\upsilon}^{\mathrm{T}}(\bar{k}))\bar{\lambda}_a(\bar{k}) + \bar{C}_\eta^{\mathrm{T}}(\bar{k}) + \upsilon(\bar{k})\bar{C}_{\eta\upsilon}^{\mathrm{T}}(\bar{k}))\bar{r}_e(\bar{k}) \\
\bar{w}_a(\bar{k}) = (\bar{B}_\eta^{\mathrm{T}}(\bar{k}) + \upsilon(\bar{k})\bar{B}_{\eta\upsilon}^{\mathrm{T}}(\bar{k}))\bar{\lambda}_a(\bar{k}) + (\bar{D}_\eta^{\mathrm{T}}(\bar{k}) + \upsilon(\bar{k})\bar{D}_{\eta\upsilon}^{\mathrm{T}}(\bar{k}))\bar{r}_e(\bar{k})
\end{cases}
$$

$$(3.2.14)$$

where

$$
\begin{aligned}
\bar{A}_\eta(\bar{k}) &= A_\eta(N - \bar{k}), & \bar{A}_{\eta\upsilon}(\bar{k}) &= A_{\eta\upsilon}(N - \bar{k}) \\
\bar{B}_\eta(\bar{k}) &= B_\eta(N - \bar{k}), & \bar{B}_{\eta\upsilon}(\bar{k}) &= B_{\eta\upsilon}(N - \bar{k}) \\
\bar{C}_\eta(\bar{k}) &= C_\eta(N - \bar{k}), & \bar{C}_{\eta\upsilon}(\bar{k}) &= C_{\eta\upsilon}(N - \bar{k}) \\
\bar{D}_\eta(\bar{k}) &= D_\eta(N - \bar{k}), & \bar{D}_{\eta\upsilon}(\bar{k}) &= D_{\eta\upsilon}(N - \bar{k}) \\
\bar{\lambda}_a(\bar{k}) &= \lambda_a(N - \bar{k}), & \bar{\gamma}_e(\bar{k}) &= \lambda_e(N - \bar{k}) \\
\bar{\omega}_a(\bar{k}) &= \omega_a(N - \bar{k}), & \bar{\lambda}_a(0) &= 0.
\end{aligned}
$$

By applying Theorem 3.2.1, if there exist a positive scalar ζ and a solution $P(\bar{k}) > 0$ to the following equation,

$$
\begin{cases}
P(\bar{k}) = \bar{A}_\eta(\bar{k})P(\bar{k} + 1)\bar{A}_\eta^{\mathrm{T}}(\bar{k}) + \varepsilon \bar{B}_{\eta\upsilon}(\bar{k})\bar{B}_{\eta\upsilon}^{\mathrm{T}}(\bar{k}) + \varepsilon \bar{A}_{\eta\upsilon}(\bar{k})P(\bar{k} + 1)\bar{A}_{\eta\upsilon}^{\mathrm{T}}(\bar{k}) \\
\qquad + \bar{B}_\eta(\bar{k})\bar{B}_\eta^{\mathrm{T}}(\bar{k}) + F(\bar{k})\bar{\Theta}^{-1}(\bar{k})F^{\mathrm{T}}(\bar{k}) + \zeta I \\
P(N + 1) = S^{-1}
\end{cases}
$$

$$(3.2.15)$$

where

$$
F(\bar{k}) = (\bar{C}_\eta(\bar{k})P(\bar{k} + 1)\bar{A}_\eta^{\mathrm{T}}(\bar{k}) + \varepsilon \bar{D}_{\eta\upsilon}(\bar{k})\bar{B}_{\eta\upsilon}^{\mathrm{T}}(\bar{k}) + \varepsilon \bar{C}_{\eta\upsilon}(\bar{k})P(\bar{k} + 1)\bar{A}_{\eta\upsilon}^{\mathrm{T}}(\bar{k})
$$
$$
\qquad + \bar{D}_\eta(\bar{k})\bar{B}_\eta^{\mathrm{T}}(\bar{k}))^{\mathrm{T}}
$$
$$
\bar{\Theta}(\bar{k}) = \gamma^2 I - \bar{C}_\eta(\bar{k})P(\bar{k} + 1)\bar{C}_\eta^{\mathrm{T}}(\bar{k}) - \bar{D}_\eta(\bar{k})\bar{D}_\eta^{\mathrm{T}}(\bar{k}) - \varepsilon \bar{C}_{\eta\upsilon}(\bar{k})P(\bar{k} + 1)\bar{C}_{\eta\upsilon}^{\mathrm{T}}(\bar{k})
$$
$$
\qquad - \varepsilon \bar{D}_{\eta\upsilon}(\bar{k})\bar{D}_{\eta\upsilon}^{\mathrm{T}}(\bar{k}) > 0,
$$

then system (3.2.14) is exponential stable in mean square and satisfies the following H_∞ performance:

$$
\sup_{\|r_e(\bar{k})\|_{2,\mathrm{E}} \neq 0} \frac{\|\bar{w}_a(\bar{k})\|_{2,\mathrm{E}}^2 + \mathrm{E}\{\bar{\lambda}^{\mathrm{T}}(N + 1)S^{-1}\lambda^{\mathrm{T}}(N + 1)\}}{\|r_e(\bar{k})\|_{2,\mathrm{E}}^2} < \gamma^2. \qquad (3.2.16)
$$

Notice that the H_∞ performance (3.2.16) for system (3.2.14) and H_∞ performance (3.2.4) for system (3.2.3) are induced norms in $G^\sim$ and G, respectively. Thus, from Theorem 3.9-2 in the work of Kreyszig [29], we know that the H_∞ performance (3.2.16) and (3.2.4) are equivalent. Let $Q(k) = P(N + 1 - k)$, then (3.2.15) will reduce to the forward equation (3.2.11), which completes the proof. $\square$

Remark 3.2.2. If system (3.2.1) is LTI, then for $k \to \infty$, Theorem 3.2.2 is the Riccati equation version of Lemma 3.2.1 proposed in the work of Ma et al.

[17] for the time-invariant robust FDF design when multiplicative noise exists. If system (3.2.1) is LTV with $A_{\eta v}(k) = 0$, $B_{\eta v}(k) = 0$, $D_{\eta}(k) = 0$, and $D_{\eta v}(k) = 0$ without considering the exponential stability in mean square in finite horizon, Theorem 3.2.2 is identical to Lemma 3.2.1 given in the work of Zhong et al. [30] for the deterministic H_{∞} fault estimation problem.

3.2.3 Design of parameter matrices

Based on Theorem 3.2.2, we are now ready to give a solution to the H_{∞}-FDF design problem. First, the determination of the parameter matrices will be converted into a quadratic optimization problem and an analytical solution will be derived by solving this problem.

Let

$$Q(k) = \begin{bmatrix} Q_{11}(k) & Q_{12}(k) \\ Q_{21}(k) & Q_{22}(k) \end{bmatrix}$$

$$
\begin{aligned}
\Xi(k) = &-W(k)C(k)Q_{22}(k)C^{\mathrm{T}}(k)W^{\mathrm{T}}(k) - W(k)D_d(k)D_d^{\mathrm{T}}(k)W^{\mathrm{T}}(k) \\
&- \varepsilon W(k)D_{fv}(k)D_{fv}^{\mathrm{T}}(k)W^{\mathrm{T}}(k) - \varepsilon V(k)D_{dv}(k)D_{dv}^{\mathrm{T}}(k)W^{\mathrm{T}}(k) \\
&- \varepsilon W(k)C_v(k)Q_{11}(k)C_v^{\mathrm{T}}(k)W^{\mathrm{T}}(k) + \gamma^2 I \\
&- (W(k)D_f(k) - I)(W(k)D_f(k) - I)^{\mathrm{T}}
\end{aligned} \tag{3.2.17}
$$

From (3.2.11), it concludes that $Q_{11}(k+1)$ is independent of $L(k)$, whereas for $Q_{22}(k+1)$, we have

$$
\begin{aligned}
Q_{22}(k+1) = &\; \varepsilon(A_v(k) - L(k)C_v(k))Q_{11}(k)(A_v(k) - L(k)C_v(k))^{\mathrm{T}} \\
&+ (A(k) - L(k)C(k))Q_{22}(k)(A(k) - L(k)C(k))^{\mathrm{T}} \\
&+ (B_d(k) - L(k)D_d(k))(B_d(k) - L(k)D_d(k))^{\mathrm{T}} \\
&+ (B_f(k) - L(k)D_f(k))(B_f(k) - L(k)D_f(k))^{\mathrm{T}} \\
&+ \varepsilon(B_{fv}(k) - L(k)D_{fv}(k))(B_{fv}(k) - L(k)D_{fv}(k))^{\mathrm{T}} \\
&+ \varepsilon(B_{dv}(k) - L(k)D_{dv}(k))(B_{dv}(k) - L(k)D_{dv}(k)) \\
&+ \Gamma(k)\Xi^{-1}(k)\Gamma^{\mathrm{T}}(k) + \zeta I,
\end{aligned} \tag{3.2.18}
$$

where

$$
\begin{aligned}
\Gamma(k) = &\; (A(k) - L(k)C(k))Q_{22}(k)C^{\mathrm{T}}(k)W^{\mathrm{T}}(k) + B_d(k) - L(k)D_d(k)D_d^{\mathrm{T}}(k)W^{\mathrm{T}}(k) \\
&+ (\varepsilon(B_{fv}(k) - L(k)D_{fv}(k))D_{fv}^{\mathrm{T}}W^{\mathrm{T}}(k) \\
&+ \varepsilon(A_v(k) - L(k)C_v(k))Q_{21}(k)C_v^{\mathrm{T}}(k)W^{\mathrm{T}}(k) \\
&+ (B_f(k) - L(k)D_f(k))(W(k)D_{f1}(k) - I)^{\mathrm{T}} \\
&+ \varepsilon(B_{dv}(k) - L(k)D_{dv}(k))D_{dv}^{\mathrm{T}}W^{\mathrm{T}}(k).
\end{aligned}
$$

Let $L_f(k)$ denote a feasible solution of $L(k)$ and $W_f(k)$ denote a feasible solution of $W(k)$. Motivated by Yu et al. [31], $L_f(k)$ and $W_f(k)$ can be derived such that makes γ as small as possible meanwhile guarantees $\Xi(k) > 0$. Thus, $L_f(k)$ and $W_f(k)$ are supposed to satisfy the following inequality:

$$\varsigma^T(k)\Xi(W_f(k), L_f(k-1), \gamma)\varsigma(k) \geq \varsigma^T(k)\Xi(W(k), L(k-1), \gamma)\varsigma(k),$$
(3.2.19)

where $\varsigma(k)$ is any nonzero vector with appropriate dimension. Based on this idea, for obtaining $L_f(k)$, it follows from

$$\frac{\partial \varsigma^T(k)Q_{22}(k+1)\varsigma(k)}{\partial(L^T(k)\varsigma(k))} = 0$$

that

$$0 = (H(k) + K(k)\Xi^{-1}(k)K^T(k))L^T(k) - K(k)\Xi^{-1}(k)G^T(k) - T(k) \quad (3.2.20)$$

where

$$K(k) = D_d(k)D_d^T(k)W^T(k) + \varepsilon D_{dv}(k)D_{dv}^T(k)W^T(k) + \varepsilon D_{fv}(k)D_{fv}^T(k)W^T(k)$$
$$\quad + D_f(k)(W(k)D_f(k) - I)^T + C(k)Q_{22}(k)C^T(k)W^T(k)$$
$$\quad + \varepsilon C_v(k)Q_{21}(k)C_v^T(k)W^T(k)$$

$$H(k) = D_d(k)D_d^T(k) + \varepsilon D_{dv}(k)D_{dv}^T(k) + D_f(k)D_f^T(k) + \varepsilon C_v(k)Q_{11}(k)C_v^T(k)$$
$$\quad + C(k)Q_{22}(k)C^T(k) + \varepsilon D_{fv}(k)D_{fv}^T(k)$$

$$G(k) = \varepsilon A_v(k)Q_{21}(k)C_v^T(k)W^T(k) + B_d(k)D_d^T(k)W^T(k)$$
$$\quad + B_f(k)(W(k)D_f(k) - I)^T + A(k)Q_{22}(k)C^T(k)W^T(k)$$
$$\quad + \varepsilon B_{dv}(k)D_{dv}^T(k)W^T(k) + \varepsilon B_{fv}(k)D_{fv}^T(k)W^T(k)$$

$$T(k) = C(k)Q_{22}(k)A^T(k) + \varepsilon D_{dv}(k)B_{dv}^T(k) + \varepsilon C_v(k)Q_{11}(k)A_v^T(k)$$
$$\quad + D_f(k)B_f^T(k) + D_d(k)B_d^T(k)^T + \varepsilon D_{fv}(k)B_{fv}^T(k).$$

Simultaneously, for deriving $W_f(k)$, it follows from

$$\frac{\partial \varsigma^T(k)\Xi(k)\varsigma(k)}{\partial(W^T(k)\varsigma(k))} = 0$$

that

$$0 = -W(k)H(k) + D_f^T(k).$$

Furthermore, under the assumption that $C(k)$ is full row rank for all k, we have

$$\frac{\partial^2 \varsigma^T(k)Q_{22}(k+1)\varsigma(k)}{\partial(L^T(k)\varsigma(k))^2} = H(k) + K(k)\Xi^{-1}(k)K^T(k) > 0$$

and

$$\frac{\partial^2 \varsigma^T(k)\Xi(k)\varsigma(k)}{\partial (W^T(k)\varsigma(k))^2} = -H(k) < 0,$$

and thus $L_f(k)$ and $W_f(k)$ can be chosen as

$$L_f(k) = \Omega^{\mathrm{T}}(k)(H(k) + K(k)\Xi^{-1}(k)K^{\mathrm{T}}(k))^{-1} \quad (3.2.21)$$

$$W_f(k) = D_f^{\mathrm{T}}(k)H^{-1}(k) \quad (3.2.22)$$

where

$$\Omega(k) = T(k) + K(k)\Xi^{-1}(k)N^T(k),$$

which implies (3.2.19) holds such that γ satisfying $\Xi(k) > 0$ can be obtained as small as possible. Substituting $W_f(k)$ back into (3.2.17), $\Xi(k)$ can be calculated as follows:

$$\Xi(k) = (\gamma^2 - 1)I + D_f^{\mathrm{T}}(k)H^{-1}(k)D_f(k) \quad (3.2.23)$$

Theorem 3.2.3. *Let*

$$Q(k) = \begin{bmatrix} Q_{11}(k) & Q_{12}(k) \\ Q_{21}(k) & Q_{22}(k) \end{bmatrix}$$

given γ; if there exit a positive scalar ζ and a solution $Q(k) > 0$ to the Riccati equation (3.2.11) with the resulted in (3.2.23) is positive definite, then (3.2.3) is exponentially stable in mean square and H_∞ performance (3.2.4) is satisfied. The design parameter matrices $L(k)$ and $W(k)$ are given in (3.2.21) and (3.2.22), respectively.

Remark 3.2.3. The design of post-filter $W(k)$ brings more freedom such that a smaller attenuation level γ can be achieved. Comparing (3.2.23) with (3.2.17), it can be seen that when $W(k) = I$, $\Xi(k)$ is inclined to be negative when γ tends to be small. Thus, the H_∞ performance (3.2.4) can be optimized by using $W_f(k)$ in some sense in contrast with designing $L(k)$ only.

3.2.4 Numerical examples

To illustrate the result achieved in this chapter, consider the following stochastic LDTV system with

$$A(k) = \begin{bmatrix} -0.1e^{-k/100} & 0.9^k \\ -0.85 & -0.1 \end{bmatrix}, \quad A_v(k) = \begin{bmatrix} 0 & 0 \\ 0 & 0.5 \end{bmatrix}$$

$$B_f(k) = \begin{bmatrix} 0.6\sin(k) \\ 0.4 \end{bmatrix}, \quad B_{fv}(k) = \begin{bmatrix} 0.5 \\ 0 \end{bmatrix}, B_d(k) = \begin{bmatrix} 0.6 \\ 0.2 \end{bmatrix}, \quad B_{dv}(k) = \begin{bmatrix} 0.3 \\ 0.1 \end{bmatrix}$$

$$C(k) = \begin{bmatrix} -0.1 & 0.3 \end{bmatrix}, \quad C_v(k) = \begin{bmatrix} 0.1 & 0 \end{bmatrix}$$

$$D_f(k) = 0.75, \quad D_{fv}(k) = 0.1, D_d(k) = 0.3, \quad D_{dv}(k) = 0.1.$$

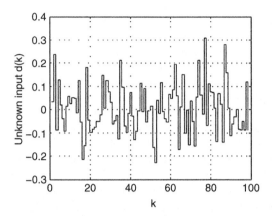

FIGURE 3.1 Unknown input $d(k)$.

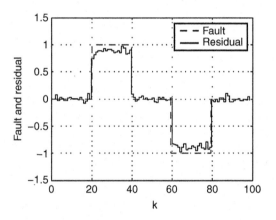

FIGURE 3.2 Stepwise fault $f(k)$ and the residual $r(k)$.

$\{v(k)\}$ is a zero-mean scalar white noise sequence with unit variance, $x(0) = [0.2 \ 0]^T$, $\hat{x}(0) = [0 \ 0]^T$, $Q(0) = I$, $\zeta = 0.01$, and $\gamma = 0.8$. The unknown input $d(k)$ is simulated as shown in Fig. 3.1. By applying Theorem 3.2.3, a stepwise fault and the corresponding generated residual are shown in Fig. 3.2. In addition, a sine wave fault and its corresponding generated residual are shown in Fig. 3.3. It can be seen from the simulation results that the generated residual can approach to fault well when there exists multiplicative noise in systems.

3.3 Robust H_∞ fault detection for LDTV systems with measurement packet loss

3.3.1 Problem formulation

Consider the following LDTV systems:

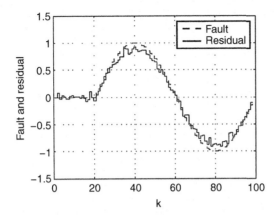

FIGURE 3.3 Sine wave fault $f(k)$ and residual $r(k)$.

$$\begin{cases} x(k+1) = A(k)x(k) + B_f(k)f(k) + B_d(k)d(k) \\ \qquad y(k) = C(k)x(k) + D_f(k)f(k) + D_d(k)d(k) \end{cases} \qquad (3.3.1)$$

where $x(k) \in R^n$, $y(k) \in R^{n_y}$, $d(k) \in R^{n_d}$, and $f(k) \in R^{n_f}$ are the state, measurement output, unknown input, and fault to be detected on system (3.3.1), respectively; without losing generality, under the assumption that $d(k)$, $f(k)$ are l_2 norm-bounded signal and $C(k)$ is full of rank; $A(k), B_f(k), B_d(k), C(k), D_f(k)$, and $D_d(k)$ are known time-varying matrices of appropriate dimensions.

When there is a data packet loss phenomenon in the transmission process of the measurement output, it is assumed that the actually obtained measurement signal $\psi(k) \in R^q$ is

$$\psi(k) = \theta(k)y(k) + (1 - \theta(k))\psi(k-1), \qquad (3.3.2)$$

where $\theta(k)$ is an independent identically distributed Bernoulli random variable and satisfies

$$\begin{cases} \Pr\{\theta(k) = 1\} = \mathrm{E}\{\theta(k)\} = \rho \\ \Pr\{\theta(k) = 0\} = 1 - \mathrm{E}\{\theta(k)\} = 1 - \rho \end{cases} \qquad (3.3.3)$$

$\rho \in (0, 1]$ is the known scalar.

Define $\alpha(k) = \theta(k) - \rho$, $\alpha(k)$ has the following statistical characteristics from (3.3.3):

$$\begin{cases} \mathrm{E}\{\alpha(k)\} = 0 \\ \mathrm{E}\{\alpha^2(k)\} = \rho - \rho^2 := \varepsilon \end{cases} \qquad (3.3.4)$$

Introducing the augmented vector $\xi(k) = [x^T(k) \quad \psi^T(k-1)]^T$ and combining (3.3.1) and (3.3.2), we have

$$
\begin{cases}
\xi(k+1) = (A_1(k) + \alpha(k)A_\alpha(k))\xi(k) + (B_{f1}(k) + \alpha(k)B_{f\alpha}(k))f(k) \\
\qquad + (B_{d1}(k) + \alpha(k)B_{d\alpha}(k))d(k) \\
\psi(k) = (C_1(k) + \alpha(k)C_\alpha(k))\xi(k) + (D_{f1}(k) + \alpha(k)D_{f\alpha}(k))f(k) \\
\qquad + (D_{d1}(k) + \alpha(k)D_{d\alpha}(k))d(k)
\end{cases}
$$

$$(3.3.5)$$

where

$$
A_1(k) = \begin{bmatrix} A(k) & 0 \\ \rho C(k) & (1-\rho)I_q \end{bmatrix}, \quad B_{f1} = \begin{bmatrix} B_f(k) \\ \rho D_f(k) \end{bmatrix}, \quad B_{d1} = \begin{bmatrix} B_d(k) \\ \rho D_d(k) \end{bmatrix}
$$

$$
C_1 = \begin{bmatrix} \rho C(k) & (1-\rho)I_q \end{bmatrix}, \quad D_{f1} = \rho D_f(k), \quad D_{d1} = \rho D_d(k)
$$

$$
A_\alpha(k) = \begin{bmatrix} 0 & 0 \\ C(k) & -I \end{bmatrix}, \quad B_{f\alpha}(k) = \begin{bmatrix} 0 \\ D_f(k) \end{bmatrix}, \quad B_{d\alpha}(k) = \begin{bmatrix} 0 \\ D_d(k) \end{bmatrix}
$$

$$
C_\alpha(k) = \begin{bmatrix} C(k) & -I \end{bmatrix}, \quad D_{f\alpha}(k) = D_f(k), \quad D_{d\alpha}(k) = D_d(k).
$$

Residual generation is the crucial segment in the design of the FDI system. Therefore, the following observer-based FDF is considered as the residual generator:

$$
\begin{cases}
\hat{\xi}(k+1) = A_1(k)\hat{\xi}(k) + L(k)(\psi(k) - C_1(k)\hat{\xi}(k)) \\
r(k) = V(k)(\psi(k) - C_1(k)\hat{x}(k)) \\
\hat{\xi}(0) = \hat{\xi}_0
\end{cases}
$$

$$(3.3.6)$$

where $\hat{\xi}(k)$ is the estimation of $\xi(k)$ and $\hat{\xi}_0$ is the initial value of the designed filter, $r(k) \in R^r$ is the residual, and observer gain matrix $L(k)$ and post-filter $V(k)$ are the parameters to be designed.

Let

$$
e(k) = x(k) - \hat{x}(k), \quad \eta(k) = \begin{bmatrix} \xi^T(k) & e^T(k) \end{bmatrix}^T
$$

$$
r_e(k) = r(k) - f(k), \quad w(k) = [f^T(k) \quad d^T(k)]^T
$$

from Eqs. (3.3.5) and (3.3.6), and we have

$$
\begin{cases}
\eta(k+1) = (A_\eta(k) + \alpha(k)A_{\eta\alpha}(k))\eta(k) + (B_\eta(k) + \alpha(k)B_{\eta\alpha}(k))w(k) \\
r_e(k) = (C_\eta(k) + \alpha(k)C_{\eta\alpha}(k))\eta(k) + (D_\eta(k) + \alpha(k)D_{\eta\alpha}(k))w(k)
\end{cases}
$$

$$(3.3.7)$$

where

$$
A_\eta(k) = \begin{bmatrix} A_1(k) & 0 \\ 0 & A_1(k) - L(k)C_1(k) \end{bmatrix}, \quad A_{\eta\alpha}(k) = \begin{bmatrix} A_\alpha(k) & 0 \\ A_\alpha(k) - L(k)C_\alpha(k) & 0 \end{bmatrix}
$$

$$
B_\eta(k) = \begin{bmatrix} B_{f1}(k) & B_{d1}(k) \\ B_{f1}(k) - L(k)D_{f1}(k) & B_{d1}(k) - L(k)D_{d1}(k) \end{bmatrix}
$$

$$
B_{\eta\alpha}(k) = \begin{bmatrix} B_{f\alpha}(k) & B_{d\alpha}(k) \\ B_{f\alpha}(k) - L(k)D_{f\alpha}(k) & B_{d\alpha}(k) - L(k)D_{d\alpha}(k) \end{bmatrix}
$$

$$
C_\eta(k) = \begin{bmatrix} 0 & V(k)C_1(k) \end{bmatrix}, \quad C_{\eta\alpha}(k) = \begin{bmatrix} V(k)C_\alpha(k) & 0 \end{bmatrix}
$$

$$
D_\eta(k) = \begin{bmatrix} V(k)D_{f1}(k) - I & V(k)D_{d1}(k) \end{bmatrix}, \quad D_{\eta\alpha}(k) = \begin{bmatrix} V(k)D_{f\alpha}(k) & V(k)D_{d\alpha}(k) \end{bmatrix}
$$

Note that system (3.3.7) is a time-varying system containing a random variable $\alpha(k)$, and the following definition is presented, first given in the work of Morozan [26].

Definition 3.3.1. System (3.3.7) is said to be mean square exponential stable if $c \geq 0$ and $q \in (0, 1)$ exist under the condition of zero input (i.e., $w(k) = 0$) so that the following relation holds:

$$E\{\|\eta(k)\|^2\} \leq cq^k \|\eta(0)\|^2$$

To sum up, the FDF design problems to be solved in this section can be summarized as follows.

Problem 3.3.1. Given $\gamma > 0$, design parameter matrices $L(k)$ and $V(k)$ so as to make system (3.3.7) mean square index stable and meet the following performance indexes,

$$\sup_{\|w(k)\|_{2,N} \neq 0} \frac{\|r_e(k)\|_{2,E}^2}{\eta^T(0)S\eta(0) + \|w(k)\|_2^2} < \gamma^2 \tag{3.3.8}$$

where $S > 0$ is the weighting matrix for initial state.

Remark 3.3.1. Systems with packet loss characteristics described by Bernoulli random variables can be roughly divided into two categories. One is the case with multistep measurement of packet loss in the form of (3.3.2); when the parameter matrix of system (3.3.1) is constant, the FDF design problem mentioned earlier is the problem studied by the document [23]. The second is the case of single-step data packet loss measurement as given in the work of Gao et al. [32] and Zhao et al. [33]. This section mainly studies the FDF design of an LDTV system with multistep data packet loss measurement. The proposed algorithm can also be applied to an LDTV system with single-step data packet loss measurement.

3.3.2 Main results

Note that the form of system (3.3.7) is similar to the LDTV system (3.2.3) with multiplicative noise in the previous section. It is known that the statistical characteristics of the random variable $\alpha(k)$ are also similar to the multiplicative noise $v(k)$ according to (3.3.4).Therefore, the relevant Theorem 3.2.2 in Section 3.2.2 can be extended to system (3.3.7) to obtain the necessary and sufficient conditions for FDF to exist, which can be summarized as the following theorem.

Theorem 3.3.1. *For system (3.3.7), given $\gamma > 0$, if the constant $\beta > 0$ and positive definite matrix $Q(k)$ make the following Riccati equation hold so that system (3.3.7) is mean square exponential stable and meets the H_∞ performance* *index (3.3.8):*

$$\begin{cases} Q(k+1) = A_\eta(k)Q(k)A_\eta^T(k) + \varepsilon B_{\eta\alpha}(k)B_{\eta\alpha}^T(k) + \varepsilon A_{\eta\alpha}(k)Q(k)A_{\eta\alpha}^T(k) \\ \qquad + B_\eta(k)B_\eta^T(k) + M(k)\Theta^{-1}(k)M^T(k) + \beta I \\ Q(0) = S^{-1} \end{cases}$$

$$(3.3.9)$$

where

$$M(k) = (C_\eta(k)Q(k)A_\eta^T(k) + \varepsilon D_{\eta\alpha}(k)B_{\eta\alpha}^T(k) + \varepsilon C_{\eta\alpha}(k)Q(k)A_{\eta\alpha}^T(k) + D_\eta(k)B_\eta^T(k))^T$$

$$\Theta(k) = \gamma^2 I - C_\eta(k)Q(k)C_\eta^T(k) - D_\eta(k)D_\eta^T(k) - \varepsilon C_{\eta\alpha}(k)Q(k)C_{\eta\alpha}^T(k)$$

$$\quad - \varepsilon D_{\eta\alpha}(k)D_{\eta\alpha}^T(k) > 0.$$

Based on the sufficient conditions for the existence of H_∞-FDF given by Theorem 3.3.1, the calculation of parameter matrices $L(k)$, $V(k)$ is transformed into a quadratic optimization problem. The main idea is to obtain a set of feasible solutions of $L(k)$, $V(k)$ by solving Riccati equation (3.3.9), making $Theta(k)$ satisfies the positive qualitative condition for the smallest possible γ and further optimizes the performance index (3.3.8) to a certain extent [34].

Define

$$Q(k) = \begin{bmatrix} Q_{11}(k) & Q_{12}(k) \\ Q_{21}(k) & Q_{22}(k) \end{bmatrix}$$

and we have

$$\Theta(k) = -V(k)C_1(k)Q_{22}(k)C_1^T(k)V^T(k) - \varepsilon V(k)C_\alpha(k)Q_{11}(k)C_\alpha^T(k)V^T(k)$$

$$\quad - \varepsilon V(k)D_{f\alpha}(k)D_{f\alpha}^T(k)V^T(k) - \varepsilon V(k)D_{d\alpha}(k)D_{d\alpha}^T(k)V^T(k) + \gamma^2 I$$

$$\quad - (V(k)D_{f1}(k) - I)(V(k)D_{f1}(k) - I)^T - V(k)D_{d1}(k)D_{d1}^T(k)V^T(k).$$

$$(3.3.10)$$

Notice that $\Theta(k)$ is a quadratic function of $V(k)$, then a feasible solution $V_f(k)$ of $V(k)$ can be defined as follows: at time instant k, given $Q(k)$, the following relation holds for any nonzero column vector of appropriate dimension $\varsigma(k)$ as follows:

$$\varsigma^T(k)\Theta(V_f(k), k)\varsigma(k) \geq \varsigma^T(k)\Theta(V(k), k)\varsigma(k) > 0.$$

Associating the following equation,

$$\frac{\partial \varsigma^T(k)\Theta(V(k), k)\varsigma(k)}{\partial(V^T(k)\varsigma(k))} = 0.$$

we have

$$- \varsigma^T(k)V(k)(C_1(k)Q_{22}(k)C_1^T(k) + \varepsilon C_\alpha(k)Q_{11}(k)C_\alpha^T(k) + \varepsilon D_{f\alpha}(k)D_{f\alpha}^T(k)$$

$$+ \varepsilon D_{d\alpha}(k)D_{d\alpha}^T(k) + D_{f1}(k)D_{f1}^T(k) + D_{d1}(k)D_{d1}^T(k)) + \varsigma^T(k)D_{f1}^T(k) = 0.$$

$$(3.3.11)$$

Furthermore, from the prior condition that $C(k)$ is a row-full rank, we get

$$\frac{\partial^2 \varsigma^T(k)\Theta(V(k),k)v(k)}{\partial(V^T(k)\varsigma(k))^2} = -(C_1(k)Q_{22}(k)C_1^T(k) + \varepsilon C_\alpha(k)Q_{11}(k)C_\alpha^T(k)$$
$$+ \varepsilon D_{f\alpha}(k)D_{f\alpha}^T(k) + \varepsilon D_{d\alpha}(k)D_{d\alpha}^T(k) + D_{f1}(k)D_{f1}^T(k)$$
$$+ D_{d1}(k)D_{d1}^T(k))$$
$$< 0,$$

then

$$V_f(k) = D_{f1}^T(k)(C_1(k)Q_{22}(k)C_1^T(k) + \varepsilon C_\alpha(k)Q_{11}(k)C_\alpha^T(k) + \varepsilon D_{f\alpha}(k)D_{f\alpha}^T(k)$$
$$+ \varepsilon D_{d\alpha}(k)D_{d\alpha}^T(k) + D_{f1}(k)D_{f1}^T(k) + D_{d1}(k)D_{d1}^T(k))^{-1} \qquad (3.3.12)$$

Combining (3.3.12) and (3.3.10), we get

$$\Theta(k) = (\gamma^2 - 1)I + D_{f1}^T(k)(C_1(k)Q_{22}(k)C_1^T(k) + \varepsilon C_\alpha(k)Q_{11}(k)C_\alpha^T(k)$$
$$+ \varepsilon D_{f\alpha}(k)D_{f\alpha}^T(k) + \varepsilon D_{d\alpha}(k)D_{d\alpha}^T(k) + D_{f1}(k)D_{f1}^T(k)$$
$$+ D_{d1}(k)D_{d1}^T(k))D_{f1}(k). \qquad (3.3.13)$$

Referring (3.3.7) and (3.3.9), the matrix $Q_{11}(k+1)$ and $L(k)$ are independent of each other, and the following relation holds for $Q_{22}(k+1)$:

$$Q_{22}(k+1) = (A_1(k) - L(k)C_1(k))Q_{22}(k)(A_1(k) - L(k)C_1(k))^T$$
$$+ \varepsilon(A_\alpha(k) - L(k)C_\alpha(k))Q_{11}(k)(A_\alpha(k) - L(k)C_\alpha(k))^T$$
$$+ (B_{f1}(k) - L(k)D_{f1}(k))(B_{f1}(k) - L(k)D_{f1}(k))^T$$
$$+ (B_{d1}(k) - L(k)D_{d1}(k))(B_{d1}(k) - L(k)D_{d1}(k))^T$$
$$+ \varepsilon(B_{f\alpha}(k) - L(k)D_{f\alpha}(k))(B_{f\alpha}(k) - L(k)D_{f\alpha}(k))^T$$
$$- L(k)D_{d\alpha}(k))(B_{d\alpha}(k) - L(k)D_{d\alpha}(k))^T$$
$$+ \Gamma(k)\Theta^{-1}(k)\Gamma^T(k) + \beta I,$$

where

$$\Gamma(k) = (A_1(k) - L(k)C_1(k))Q_{22}(k)C_1^T(k)V^T(k)$$
$$+ \varepsilon(A_\alpha(k) - L(k)C_\alpha(k))Q_{11}(k)C_\alpha^T(k)V^T(k)$$
$$+ (B_{f1}(k) - L(k)D_{f1}(k))(V(k)D_{f1}(k) - I)^T$$
$$+ (B_{d1}(k) - L(k)D_{d1}(k))D_{d1}^T(k)V^T(k)$$
$$+ \varepsilon(B_{f\alpha}(k) - L(k)D_{f\alpha}(k))(V(k)D_{f\alpha}(k) - I)^T$$
$$+ \varepsilon(B_{d\alpha}(k) - L(k)D_{d\alpha}(k))D_{d\alpha}^T(k)V^T(k).$$

Similar to the preceding analysis for finding $V(k)$, a feasible solution $L_f(k)$ of $L(k)$ can be defined as follows: at time instant k, given $V(k)$, $Q(k)$, the following relation holds for any nonzero column vector of appropriate dimension $\varsigma(k)$:

$$\varsigma^T(k)Q_{22}(k+1)\varsigma(k)|_{L(k)=L_f(k)} \leq \varsigma^T(k)Q_{22}(k+1)\varsigma(k)|_{L(k)\neq L_f(k)}$$

so that the following inequality holds:

$$\varsigma^T(k+1)\Theta(L_f(k+1), k+1)\varsigma(k+1)$$
$$\geq \varsigma^T(k+1)\Theta(L(k+1), k+1)\varsigma(k+1) > 0.$$

Based on

$$\frac{\partial \varsigma^T(k)Q_{22}(k+1)\varsigma(k)}{\partial(L^T(k)\varsigma(k))} = 0,$$

we get

$$0 = (H(k) + G(k)\Theta^{-1}(k)G^T(k))L^T(k) - G(k)\Theta^{-1}(k)N^T(k) - T(k),$$

where

$$H(k) = C_1(k)Q_{22}(k)C_1^T(k) + \varepsilon C_\alpha(k)Q_{11}(k)C_\alpha^T(k)$$
$$+ D_{f1}(k)D_{f1}^T(k) + D_{d1}(k)D_{d1}^T(k)$$
$$+ \varepsilon D_{f\alpha}(k)D_{f\alpha}^T(k)^T + D_{d1}(k)D_{d1}^T(k) + \varepsilon D_{d\alpha}(k)D_{d\alpha}^T(k)$$

$$G(k) = C_1(k)Q_{22}(k)C_1^T(k)V^T(k) + \varepsilon C_\alpha(k)Q_{11}(k)C_\alpha^T(k)V^T(k)$$
$$+ D_{d1}(k)D_{d1}^T(k)V^T(k) + \varepsilon D_{f\alpha}(k)D_{f\alpha}^T(k)V^T(k)$$
$$+ \varepsilon D_{d\alpha}(k)D_{d\alpha}^T(k)V^T(k) + D_{f1}(k)(V(k)D_{f1}(k) - I)^T$$

$$N(k) = A_1(k)Q_{22}(k)C_1^T(k)V^T(k) + \varepsilon A_\alpha(k)Q_{11}(k)C_\alpha^T(k)V^T(k)$$
$$+ B_{d1}(k)D_{d1}^T(k)V^T(k) + B_{f1}(k)(V(k)D_{f1}(k) - I)^T$$
$$+ \varepsilon B_{f\alpha}(k)D_{d\alpha}^T(k) + \varepsilon B_{d\alpha}(k)D_{d\alpha}^T(k)V^T(k)$$

$$T(k) = C_1(k)Q_{22}(k)A_1^T(k) + \varepsilon C_\alpha(k)Q_{11}(k)A_\alpha^T(k) + D_{d1}(k)B_{d1}^T(k)$$
$$+ \varepsilon D_{f\alpha}(k)B_{d\alpha}^T(k) + \varepsilon D_{d\alpha}(k)B_{d\alpha}^T(k) + D_{f1}(k)B_{f1}^T(k).$$

Furthermore, combining the known condition that $C(k)$ is the full row rank, we obtain

$$\frac{\partial^2 \varsigma^T(k)Q_{22}(k+1)\varsigma(k)}{(\partial L^T(k)\varsigma(k))^2} = H(k) + G(k)\Theta^{-1}(k)G^T(k) > 0,$$

then

$$L_f(k) = \Omega^T(k)(H(k) + G(k)\Theta^{-1}(k)G^T(k))^{-1} \qquad (3.3.14)$$

where

$$\Omega(k) = T(k) + G(k)\Theta^{-1}(k)N^T(k).$$

$\Theta(k)$ could be obtained following (3.3.13).

Based on the preceding analysis, the following theorem is obtained.

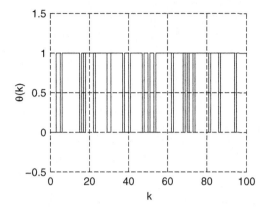

FIGURE 3.4 The rate of change $\theta(k)$.

Theorem 3.3.2. *For system (3.3.7), given $\gamma > 0$, if the constant $\beta > 0$ and symmetric matrix $Q(k)$ make Eq. (3.3.9) hold and $\Theta(k) > 0$ in (3.3.13) so that system (3.3.7) is mean square exponential stable and meets the H_∞ performance index (3.3.8), where the parameter matrices $L(k)$, $V(k)$ can be obtained by the expressions (3.3.14) and (3.3.12), respectively.*

$$Q(k) = \begin{bmatrix} Q_{11}(k) & Q_{12}(k) \\ Q_{21}(k) & Q_{22}(k) \end{bmatrix}$$

Remark 3.3.2. Comparing the form of the FDF parameter matrix of the LDTV system with that affected by multiplicative noise in Section 2 under the condition of multistep measurement data packet loss in this section, it can be seen that for a specific form of system (3.2.3) or (3.3.7), the feasible solution of parameter matrix can be obtained by adopting the matrix optimization idea given in this section to solve the Riccati equation and the method of minimizing the upper bound of variance in Section 2.

3.3.3 Numerical examples

To illustrate the effectiveness of FDF designed by Theorem 3.3.2, a set of system (3.3.1) parameter matrices are selected as follows:

$$A(k) = \begin{bmatrix} -0.1e^{-k/100} & 0.9^k \\ -0.85 & -0.1 \end{bmatrix}, \quad B_f(k) = \begin{bmatrix} 0.6\sin(k) \\ 0.4 \end{bmatrix} B_d(k) = \begin{bmatrix} 0.6 \\ 0.2 \end{bmatrix}$$

$$C(k) = \begin{bmatrix} -0.1 & 0.3 \end{bmatrix}, \quad D_f(k) = 0.8, D_d(k) = 0.3.$$

Set packet loss probability $\rho = 0.85$, $Q(0) = 0.1I$, $x(0) = [0.2 \ 0]^T$, $\hat{x}(0) = [0 \ 0]^T$, $Q(0) = I$, $\beta = 0.01$, and $\gamma = 1.1$. The rate of change $\theta(k)$ and the unknown input signal is shown in Fig. 3.4 and Fig. 3.5 respectively. Fig. 3.6 and Fig. 3.7 respectively give the square wave fault and sine wave fault and the

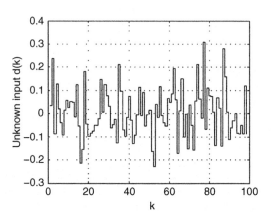

FIGURE 3.5 The unknown input signal $d(k)$.

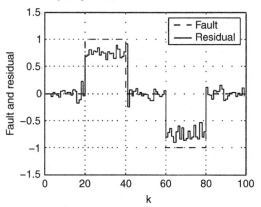

FIGURE 3.6 Square wave fault $f(k)$ and the residual $r(k)$.

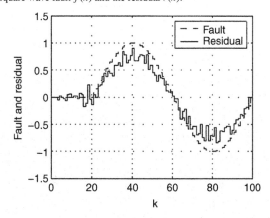

FIGURE 3.7 Sine wave fault $f(k)$ and the residual $r(k)$.

corresponding residual signal. From Fig. 3.6 and Fig. 3.7, it can be seen that the robust H_∞-FDF designed based on Theorem 3.3.2 can obtain an effective residual signal when a fault occurs.

3.4   Fixed-lag H_∞ fault estimator design for LDTV systems under an unreliable communication link

3.4.1   Problem formulation and preliminaries

Consider the following LDTV system:

$$\begin{cases} x(k+1) = A(k)x(k) + B_f(k)f(k) + D(k)d(k) \\ y(k) = \theta(k)C(k)x(k) + v(k) \\ x(0) = x_0 \end{cases} \tag{3.4.1}$$

where $x(k) \in R^n$, $y(k) \in R^q$, $d(k) \in R^{n_d}$, $v(k) \in R^{n_v}$, and $f(k) \in R^{n_f}$ denote the state, sensor measurement, process noise, observation noise, and fault, respectively. $f(k)$, $d(k)$, and $v(k)$ belong to $l_2[0, N]$. $A(k)$, $B_f(k)$, $C(k)$, and $D(k)$ are known time-varying matrices with appropriate dimensions. $\theta(k)$ is a Bernoulli distributed binary stochastic variable to describe the measurement packet dropouts, which satisfies

$$\begin{cases} \text{Prob}\{\theta(k) = 1\} = E\{\theta(k)\} = \rho \\ \text{Prob}\{\theta(k) = 0\} = 1 - E\{\theta(k)\} = 1 - \rho \end{cases} \tag{3.4.2}$$

with ρ as a known constant. The value of ρ can be obtained by empirical observations, experimentations, and statistical analysis [35].

The main purpose of this chapter is as follows. Given a prescribed disturbance attenuation level γ, by collecting the observations $y(0), \ldots, y(k)$, find $\check{f}(k-l|k)$ as a suitable estimation of the fault signal $f(k)$ such that the following l-step delayed H_∞ performance index is fulfilled with l as a positive integer:

$$\sup_{(x_0, f_k, d_k, v_k) \neq 0} \frac{E\left\{\sum_{k=l}^{N}(\check{f}(k-l|k) - f(k-l))^{\mathrm{T}}(\check{f}(k-l|k) - f(k-l))\right\}}{x_0^{\mathrm{T}}P_0^{-1}x_0 + \sum_{k=0}^{N}f^{\mathrm{T}}(k)f(k) + \sum_{k=0}^{N-1}d^{\mathrm{T}}(k)d(k) + \sum_{k=0}^{N}v^{\mathrm{T}}(k)v(k)} < \gamma^2 \tag{3.4.3}$$

where $f_k = [f^{\mathrm{T}}(0) \cdots f^{\mathrm{T}}(k)]^{\mathrm{T}}$, $d_k = [d^{\mathrm{T}}(0) \cdots d^{\mathrm{T}}(k)]^{\mathrm{T}}$, and $v_k = [v^{\mathrm{T}}(0) \cdots v^{\mathrm{T}}(k)]^{\mathrm{T}}$.

Due to the denominator of the left side of (3.4.3) being positive, (3.4.3) can be rewritten as

$$J_0 = x_0^{\mathrm{T}}P_0^{-1}x_0 + \sum_{k=0}^{N}f^{\mathrm{T}}(k)f(k) + \sum_{k=0}^{N-1}d^{\mathrm{T}}(k)d(k) + \sum_{k=0}^{N}v^{\mathrm{T}}(k)v(k)$$

$$- E\left\{\gamma^{-2}\sum_{k=l}^{N}v_s^{\mathrm{T}}(k)v_s(k)\right\} > 0, \tag{3.4.4}$$

where $v_s(k) = \check{f}(k - l|k) - f(k - l)$. Consequently, according to Stoorvogel et al. [36], the H_∞ fixed-lag fault estimation problem can be restated as follows. Given a constant $\gamma > 0$, design an estimator in the following way:

$$\check{f} = \Psi(y) = \bar{\Psi}(f, d, v),$$

where Ψ denotes a stable operator that generates a bounded operator $\bar{\Psi}$ mapping from f, d, v to $\check{f}$ such that the indefinite cost function (3.4.4) has a positive minimum with respect to f, d, and v.

Remark 3.4.1. In the existing results (e.g., [19–21, 37-40]), the Bernoulli distributed random variables are introduced to describe the packet dropping or finite step measurement time-delay phenomenon. It is noteworthy that the designed estimators only depend on the probability (i.e. ρ) rather than $\theta(k)$. This indicates that the desired fault estimator does not require the timestamp of the data packet.

Remark 3.4.2. Notice that when $y(k)$ is affected by the so-called sensor fault with the following form,

$$y(k) = \theta(k)C(k)x(k) + D_f(k)f(k) + v(k),$$

the existing BRL-based H_∞ fault estimation algorithm in the work of Li et al. [39] is applicable in a *filter* manner. In the case that $D_f(k) = 0$, the estimator is supposed to be designed as a *smoother* with the proposed performance index (3.4.4). In this scenario, the methodology of Li et al. [39] may induce computational burden via a state augmentation approach and the gain matrices of the estimator are arduous to be derived due to some coupled product terms. In what follows, a Krein space–based fault estimator design scheme will be addressed to overcome the aforementioned defects.

In this section, inspired by the work of Zhao et al. [41] and Lu et al. [42], an equivalent Krein space stochastic system and a corresponding H_∞ performance index are first introduced. Then, by exploiting the reorganized innovation analysis and the projection theory in Krein space, the H_∞ fault estimator is derived.

3.4.2 Krein space model design

Before we proceed, we would like to propose the following lemma to construct an auxiliary stochastic system in Krein space.

Lemma 3.4.1. *Given a scalar $\gamma > 0$ and an integer $l > 0$, the H_∞ performance (3.4.4) is fulfilled if and only if there exists a fault estimator $\check{f}(k - l|k)$ such that* *the following inequality holds:*

$$J = x_0^T P_0^{-1} x_0 + \sum_{k=0}^{N} f^T(k)f(k) + \sum_{k=0}^{N} v_0^T(k)v_0(k) + \sum_{k=0}^{N-1} d^T(k)d(k)$$

$$+ \sum_{k=0}^{N} v_z^T(k)v_z(k) - \gamma^{-2} \sum_{k=l}^{N} v_s^T(k)v_s(k) > 0 \tag{3.4.5}$$

subject to the following dynamic constraints

$$\begin{cases} x(k+1) = A(k)x(k) + B_f(k)f(k) + D(k)d(k) \\ y_0(k) = \rho C(k)x(k) + v_0(k) \\ y_z(k) = \sqrt{\rho(1-\rho)}C(k)x(k) + v_z(k) \\ \check{f}(k-l|k) = f(k-l) + v_s(k) \\ x(0) = x_0 \end{cases} \tag{3.4.6}$$

where $y_0(k)$ and $y_z(k)$ are the fictitious observations with their corresponding observation noises $v_0(k)$ and $v_z(k)$, respectively. The instantaneous value of y_0 at each time instant k is equal to $y(k)$ along with $y_z(k) \equiv 0$.

Proof. *Necessity*: From (3.4.1), the state transition matrix Φ is defined as

$$\Phi(k,j) = \begin{cases} A(k-1)\cdots A(j), & 0 < k < j \\ I, & k = j \end{cases}$$

and hence we have

$$x(k) = \Phi(k,0)x_0 + \sum_{i=0}^{k-1} \Phi(k,i+1)B_f(i)f(i) + \sum_{i=0}^{k-1} \Phi(k,i+1)D(i)d(i). \tag{3.4.7}$$

Define

$$y_k = [y^T(0)\cdots y^T(k)]^T, \quad v_{s,k} = [v_s^T(0)\cdots v_s^T(k)]^T$$
$$\check{f}_k = [\check{f}^T(0|l)\cdots \check{f}^T(k-l|k)]^T,$$

then, in view of (3.4.7), we have

$$\begin{cases} y_N = \Xi(k)G_x x_0 + \Xi(k)G_f f_N + \Xi(k)G_d d_N + v_N \\ \check{f}_N = f_{N-l} + v_{s,N} \end{cases} \tag{3.4.8}$$

where

$$\Xi(k) = diag\{\theta(1),\ldots,\theta(k)\}, \quad G_f(k,i) = C(k)\Phi(k,i+1)B_f(i)$$
$$G_d(k,i) = C(k)\Phi(k,i+1)D(i)$$

$$G_x = \begin{bmatrix} C(0)\Phi(0,0) \\ C(1)\Phi(1,0) \\ \vdots \\ C(N)\Phi(N,0) \end{bmatrix}, \quad G_f = \begin{bmatrix} 0 & \cdots & & \cdots & 0 \\ G_f(1,0) & 0 & & \cdots & 0 \\ \vdots & & \ddots & \ddots & \\ G_f(N,0) & G_f(N,1) & \cdots & & 0 \end{bmatrix}$$

$$
G_d = \begin{bmatrix} 0 & \cdots & \cdots & 0 \\ G_d(1,0) & 0 & \cdots & 0 \\ \vdots & \ddots & \ddots & \\ G_d(N,0) & G_d(N,1) & \cdots & 0 \end{bmatrix}
$$

Thus, by substituting (3.4.8) into (3.4.4) and taking (3.4.2) into consideration, we have

$$
\begin{aligned}
J_0 = \mathrm{E} \Bigg\{ & x_0^{\mathrm{T}} P_0^{-1} x_0 + \sum_{k=0}^{N} f^{\mathrm{T}}(k)f(k) + \sum_{k=0}^{N-1} d^{\mathrm{T}}(k)d(k) \\
& - (y_N - \Xi(k)G_x x_0 - \Xi(k)G_f f_N - \Xi(k)G_d d_N)^{\mathrm{T}} \\
& \times (y_N - \Xi(k)G_x x_0 - \Xi(k)G_f f_N - \Xi(k)G_d d_N) \\
& - \gamma^{-2} \sum_{k=l}^{N} (\check{f}(k-l|k) - f(k-l))^{\mathrm{T}} (\check{f}(k-l|k) - f(k-l)) \Bigg\} \\
= & x_0^{\mathrm{T}} P_0^{-1} x_0 + \sum_{k=0}^{N} f^{\mathrm{T}}(k)f(k) + \sum_{k=0}^{N-1} d^{\mathrm{T}}(k)d(k) \\
& + (y_{0,N} - \bar{\Xi}G_x x_0 - \bar{\Xi}G_f f_N - \bar{\Xi}G_d d_N)^{\mathrm{T}} (y_{0,N} - \bar{\Xi}G_x x_0 - \bar{\Xi}G_f f_N - \bar{\Xi}G_d d_N) \\
& + (y_{z,N} - \tilde{\Xi}G_x x_0 - \tilde{\Xi}G_f f_N - \tilde{\Xi}G_d d_N)^{\mathrm{T}} (y_{z,N} - \tilde{\Xi}G_x x_0 - \tilde{\Xi}G_f f_N - \tilde{\Xi}G_d d_N) \\
& - \gamma^{-2} \sum_{k=l}^{N} (\check{f}(k-l|k) - f(k-l))^{\mathrm{T}} (\check{f}(k-l|k) - f(k-l)), \qquad (3.4.9)
\end{aligned}
$$

where

$$
\begin{aligned}
& y_{0,k} = [y_0^{\mathrm{T}}(0) \cdots y_0^{\mathrm{T}}(k)]^{\mathrm{T}}, \ y_{z,k} = [y_z^{\mathrm{T}}(0) \cdots y_z^{\mathrm{T}}(k)]^{\mathrm{T}} \\
& y_0(i) = y(i), \ y_z(i) = 0 \ (i = 0, \ldots, k) \\
& \bar{\Xi} = \rho I, \ \tilde{\Xi} = \sqrt{\rho(1-\rho)} I.
\end{aligned}
$$

Therefore, if the H_∞ performance index (3.4.4) is satisfied, then following the same line with the correlation between (3.4.1) and (3.4.4), we have $J > 0$ subject to the dynamics (3.4.6) over x_0, f_k, and d_k.

Sufficiency. For (3.4.6), since the value of $y_0(k)$ is equivalent to $y(k)$ and $y_z(k) \equiv 0$, in light of (3.4.9), it is easy to find out that for a given constant $\gamma > 0$ and an integer $l > 0$, $J_0 = J$, which indicates that if $J > 0$ holds, then the H_∞ performance (3.4.4) is satisfied. Combing the sufficiency and necessity part, the proof is complete. □

In virtue of Lemma 3.4.1, the auxiliary performance index J in (3.4.5) can be converted into the following compact form:

$$
J = \begin{bmatrix} x_0 \\ d_N \\ f_N \\ v_{a,N} \end{bmatrix}^{\mathrm{T}} \begin{bmatrix} I & 0 & 0 & 0 \\ 0 & I & 0 & 0 \\ 0 & 0 & I & 0 \\ 0 & 0 & 0 & Q_{a,N} \end{bmatrix}^{-1} \begin{bmatrix} x_0 \\ d_N \\ f_N \\ v_{a,N} \end{bmatrix} \qquad (3.4.10)
$$

where

$$v_a(k) = \begin{cases} v_1(k) = \begin{bmatrix} v_0(k) \\ v_z(k) \end{bmatrix}, & 0 \le k < l \\[3mm] v_2(k) = \begin{bmatrix} v_0(k) \\ v_z(k) \\ v_s(k) \end{bmatrix}, & k \ge l \end{cases} \tag{3.4.11}$$

$$v_{a,N} = \begin{bmatrix} v_a^{\mathrm{T}}(0) \dots v_a^{\mathrm{T}}(N) \end{bmatrix}^{\mathrm{T}}$$

$$Q_a(k) = \begin{cases} Q_{v_1}(k) = diag\{I, I\}, \ 0 \le k < l \\ Q_{v_2}(k) = diag\{I, I, -\gamma^2 I\}, \ k \ge l \end{cases}$$

$$Q_{a,N} = diag\{Q_a(0), \dots, Q_a(N)\}. \tag{3.4.12}$$

From (3.4.6) and (3.4.11), we have

$$y_f(k) = \begin{bmatrix} y(k) \\ y_z(k) \end{bmatrix} = C_1(k)x(k) + v_1(k) \tag{3.4.13}$$

$$y_a(k) = \begin{cases} y_f(k), \ 0 \le k < l \\[3mm] \begin{bmatrix} y_f(k) \\ \check{f}(k - l | k) \end{bmatrix} = C_2(k)x(k) + Hf(k - l) + v_2(k), \ k \ge l \end{cases}$$

$$\tag{3.4.14}$$

where

$$C_1(k) = \begin{bmatrix} \rho C(k) \\ \sqrt{\rho(1 - \rho)} C(k) \end{bmatrix}, \ C_2(k) = \begin{bmatrix} C_1(k) \\ 0 \end{bmatrix}$$

$$H = \begin{bmatrix} 0 & 0 & I \end{bmatrix}^{\mathrm{T}}.$$

Thus, according to Xie and Zhang [43] and Hassibi et al. [44], we introduce the following Krein space system associated with (3.4.6), (3.4.10), (3.4.12), (3.4.13) and (3.4.14)

$$\begin{cases} \mathbf{x}(k + 1) = A(k)\mathbf{x}(k) + B_f(k)\mathbf{f}(k) + D(k)\mathbf{d}(k) \\[3mm] \mathbf{y}_a(k) = \begin{cases} \mathbf{y}_f(k) = C_1(k)\mathbf{x}(k) + \mathbf{v}_1(k), \ 0 \le k < l \\[3mm] \begin{bmatrix} \mathbf{y}_f(k) \\ \check{\mathbf{f}}(k - l | k) \end{bmatrix} = C_2(k)\mathbf{x}(k) + H\mathbf{f}(k - l) + \mathbf{v}_2(k), \ k \ge l \end{cases} \\[3mm] \mathbf{x}(0) = \mathbf{x}_0 \end{cases} \tag{3.4.15}$$

where $\mathbf{x}_0(i)$, $\mathbf{d}(i)$, $\mathbf{f}(i)$, $\mathbf{v}_1(i)$, and $\mathbf{v}_2(i)$ are uncorrelated white noises in Krein space that satisfying

$$\left\langle \begin{bmatrix} \mathbf{x}_0 \\ \mathbf{d}(i) \\ \mathbf{f}(i) \\ \mathbf{v}_a(i) \end{bmatrix}, \begin{bmatrix} \mathbf{x}_0 \\ \mathbf{d}(i) \\ \mathbf{f}(i) \\ \mathbf{v}_a(i) \end{bmatrix} \right\rangle = \begin{bmatrix} I\delta_{ij} & 0 & 0 & 0 \\ 0 & I\delta_{ij} & 0 & 0 \\ 0 & 0 & I\delta_{ij} & 0 \\ 0 & 0 & 0 & Q_a(i)\delta_{ij} \end{bmatrix} \tag{3.4.16}$$

and

$$
\mathbf{v}_a(k) = \begin{cases} \mathbf{v}_1(k) = \begin{bmatrix} \mathbf{v}_0(k) \\ \mathbf{v}_z(k) \end{bmatrix}, & 0 \le k < l \\[2ex] \mathbf{v}_2(k) = \begin{bmatrix} \mathbf{v}_0(k) \\ \mathbf{v}_z(k) \\ \mathbf{v}_s(k) \end{bmatrix}, & k \ge l \end{cases}
$$

with $\mathbf{v}_0(k)$, $\mathbf{v}_z(k)$, and $\mathbf{v}_s(k)$ are fictitious noise in Krein space corresponding to (3.4.11).

Consequently, on the basis of Lemma 4.2.1 in the work of Xie and Zhang [43], we have the following lemma.

Lemma 3.4.2. *For (3.4.6), given a scalar $\gamma > 0$ and an integer $l > 0$, the H_∞ performance (3.4.5) has a minimum over x_0, f, d if and only if $Q_a(k)$ and $Q_w(k)$ have the same inertia, where $Q_w(k) = \langle \mathbf{w}(k), \mathbf{w}(k) \rangle$ is the covariance matrix of innovation sequence $\mathbf{w}(k)$ given by*

$$
\mathbf{w}(k) = \mathbf{y}_a(k) - \hat{\mathbf{y}}_a(k), \tag{3.4.17}
$$

where $\hat{\mathbf{y}}_a(k)$ is the projection of $\mathbf{y}_a(k)$ onto $\mathcal{L}\{\{\mathbf{y}_a(j)\}_{j=0}^{k-1}\}$. Furthermore, the minimum value of J is

$$
J_{min} = \sum_{k=l}^{N} \begin{bmatrix} y_f(k) - C_1(k)\hat{x}(k) \\ \check{f}(k-l|k) - \hat{f}(k-l|k-1) \end{bmatrix}^{\mathrm{T}} Q_w^{-1}(k) \begin{bmatrix} y_f(k) - C_1(k)\hat{x}(k) \\ \check{f}(k-l|k) - \hat{f}(k-l|k-1) \end{bmatrix}
$$
$$
+ \sum_{k=0}^{l-1} [y_f(k) - C_1(k)\hat{x}(k)]^{\mathrm{T}} Q_w^{-1}(k) [y_f(k) - C_1(k)\hat{x}(k)] \tag{3.4.18}
$$

with $\hat{x}(k)$ and $\hat{f}(k-l|k-1)$ are respectively calculated from the Krein space projections of $\mathbf{x}(k)$ and $\mathbf{f}(k-l)$ onto $\mathcal{L}\{\{\mathbf{y}_a(j)\}_{j=0}^{k-1}\}$.

Remark 3.4.3. According to Lemma 3.4.1 and Lemma 3.4.2, the purpose of establishing the dynamic model (3.4.6) associated with (3.4.5) is to derive a positive minimum of the cost function (3.4.4) by applying the projection theory in Krein space. Notice that although the measurement $\{y(k)\}_{k=0}^{N}$ is a substantially stochastic sequence, the instantaneous values of $y(k)$ and $\check{f}(k-l|k)$ at each instant are available for the estimator. Thus, the equivalent cost function (3.4.5) and its corresponding dynamic constraint are constructed in a *conditional expectation* sense by gathering up $\{y(k)\}_{k=0}^{N}$ (cf. Equation (3.4.9) in the proof of Lemma 3.4.1).

3.4.3 Kalman filtering in krein space

From the preceding analysis, the key step to achieve our goal is to find a suitable $\hat{x}(k)$ and $\hat{f}(k-l|k-1)$. To this end, let

$$\mathbf{y}_1(k) = \mathbf{y}_f(k)$$

$$\mathbf{y}_2(k) = \begin{bmatrix} \mathbf{y}_f(k) \\ \check{\mathbf{f}}(k|k+l) \end{bmatrix},$$

then

$$\mathbf{y}_1(k-l+i) = C_1(k-l+i)\mathbf{x}(k-l+i) + \tilde{\mathbf{v}}_1(k-l+i), \; i = 1, \dots, l$$

$$\mathbf{y}_2(i) = C_2(i)\mathbf{x}(i) + Hf(i) + \tilde{\mathbf{v}}_2(i), \; i = 0, \dots, k-l,$$

where $\tilde{\mathbf{v}}_1(k) = \mathbf{v}_1(k)$ and $\tilde{\mathbf{v}}_2 = [\mathbf{v}_1^T(k) \; \mathbf{v}_s^T(k+l)]^T$ are zero-mean white noises with the following covariance matrices, respectively:

$$Q_{\tilde{v}_1}(k) = diag\{I, \; I\}$$

$$Q_{\tilde{v}_2}(k) = diag\{I, I, -\gamma^2 I\}.$$

It is easy to check out that $\{\mathbf{y}_2(0), \dots, \mathbf{y}_2(k-l); \mathbf{y}_1(k-l+1), \dots, \mathbf{y}_1(k)\}$ span the same linear space as $\mathcal{L}\{\{\mathbf{y}_a(j)\}_{j=0}^{k}\}$.

To proceed, the following definition is introduced.

Definition 3.4.1 [42]. For $t > k - l$, the estimator $\hat{\eta}(t, 1)$ is the optimal estimation of $\eta(t)$ on the observation $\mathcal{L}\{\{\mathbf{y}_2(t)\}_{t=0}^{k-l-1}; \{\mathbf{y}_1(t)\}_{t=k-l}^{t=k-1}\}$. For $0 < t \leq k - l$, the estimator $\hat{\eta}(t, 2)$ is the optimal estimation of $\eta(t)$ on the observation $\mathcal{L}\{\{\mathbf{y}_2(t)\}_{t=0}^{t=k-1}\}$.

In accordance with (3.4.17), the innovation sequence is defined as follows:

$$\mathbf{w}_1(k-l+i) = C_1(k-l+i)\mathbf{e}_1(k-l+i) + \tilde{\mathbf{v}}_1(k-l+i), \; i = 0, \dots, l$$

$$(3.4.19)$$

$$\mathbf{w}_2(i) = C_2(i)\mathbf{e}_2(i) + Hf(i) + \tilde{\mathbf{v}}_2(i), \; i = 0, \dots, k-l, \qquad (3.4.20)$$

where

$$\mathbf{e}_1(k-l+i) = \mathbf{x}(k-l+i) - \hat{\mathbf{x}}(k-l+i, 1), \; i = 0, \dots, l$$

$$\mathbf{e}_2(i) = \mathbf{x}(i) - \hat{\mathbf{x}}(i, 2), \; i = 0, \dots, k-l$$

with the corresponding covariance matrices given as

$$P_1(k-l+i) = \langle \mathbf{e}_1(k-l+i), \mathbf{e}_1(k-l+i) \rangle, \; i = 0, \dots, l$$

$$P_2(i) = \langle \mathbf{e}_2(i), \mathbf{e}_2(i) \rangle, \; i = 0, \dots, k-l.$$

In light of Lemma 2.2.1 in the work of Xie and Zhang [43], the innovation sequences $\mathcal{L}\{\{\mathbf{w}_2(t)\}_{t=0}^{k-l-1}; \{\mathbf{w}_1(t)\}_{t=k-l}^{t=k-1}\}$ are uncorrelated white noises and span the same linear space as $\mathcal{L}\{\{\mathbf{y}_a(j)\}_{j=0}^{k}\}$.

For deriving $\hat{\mathbf{x}}(k-l, 2)$ $(k = l+1, l+2, \dots)$, applying the Krein space–based projection formula in the work of Hassibi et al. [44] by taking (3.4.15) and (3.4.16) into account, we have that

$$\begin{cases} \hat{\mathbf{x}}(k-l, 2) = A(k-l-1)\hat{\mathbf{x}}(k-l-1, 2) + \langle \mathbf{x}(k-l), \mathbf{w}_2(k-l-1) \rangle \\ \qquad \times \langle \mathbf{w}_2(k-l-1), \mathbf{w}_2(k-l-1) \rangle^{-1} \mathbf{w}_2(k-l-1) \\ \qquad = A(k-l-1)\hat{\mathbf{x}}(k-l-1, 2) + K_2(k-l-1)\mathbf{w}_2(k-l-1) \\ \hat{\mathbf{x}}(0) = 0 \end{cases} \quad (3.4.21)$$

where

$$K_2(k - l - 1) = (A(k - l - 1)P_2(k - l - 1)C_2^T(k - l - 1)$$
$$+ B_f(k - l - 1)H)Q_2^{-1}(k - l - 1)$$

with $Q_2(k-l-1) = C_2(k-l-1)P_2(k-l-1)C_2^T(k-l-1)+HH^T+Q_{\tilde{v}_2}(k-l-1)$.

In addition, following the definition of $P_2(i)$ and (3.4.21), $P_2(i)$ ($i = 0, 1, \ldots, k - l - 1$) is the solution to the following standard Riccati equation

$$\begin{cases} P_2(i + 1) = A(i)P_2(i)A^T(i) + B_f(i)B_f^T(i) + D(i)D^T(i) \\ \qquad\qquad - K_2(i)Q_2^{-1}(i)K_2^T(i) \\ P_2(0) = P_0 \end{cases} \qquad (3.4.22)$$

For calculating $\hat{\mathbf{x}}(k - l + i, 1)$ ($i = 1, \ldots, l$) with the initial condition $\hat{\mathbf{x}}(k - l, 1) = \hat{\mathbf{x}}(k - l, 2)$, we apply the projection formula once again such that

$$\hat{\mathbf{x}}(k - l + i + 1, 1) = A(k - l + i)\hat{\mathbf{x}}(k - l + i, 1)$$
$$+ A(k - l + i)\langle \mathbf{x}(k - l + i), \mathbf{w}_1(k - l + i)\rangle$$
$$\times \langle \mathbf{w}_1(k - l + i), \mathbf{w}_1(k - l + i)\rangle^{-1}\mathbf{w}_1(k - l + i)$$
$$= A(k - l + i)\hat{\mathbf{x}}(k - l + i, 1)$$
$$+ K_1(k - l + i)\mathbf{w}_1(k - l + i), \qquad (3.4.23)$$

where

$$K_1(k - l - 1) = A(k - l + i)P_1(k - l + i)C_1^T(k - l + i)Q_1^{-1}(k - l + i)$$

with $Q_1(k - l + i) = C_1(k - l + i)P_1(k - l + i)C_1^T(k - l + i) + Q_{\tilde{v}_1}(k - l + i)$, and $P_1(k - l + i)$ is computed recursively in the following form:

$$\begin{cases} P_1(k - l + i + 1) = A(k - l + i)P_1(k - l + i)A^T(k - l + i) \\ \qquad\qquad + B_f(k - l + i)B_f^T(k - l + i) \\ \qquad\qquad + D(k - l + i)D^T(k - l + i) \\ \qquad\qquad - K_1(k - l + i)Q_2^{-1}(k - l + i)K_1^T(k - l + i) \\ P_1(k - l) = P_2(k - l) \end{cases} \qquad (3.4.24)$$

Similarly, the projection formula is reutilized to compute $\hat{\mathbf{f}}(k - l|k - 1)$—that is,

$$\hat{\mathbf{f}}(k - l|k - 1) = \sum_{i=0}^{l-1} \langle \mathbf{f}(k - l), \mathbf{w}_1(k - l + i)\rangle Q_1^{-1}(k - l + i)\mathbf{w}_1(k - l + i)$$

$$= \sum_{i=0}^{l-1} \Omega_{k-l+i}^{k-l} C_1^T(k - l + i)Q_1^{-1}(k - l + i)\mathbf{w}_1(k - l + i),$$

$$i = 1, \ldots, l - 1, \qquad (3.4.25)$$

where Ω_{k-l+i}^{k-l}, $i = 1, \ldots, l - 1$ is obtained recursively in terms of

$$\begin{cases} \Omega_{k-l+i}^{k-l} = \Omega_{k-l+i-1}^{k-l}[A(k - l + i - 1) - K_1(k - l + i - 1)C_1(k - l + i - 1)]^T \\ \Omega_{k-l+1}^{k-l} = B_f^T(k - l). \end{cases} \qquad (3.4.26)$$

Finally, to calculate $Q_w(k)$ that is associated with J_{min} and $\check{f}(k-l)|k$, define $\tilde{\mathbf{f}}(k-l) = \mathbf{f}(k-l) - \hat{\mathbf{f}}(k-l|k-1)$, and then, from (3.4.25), we know that

$$\langle \tilde{\mathbf{f}}(k-l), \tilde{\mathbf{f}}(k-l) \rangle = I - \sum_{i=0}^{l-1} \Omega_{k-l+i}^{k-l} C_1^{\mathrm{T}}(k-l+i) Q_1^{-1}(k-l+i)$$

$$\times \left(\Omega_{k-l+i}^{k-l} C_1^{\mathrm{T}}(k-l+i) \right)^{\mathrm{T}}. \tag{3.4.27}$$

Combing (3.4.19), (3.4.20) and (3.4.27), we have

$$Q_w(k) = \begin{cases} C_1(k)P_1(k)C_1^{\mathrm{T}}(k) + I, & 0 < k < l \\ \begin{bmatrix} C_1(k)P_1(k)C_1^{\mathrm{T}}(k) + I & C_1(k)(\Omega_k^{k-l})^{\mathrm{T}} \\ \Omega_k^{k-l}C_1^{\mathrm{T}}(k) & -\gamma^2 I + I - \langle \tilde{f}(k-l), \tilde{f}(k-l) \rangle \end{bmatrix}, \\ k \geq l \end{cases}$$

$$\tag{3.4.28}$$

where $P_1(k)$ and Ω_{l-l+i}^{k-l} are the same as in (3.4.24) and (3.4.26).

3.4.4 H_∞ fault estimator design

From analysis and lemmas presented previously, we are now in the position to give our main results for designing the fault estimator, which is summarized in the following theorem.

Theorem 3.4.1. *For (3.4.6), given a scalar $\gamma > 0$ and an integer $l > 0$, the H_∞ fixed-lag fault estimator that satisfies (3.4.5) exists if and only if*

$$\Lambda_1(k) = C_1(k)P_1(k)C_1^{\mathrm{T}}(k) + I > 0 \tag{3.4.29}$$

and

$$\Lambda_3(k) = -\gamma^2 I + I - \sum_{i=0}^{l-1} \Omega_{k-l+i}^{k-l} C_1^{\mathrm{T}}(k-l+i)$$

$$\times Q_1^{-1}(k-l+i)(\Omega_{k-l+i}^{k-l} C_1^{\mathrm{T}}(k-l+i))^{\mathrm{T}}$$

$$-\Omega_k^{k-l} C_1^{\mathrm{T}}(k)\Lambda_1^{-1}(k)(\Omega_k^{k-l} C_1^{\mathrm{T}}(k))^{\mathrm{T}} < 0. \tag{3.4.30}$$

In this case, a feasible fault estimator is given by

$$\check{f}(k-l|k) = \sum_{i=0}^{l} \Omega_{k-l+i}^{k-l} C_1^{\mathrm{T}}(k-l+i) Q_1^{-1}(k-l+i)$$

$$\times [y_f(k-l+i) - C_1(k-l+i)\hat{x}(k-l+i, 1)], \tag{3.4.31}$$

where $\hat{x}(k-l+i, 1)$, $Q_1(k-l+i)$, and Ω_{k-l+i}^{k-l} are calculated by (3.4.21), (3.4.22), (3.4.23), (3.4.24), and (3.4.26).

Proof. For $k \geq l$, applying the block triangular factorization technique to $Q_w(k)$ in (3.4.28), we have

$$
Q_w(k) = \begin{bmatrix} I & 0 \\ \Lambda_2(k)^{\mathrm{T}}\Lambda_1(k)^{-1} & I \end{bmatrix} \begin{bmatrix} \Lambda_1(k) & 0 \\ 0 & \Lambda_3(k) \end{bmatrix} \begin{bmatrix} I & 0 \\ \Lambda_2(k)^{\mathrm{T}}\Lambda_1(k)^{-1} & I \end{bmatrix}^{\mathrm{T}},
$$

$$(3.4.32)$$

where $\Lambda_2(k) = \Omega_k^{k-l} C_1^{\mathrm{T}}(k)$. Thus, from Lemma 3.4.2, we know that $Q_a(k)$ and $Q_w(k)$ have the same inertia if and only if $\Lambda_1(k) > 0$ as well as $\Lambda_3(k) < 0$. Furthermore, based on (3.4.18) and (3.4.32), J has a minimum J_{min} if (3.4.29) and (3.4.30) are satisfied, where

$$
J_{min} = \sum_{k=l}^{N} \left(\begin{bmatrix} I & 0 \\ -\Lambda_2(k)\Lambda_1^{-1}(k) & I \end{bmatrix} \begin{bmatrix} y_f(k) - C_1(k)\hat{x}(k) \\ \check{f}(k-l|k) - \hat{f}(k-l|k-1) \end{bmatrix} \right)^{\mathrm{T}}
$$

$$
\times \begin{bmatrix} \Lambda_1^{-1}(k) & 0 \\ 0 & \Lambda_3^{-1}(k) \end{bmatrix} \begin{bmatrix} I & 0 \\ -\Lambda_2(k)\Lambda_1^{-1}(k) & I \end{bmatrix}
$$

$$
\times \begin{bmatrix} y_f(k) - C_1(k)\hat{x}(k) \\ \check{f}(k-l|k) - \hat{f}(k-l|k-1) \end{bmatrix}
$$

$$
+ \sum_{k=0}^{l-1} [y_f(k) - C_1(k)\hat{x}(k)]^{\mathrm{T}} \Lambda_1^{-1}(k)[y_f(k) - C_1(k)\hat{x}(k)]. \quad (3.4.33)
$$

Since $\Lambda_3(k) < 0$, to guarantee $J_{min} > 0$, combining (3.4.25) with (3.4.33), we know that a possible choice of $\check{f}(k-l|k)$ is

$$
\check{f}(k-l|k) = \hat{f}(k-l|k-1) + \Lambda_2(k)\Lambda^{-1}(k)(y_f(k) - C_1\hat{x}(k))
$$

$$
= \sum_{i=0}^{l} \Omega_{k-l+i}^{k-l} C_1^{\mathrm{T}}(k-l+i)Q_1^{-1}(k-l+i)
$$

$$
\times [y_f(k-l+i) - C_1(k-l+i)\hat{x}(k-l+i,1)],
$$

which indicates (3.4.31). This completes the proof. $\quad\square$

Remark 3.4.4. It can be seen from Theorem 3.4.1 that the superiority of the proposed algorithm lies in three aspects:

(i) In contrast to the results elsewhere [19–21, 37, 38], the proposed algorithm can be applied to systems with time-varying $\rho(k)$.
(ii) Comparing to the result of Li and Zhong [39], the parameter matrices of the addressed estimator are given in terms of standard Riccati equations with the same dimension n of system (3.4.6), where no coupled Lyapunov equation with higher dimension is needed.
(iii) The fault can be estimated in an arbitrary fixed-lag l.

3.4.5 Numerical examples

To illustrate the effectiveness and the applicability of the proposed method, we shall implement our algorithm on a time-varying model. The following system

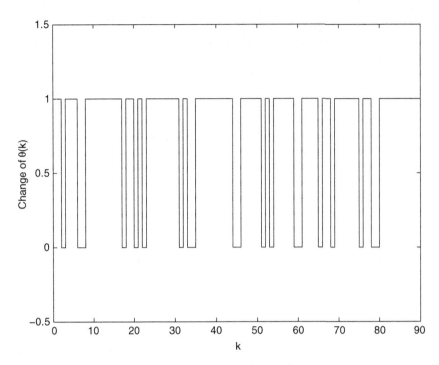

FIGURE 3.8   The change mode of $\theta(k)$.

matrices are adopted that are borrowed from Ma et al. [45] and Moayedi et al. [46]:

$$A(k) = (1 + 0.2\sin(0.02k\pi)) \times \begin{bmatrix} 0.8 & 0 \\ 0.9 & 0.2 \end{bmatrix}$$

$$B_f(k) = \begin{bmatrix} 0.5 & 0.5 \end{bmatrix}^{\mathrm{T}}, \; C(k) = \begin{bmatrix} 1 & 1 \end{bmatrix}$$

$$D(k) = \begin{bmatrix} 0.3 & 0.25 \end{bmatrix}^{\mathrm{T}}$$

The process noise $d(k)$ is uniformly randomly chosen from the interval $[-0.5, 0.5]$, and the measurement noise $v(k)$ is assumed as $v(k) = 0.5\sin(0.2k)$. The fault signal $f(k)$ is assumed to be time varying in the following sinusoidal form

$$f(k) = \begin{cases} \sin(0.5k), & k \in [30, 80] \\ 0, & \text{otherwise} \end{cases}$$

and the expectation of $\theta(k)$ is assumed as $\rho = 0.8$, where Fig. 3.8 displays the switching mode of $\theta(k)$.

Setting $l = 10$, $\gamma = 1.52$, $x_0 = [0.2 \; 0]^{\mathrm{T}}$, and $P_0 = 0.1I$, we design the fault estimator by applying Theorem 3.4.1. Fig. 3.9 displays the fault signal and its estimation simultaneously. Fig. 3.10 shows the value of $f(k - l) - \check{f}(k - l|k)$,

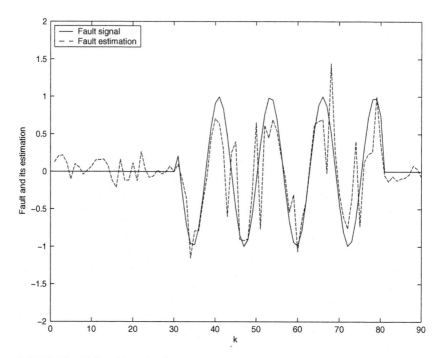

FIGURE 3.9 Fault and its estimation.

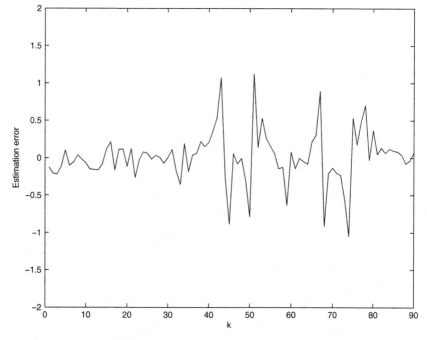

FIGURE 3.10 Fault estimation error.

which is the error between the fault and its estimation. It can be seen from the results that our algorithm can track the fault signal regardless of whether random packet dropouts occur.

3.5 Conclusion

In this chapter, we have dealt with fault estimation problem for LDTV systems with random uncertainties. It is divided into the following three aspects.

First, we have handled the problem of robust fault detection for LDTV systems subject to multiplicative noise, and l_2-norm bounded unknown disturbance is investigated. The design of the FDF is converted into the framework of H_∞-filtering. A sufficient condition on the existence of the FDF is derived in terms of a Riccati equation, and by solving the Riccati equation, an analytical solution of the parameter matrices is obtained.

Second, the FDF design problem of the LDTV system with multistep data packet dropouts is studied. The observer-based robust H_∞-FDF is used as a residual generator to transform the FDF design problem into a H_∞ filtering problem of a class of stochastic time-varying systems. By applying the derivation derived in Section 2, the sufficient conditions for the existence of H_∞-FDF based on Riccati equation are given. The solution of the parameter matrix is transformed into a quadratic optimization problem. By solving the Riccati equation, the analytic solution of the FDF parameter matrix is obtained.

Third, the problem of H_∞ fixed-lag fault estimator design for LDTV systems subject to intermittent observations has been dealt with. Special efforts have been made to handle the multiplicative uncertainty introduced by the random measurement packet dropouts. Through defining a couple of equivalent dynamic systems and the H_∞ performance index, the fault estimator has been derived by using the projection formula in Krein space based on the reorganized innovation approach. The parameter matrices of the estimator have been calculated by solving two standard Riccati equations.

Finally, the achieved results are illustrated by numerical examples.

Acknowledgments

This work was supported in part by the National Natural Science Foundation of China under grants 61973135, 91948201, and 61773242, and in part by the Shandong Provincial Key Research and Development Program (Major Scientific and Technological Innovation Project) under grant 2019JZZY10441.

References

[1] S. X. Ding. Fault identification schemes. In Model-Based Fault Diagnosis Techniques: Design Schemes, Algorithms and Tools (2nd ed.). Advances in Industrial Control. Springer, London, UK, 2013. 441–470.

[2] J.P. Cai, C.Y. Wen, H.Y. Su, Z.T. Liu, Robust adaptive failure compensation of hysteretic actuators for a class of uncertain nonlinear systems, IEEE Transactions on Automatic Control 58 (9) (2013) 2388–2394.

[3] J.L. Liu, Y. Dong, Event-based fault detection for networked systems with communication delay and nonlinear perturbation, Journal of the Franklin Institute 350 (9) (2013) 2791–2807.

[4] S.X. Ding, T. Jeinsch, A unified approach to the optimization of fault detection systems, International Journal of Adaptive Control & Signal Processing 14 (7) (2015) 725–745.

[5] Y. Zhang, H.J. Fang, Z.X. Liu, Fault detection for nonlinear networked control systems with Markov data transmission pattern, Circuits Systems & Signal Processing 31 (4) (2012) 1343–1358.

[6] Y. Zhang, Z.X. Liu, H.J. Fang, H.B. Chen, H_∞ fault detection for nonlinear networked systems with multiple channels data transmission pattern, Information Sciences: An International Journal 221 (1) (2013) 534–543.

[7] B. Shen, S.X. Ding, Z.D. Wang, Finite-horizon H_∞ fault estimation for linear discrete time-varying systems with delayed measurements, IEEE Transactions on Circuits & Systems II: Analog & Digital Signal Processing 60 (12) (2013) 902–906.

[8] M.Y. Zhong, S.X. Ding, E.L. Ding, Automatica 46 (8) (2010) 1395–1400.

[9] M.Y. Zhong, D.H. Zhou, S.X. Ding, On designing H_∞ fault detection filter for linear discrete time-varying systems, IEEE Transactions on Automatic Control 55 (7) (2010) 1689–1695.

[10] H.S. Zhang, G. Feng, C.Y. Han, Linear estimation for random delay systems, Systems & Control Letters 60 (7) (2011) 450–459.

[11] A. Bouhtouri, D. Hinrichsen, A. Pritchard, H_∞-type control for discrete-time stochastic systems, International Journal of Robust and Nonlinear Control (13) (1999) 923–948.

[12] A.M. Rami, X.Y. Zhou, Linear matrix inequalities, Riccati equations, and indefinite stochastic linear quadratic controls, International Journal of Robust and Nonlinear Control 45 (6) (2000) 1131–1143.

[13] V. Dragan, T. Morozan, A. Stoica, h_2 Optimal control for linear stochastic systems, Automatica 40 (7) (2004) 1103–1113.

[14] E. Gershon, U. Shaked, I. Yaesh, H_∞ Control and Estimation of State-Multiplicative Linear Systems, Springer, London, 2005.

[15] W.H. Zhang, Y.L. Huang, H.S. Zhang, Stochastic H_2/H_∞ control for discrete-time systems with state and disturbance dependent noise, Automatica 43 (3) (2007) 513–521.

[16] S. X. Ding, P. Zhang, E. L. Ding. Fault detection system design for linear a class of stochastically uncertain systems. IFAC Proceedings Volumes 2006;39(13):705–710.

[17] C.F. Ma, M.Y. Zhong, M. Sader, T. Jeinsch, Robust fault detection for linear systems with multiplicative noise, IFAC Proceedings Volumes 39 (13) (2006) 1228–1233.

[18] M. Tabbara, D. Nesic, A.R. Teel, Stability of wireless and wireline networked control systems, IEEE Transactions on Automatic Control 52 (9) (2007) 1615–1630.

[19] J. Yu, M. Liu, W. Yang, P. Shi, Robust fault detection for Markovian jump systems with unreliable communication links, International Journal of Systems Science 44 (11) (2013) 2015–2026.

[20] H.L. Dong, Z.D. Wang, H.J. Gao, On design of quantized fault detection filters with randomly occurring nonlinearities and mixed time-delays, Signal Processing 92 (4) (2011) 1117–1125.

[21] S.M. Alavi, M. Saif, Fault detection in nonlinear stable systems over lossy networks, IEEE Transactions on Control Systems Technology 21 (6) (2013) 2129–2142.

[22] X. He, Z.D. Wang, D.H. Zhou, Networked fault detection with random communication delays and packet losses, International Journal of Systems Science 39 (11) (2008) 1045–1054.

[23] Y.B. Ruan, W. Wang, F.W. Yang, Fault detection filter for networked systems with missing measurements, Control Theory Applications 26 (3) (2009) 291–295.

[24] Y.Q. Wang, Y. Hao, S.X. Ding, G.Z. Wang, D.H. Zhou, Residual generation and evaluation of networked control systems subject to random packet dropout, Automatica 45 (10) (2009) 2427–2434.

[25] X. He, Z.D. Wang, D.H. Zhou, Robust fault detection for networked systems with communication delay and data missing, Automatica 45 (11) (2009) 2634–2639.

[26] T. Morozan, Stabilization of some stochastic discrete–time control systems, Stochastic Analysis & Applications 1 (1) (1983) 89–116.

[27] E. Gershon, U. Shaked, I. Yaesh, H_∞ control and filtering of discrete-time stochastic systems with multiplicative noise, Automatica 37 (3) (2001) 409–417.

[28] M. Green, D.N. Limebeer, Robust Linear Control, Prentice Hall, Englewood Cliffs, NJ, 1995.

[29] E. Kreyszig, Introductory Functional Analysis with Applications., Wiley Classics Library, 1978, pp. 196–197.

[30] M.Y. Zhong, S. Liu, H.H. Zhao, Krein space-based H_∞ fault estimation for linear discrete time-varying systems, Acta Automatica Sinica 34 (12) (2008) 1529–1533.

[31] X.G. Yu, C.S. Hsu, Reduced order H_∞ filter design for discrete time-variant systems, International Journal of Robust & Nonlinear Control (8) (1997) 797–809.

[32] H. Gao, T. Chen, L. Wang, Robust fault detection with missing measurement, International Journal of Robust and Nonlinear Control 81 (5) (2008) 804–819.

[33] Y. Zhao, J. Lam, H.J. Gao, Fault detection for fuzzy systems with intermittent measurements, IEEE Transactions on Fuzzy Systems 17 (2) (2009) 398–410.

[34] J.P. Hespanha, P. Naghshtabrizi, Y.G. Xu, A survey of recent results in networked control systems, Proceedings of the IEEE 95 (2007) 138–162.

[35] J. Nilsson, Real-Time Control Systems With Delays, Department of Automatic Control, Lund Institute of Technology, Lund, Sweden (1998).

[36] A.A. Stoorvogel, H.H. Niemann, A. Saberi, P. Sannuti, Optimal fault signal estimation, International Journal of Robust & Nonlinear Control 12 (8) (2002) 697–727.

[37] X.B. Wan, H.J. Fang, F. Yang, Fault detection for a class of networked nonlinear systems subject to imperfect measurements, International Journal of Control, Automation, and Systems 10 (2) (2013) 265–274.

[38] D. Zhang, Q.G. Wang, L. Yu, H.Y. Song, Fuzzy-model-based fault detection for a class of nonlinear systems with networked measurements, IEEE Transactions on Instrumentation & Measurement 62 (12) (2013) 3148–3159.

[39] Y.Y. Li, M.Y. Zhong, On designing robust H_∞ fault detection filter for linear discrete time-varying systems with multiple packet dropouts, Acta Automatica Sinica 36 (12) (2010) 1788–1796.

[40] Y.Y. Li, S. Liu, Z.H. Wang, Fault detection for linear discrete time-varying systems with measurement packet dropping, Mathematical Problems in Engineering 2013 (pt.5) (2013) 697345.1–697345.9.

[41] H. Zhao, C. Zhang, H_∞ fixed-lag smoothing for linear discrete time-varying systems with uncertain observations, Applied Mathematics & Computation 224 (1) (2013) 387–397.

[42] X. Lu, H. Zhang, J. Yan, H_∞ deconvolution fixed-lag smoothing, International Journal of Control, Automation & Systems 8 (4) (2010) 896–902.

[43] L.H. Xie, H.S. Zhang, Control and Estimation of Systems with Input/Output Delays, Springer, Berlin, Germany, 2007.

[44] B. Hassibi, A.H. Sayed, T. Kailath, Linear estimation in Krein spacesâ¬"Part I: Theory, IEEE Transactions on Automatic Control 41 (1) (1996) 18–33.

[45] J. Ma, S.L. Sun, Optimal linear estimators for systems with random sensor delays, multiple packet dropouts and uncertain observations, IEEE Transactions on Signal Processing 59 (4) (2011) 5181–5192.

[46] M. Moayedi, Y.K. Foo, Y.C. Soh, Adaptive Kalman filtering in networked systems with random sensor delays, multiple packet dropouts and missing measurements, IEEE Transactions on Signal Processing 58 (3) (2010) 1577–1588.

Chapter 4

Fault diagnosis and failure prognosis of electrical drives

Elias G. Strangas
Michigan State University, United States

4.1 Introduction

The interest in continuous, safe, and reliable operation of electrical drives has dramatically increased in recent years. Until a few decades ago, an AC electrical machine was typically a synchronous generator or a three-phase uncontrolled motor that would operate connected directly to the power grid, or if a DC machine, its speed would be controlled through a rectifier. Although these applications remain, they have been augmented by many others.

Many developments have led to this explosion of applications. Primarily, the ubiquitous use of power electronics and microcontrollers, DSPs, and so forth, have allowed control of speed and torque, among others, as well as the utilization of new motor types, such as permanent magnet AC (PMAC) and switched and synchronous reluctance, with increased versatility, higher torque, and power density. These new power electronics-controlled machines have been finding uses in applications in manufacturing, robotics, vehicle traction and operation, airplane propulsion, wind, and wave generators, but also in medical, commercial, and consumer electronics that are becoming vital in health and other necessary devices.

The need for higher reliability and the associated fault diagnosis and failure prognosis resulted from the fact that these applications replaced older methods of control, such as pneumatic or other mechanical links, which were considered reliable, and from the desired and expected extension of life of well-established systems like generators. The tools for fault diagnosis and prognosis of failure of electrical machines and drives are not fundamentally different from those used in other complex engineering systems. What differs here are the faults themselves, the variables and parameters that are monitored, and the interaction between components. What is also important is the utilization of the results of diagnosis and prognosis. Interruption of service and request for maintenance is not the only

Fault Diagnosis and Prognosis Techniques for Complex Engineering Systems. DOI: 10.1016/B978-0-12-822473-1.00008-2
127

option; others are the employment of mitigating partial redundancies in hardware and control without interruption of service.

In the past, regularly scheduled maintenance was the standard, based on a schedule or simple evaluation of health. Condition-based maintenance processes are a precursor to fault diagnosis and prognosis, and to the resulting decisions and actions. Originally based on sensing of vibrations, overvoltages and overcurrents, and higher temperatures, it evolved to signature analysis of currents or voltages, using initially Fourier transform of signals and thresholding, to the present-day features extraction, signal processing, fault classification, and decision making. Fault diagnosis alone, to some degree, is not useful or adequate to lead to a informed decision. Prognosis, whether implicit or explicit, is what will allow the operator, human or automated, to take the drive out of operation, or use redundancies or other mitigating measures.

All of these—detection, identification, diagnosis, prognosis, decision making, and action planning—have been designed to work together under a unified control system that provides health monitoring and reliability improvement.

A fault in this context means a malfunction that allows the continuous operation of the drive, but that may lead to an eventual disruption or at least a deterioration of operation. A failure is in unplanned interruption of service. Management of the condition may include early emergency interruption of service, scheduling of maintenance, mitigation by use of redundancies, or continuous operation at lower performance. For the drive to be able to utilize any of these management systems, a fault has to be detected and identified, its severity diagnosed, and its progression monitored and predicted.

Most electrical machines, motors, or generators are a type of induction or synchronous machine. The first are typically squirrel cage or wound rotor field machines, with a number of double-fed ones. The second, synchronous machines, are typically generators with a wound field, but increasingly permanent magnet AC ones of size up to a few hundred kilowatts to megawatts. The power electronic supply consists of an inverter, usually operating at high switching frequency, and a controller, with adequate sensors and high-frequency sampling of variables.

To increase the reliability and availability of a drive, including diagnosis and prognosis, a number of steps are required:

- Design of the drive and design evaluation through risk analysis and simulation,
- Testing before assembly of components and materials to ascertain their quality and establish statistical characteristics, and
- Testing before commissioning to avoid early failures and to develop baseline operating parameters.

The qualities of diagnosis and prognosis of a drive that will allow avoidance of a catastrophic faults are as follows:

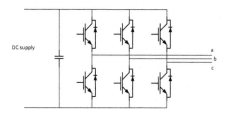

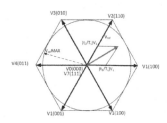

(A) Schematic of a two-level inverter   (B) Possible voltages obtained from a three-phase two-level inverter

FIGURE 4.1   A two-level inverter.

- Speed of detection, specificity, sensitivity, and confidence, and the ability to determine severity, and
- Continuously updating prognosis and confidence mitigation update of modified system remaining useful life (RUL) and reliability.

The main components of an electrical drive are an electrical machine, consisting of steel stator and rotor cores, windings, possibly magnets, bearings, and frame; a power electronics supply, consisting of power electronics switches, capacitors, inductors, and resistors and sensors: current, voltage, temperature, vibrations, magnetic field, and a controller. Fig. 4.1(A) shows the schematic of the simplest three-phase, two-leg voltage source inverter, and Fig. 4.1(B) shows the space vector of the voltages that it can produce, through all possible safe combinations of open and closed switches. Each of the three outputs can be connected to either the positive or negative DC rail and the space vector of output voltages, a complex variable defined as

$$\mathbf{v}_{\alpha\beta} = v_a + v_b e^{j\gamma} + v_c e^{-j\gamma} = v_\alpha + j v_\beta \qquad \gamma = 120^o. \qquad (4.1)$$

There are six nonzero such vectors, and two zero ones, which are arrived at when each phase is connected to the positive or negative DC rail. The desired voltages, which are not one of these eight possibilities, are created through pulse width modulated technique, connecting each output phase alternating between the two DC links during each power cycle. A commanded voltage $v_\alpha + j v_\beta$ representing a sector in Fig. 4.1(B) is synthesize by appropriately selecting duty cycles, d_1, d_2, and d_3, averaging the times of the three neighboring voltage vectors, V_r, V_l, and V_0. A pulse width modulation (PWM) inverter using fast electronic switches, insulated gate bipolar transistors (IGBTs), and MOSFETS can produce the desired voltage almost instantaneously, within a microsecond. The switching frequency, which defines the frequency of high harmonics, is typically between 5 and 25 kHz.

Two control schemes are used primarily today: most applications in synchronous and induction machines use some version of field-oriented or vector control, which is based on commanding phase currents. From these current commands, voltage commands are derived through either high gain controllers alone or in combination with feed-forward control. Both the control and the health evaluation of the drive use sensors. Some of the sensors are integral parts of a drive system, such as current and speed sensors in an AC motor drive; others, such as interferometers and magnetic field sensors, are additions that may increase the cost but also decrease reliability due to additional wiring, space, and so forth.

4.1.1   Operation under field orientation control

Vector control is a technique that allows the decoupling of the control of flux in an AC machine from the control of the torque. This is accomplished through two transformations of the terminal voltages and currents: the first is the Clarke transformation (Eq. 4.1) discussed already, which transforms the phase voltages or other three-phase quantities to two. The second is the Park transformation, which transforms these stationary quantities, with subscripts v_α and v_β, to a rotating frame of reference:

$$v_d + jv_q = (v_\alpha + jv_\beta)e^{h\theta} \tag{4.2}$$

The angle θ in Eq. 4.1 in the case of orientation to the rotor magnetic field is that of the instantaneous position of the rotor field established by the rotor and stator currents, and in the case of a PMAC machine by the magnets of the rotor.

In the case of a PMAC machine, this field position is easily measured, as it coincides with the rotor position and a simple sensing of this position suffices. In the case of an asynchronous (or induction) machine, this field moves with respect to the rotor and its location has to be estimated accurately for high-performance torque control. Equally simply, the flux due to the rotor magnets, λ_d, is usually known *a priori* through the testing of the machine commissioning or manufacturing, whereas for an induction machine, λ_d has to also be estimated, along with its position. The advantages of the decoupling through field orientation making themselves apparent in the torque equation for induction machines using field-oriented variables is

$$T = \frac{3}{2}\lambda_d i_q, \tag{4.3}$$

where λ_d is the flux linkages of the stator in the $-d$ axis and i_q is the component of the stator current perpendicular to that (i.e., the q axis). In an induction machine, the rotor field is established and controlled through the i_d component. For PMAC machines, at high speeds the field also has to be modified though the component stator current aligned with the rotor flux, i_d. For salient geometry PMAC machines, this current is used to provide an additional component of the torque.

In all of these cases, if the machine is supplied by a voltage source inverter, the commanded current is established through the use of proportional-integral

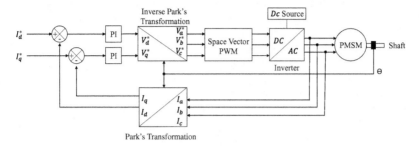

FIGURE 4.2   Schematic of a field-oriented PWM drive.

(PI) controllers, with relatively high bandwidth. Fig. 4.2 shows a typical scheme for the control of a permanent magnet synchronous machine (PMSM).

4.1.2   Operation under Direct Torque Control

Direct torque control (DTC) replaces the decoupling in field-oriented control with bang-bang control, which naturally fits the inherently discrete nature of switch-mode power inverters. DTC does not involve space vector modulation but utilizes a switching table that consists of different voltage vectors. In addition, the rotor position sensing that is essential for FOC is not necessary for DTC to operate properly even if a speed control loop is included in the DTC scheme. Instead, the desired stator flux λ' and torque T are established through comparators and voltage commands. Fig. 4.3 show a typical controller for direct torque control.

Almost all diagnosis and prognosis is based on the collection of operating data, and the design of a model—deterministic, stochastic, or entirely analytical. We should differentiate between data required to create a data-based model of the drive components and the data used to arrive at a decision for one device. The

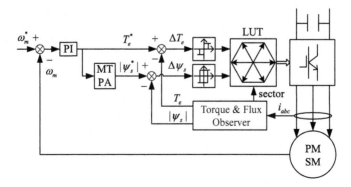

FIGURE 4.3   Schematic of a DTC PMAC drive. From Niu et al. [1]

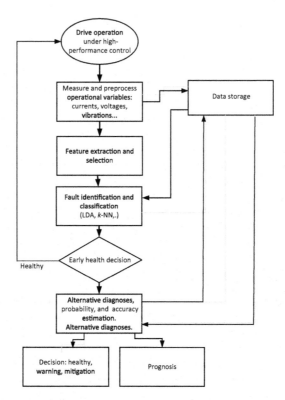

FIGURE 4.4 Fault classification and decision flowchart.

preceding is directly related to the time available to decide appropriate action and the accuracy of the decision.

4.2 What can fail and how

Not all components of a drive are prone to the same rate of failure, and of course they develop different types of faults. Although many faults involve two different components, such as an inverter and a machine, we discuss the modes of failure of each separately.

4.2.1 Electric power converters

Inverter and DC converters consist of power electronics switches and gate drivers, as well as inductors and capacitors. All of these are subjected to faults and failures, and are considered to be of the weaker components of a drive.

4.2.1.1 IGBT catastrophic failure mechanisms

Generally, catastrophic failure mechanisms of IGBTs are mostly caused by overstressed working conditions and are understood by studying the physics of the devices (a detailed review was published by Wu et al. [2]):

- *Open-circuit failure* may not be directly catastrophic, but it will disrupt the drive operation and can cause secondary faults. One is *bond wire lift-off* and can happen after a short circuit. It results from uneven thermal expansion between silicon and aluminum, together with high temperature gradients. This unevenness results in a crack around the bonding interface. One bond wire lifts off, the current in the rest increases, along with thermal gradient and stress, and the crack expands. Another type of open fault is caused by There are many possible causes of gate driver failure; a typical one is when damage in the wires connecting the drive board and IGBT are disconnected. The driver failure may result in IGBT intermittent misfiring and degraded output voltage. In addition, abnormal operating conditions may also damage the gate driver, which is more sensitive to temperature rise than the IGBT itself.

- *Short-circuit failure* can lead to destruction of the IGBT, and this can precipitate the failure of the remaining IGBTs and the motor, as it results in high current through the circuit. *High voltage breakdown* is the most common failure. It occurs as the IGBT turns off the collector current very quickly; the falling current through the small stray series inductance, not compensated for by the bar design or snubber capacitors, causes a high turn-off voltage spike. The electric field can reach a critical value. It reaches a few IGBT cells first and leads to high leakage current and high local temperature. A high value of collector-emitter and gate-emitter voltages can also lead to a short circuit during turn-on. *Latch-up* is a condition where the collector current is no longer controlled by the gate voltage. Latch-up happens when the parasitic NPN transistor of the IGBT is turned on, and works together with the main PNP transistor as a thyristor, that turns on but not off. Static latch-up happens at high collector currents, whereas dynamic latch-up happens during switching transients, usually during turn-off. *Second breakdown* is a local thermal breakdown for transistors due to high current stresses, which can also happen to IGBTs during on-state and turn-off. With the increase of current, the charge density in the collector-base junction increases, and the breakdown voltage decreases. This results in a further increase in the current density. When the area of the high current density region reduces beyond the minimum area of a stable current filament, the temperature increases rapidly and causes a short circuit. As for *energy shocks*, during short circuit at the on-state, failure may happen due to high power dissipation, an energy shock that will result in fast-rising temperature. The IGBT may not immediately fail, even when the junction temperature exceeds the rated temperature, but it may survive until more repetitive short circuits occur.

4.2.2 Electrical machines

Most of the electrical machines in use at present are squirrel cage induction and PMAC motors. They come in many variations, with distinct operational and fault characteristics and weaknesses. Switched reluctance machines are less common and have been introduced in niche applications; their faults and failures are also discussed briefly here. Many other types (permanent magnet axial flux, variable flux, etc.) are far less common and have some similarities, at least in failure modes, to the those discussed in more detail. We first discuss faults that may occur in more than one machine type, as well as specific machine types that have particular fault modes associated with them.

4.2.2.1 Bearing faults

These are of the most common and severe faults in electrical drives. Similar to faults in purely mechanical systems, they can be due to environmental conditions, high temperature, contamination, or vibrations, but there are additional fault causes in electrical machines. These faults in turn manifest themselves with increasing localized temperature, vibrations, and occasionally through high-frequency stator currents.

The following are some of the causes of bearing faults according to their origins [3]:

1. *Mechanical origin.* Mechanical origin is a mechanical bearing load applied by radial and axial forces, as well as vibrations, but also contamination. Radial and axial forces are the most frequent and most analyzed causes of bearing faults. Such loads cause increased wear and are primarily due to eccentricity and the associated radial forces or even axial forces due to rotor displacement. An important consequence of bearing deterioration for electrical machines is that magnetic pull of the rotor, especially for PMAC machines, becomes asymmetric [4], and the rotor becomes further eccentric in the stator bore, increasing static and/or dynamic eccentricity, placing more load on the bearing and causing further bearing degradation.

2. *Electric origin (electric bearing load as given by bearing currents).* A review is presented in the work of Plazenet et al. [5]. The possible damage to bearings is due to voltages building between the races of the bearings and then discharging through electric currents and has been recognized for a long time. These voltages are present under operation from a balanced, symmetric, sinusoidal system (traditional operation); additional ones are induced under operation from an inverter.

   The first include the following:
   - *Alternating voltages induced in the shaft.* Unbalanced magnetic fields caused by design—manufacturing details such as axial holes in the stator or/and rotor laminations, joints between stator segments, rotor eccentricities, and bowed rotor—can create a magnetic flux encircling the shaft. Thus, alternating voltages are induced in the shaft and may cause a

circulating current in the loop "stator frame—drive-end bearing—rotor shaft—non-drive-end bearing." If the bearing voltage increases above a threshold to break the insulating lubricant film of the bearing. As a result, design rules have been established to limit this issue.

Axial rotor flux can be generated through the shaft by residual magnetism (linked to magnetic particles and improper demagnetization) local saturation, asymmetries in the rotor field winding, and rotor eccentricities. This homopolar flux will circulate from the shaft, in the loop "stator frame—drive-end bearing—rotor shaft—non-drive-end-bearing."

- *High-frequency bearing currents.* High-frequency bearing currents are generated through capacitive and inductive coupling inside the machine. Under inverter operation, the high-frequency components of the common mode voltage excite the parasitic capacitances and inductances of the motor, producing the so-called inverter-induced bearing currents. These phenomena are in the range of 100 kHz to several megahertz. One can identify "circulating" and "noncirculating" currents for the purpose of developing models and possible design methods to limit them:

  (1) Small capacitive high-frequency bearing currents ($\approx 5 - 200mA$), "noncirculating" type that appear at low speed.

  (2) *EDM bearing currents, "noncirculating" type.* The bearing voltage mirrors the common mode voltage through a capacitive voltage divider. The bearing voltage increases, charges the lubricant until its breakdown field strength is surpassed, and causes a breakdown with the EDM current pulse ($\approx 0.5 - 3A$), which is oscillating at frequencies in the megahertz range.

  (3) *High-frequency "circulating" type of bearing current.* The parasitic capacitances between the stator winding and the frame are excited by the high dV/dt at the motor terminals, which creates a high-frequency ground current. The latter produces a circular flux around the motor shaft, inducing bearing voltages. The lubricating film breakdown high-frequency current ($\approx 0.5 - 20A$) circulates in the loop "stator frame—drive-end bearing rotor shaft—non-drive-end bearing" with a frequency of several hundred kilohertz. This type of bearing current is due to inductive coupling, and it mirrors the common mode current.

  (4) *Bearing currents, "circulating" type due to rotor ground currents.* This happens if the rotor-to-ground impedance is lower than the stator-to-ground impedance. In this case, a portion of the ground current crosses the bearings toward the shaft. These currents can reach high levels ($\approx 1 - 35A$) and prematurely damage the bearings. Small motors up to 20 kW are more sensitive to EDM bearing currents, whereas larger motors are likely to be subjected to circulating bearing currents.

- *Chemical and environmental causes have significant effects, but very little systematic study is available.* Temperature as both a cause and an effect plays an important role. Contamination and generally environmental conditions can severely shorten bearing life. Liquids (e.g., water) directly degrade the lubricant and surfaces through oxidation, and particles disrupt the lubricant films and separate the rolling body surfaces. Unlike other reasons and causes of degradation, they cannot be easily quantified and included in a model used for detection.

These causes will lead to the generation of local flaking. Studies on how a fault is initiated and developed have provided a better understanding of the fault progression and helped develop diagnosis tools. The initiation step requires the stress to exceed a threshold value, a fatigue-initiation stress criterion. Beyond stresses due to operation, stresses on bearings occur because of loading without rotation (e.g., true brinelling) or vibrations unrelated to operation (false brinelling). Healing, the smoothing of sharp edges of a crack or damage done by the rolling contact, initially reduces vibrations until the damage spreads.

As for the electric bearing load, the cause-and-effect chains of this type of load are well known. Discharge bearing currents can lead to localized pits that may translate into a gray trace, frosting, or fluting. Depending on the energy released, the discharge may lead to melting or vaporization of the bearing raceway surface. The small craters that are caused by melting are flattened by the rolling bearing balls, resulting only in a frosted raceway, and have been found to have no direct effect on the lifetime of the bearing; they can be ameliorated by frequent greasing. The large craters, however, resulting from vaporization, affect the lubricating grease, lead to corrugated patterns, and shorten the lifetime of the bearing. Similar damage patterns also result from differential-mode currents. The fault is developed further by the localized currents and by the breakdown of the grease that results when current flows where these factors are iteratively affecting each other and related to low vibration frequencies.

4.2.2.2 Winding faults

A large portion of faults in electrical machines are in the windings and are either short circuits between turns, of the same phase, between two phases, and between turns and the ground steel paths. These shorts lead to localized overcurrents that create local high temperature spots, damage the machine, and quickly result in catastrophic failure.

Armature winding faults. The causes of *short circuit* faults (turn-to-turn, phase-to-phase, phase-to-ground) are multiple [6]. Insulation degrades for a number of reasons. High temperature leads to oxidation and other forms of chemical degradation, and with it the mechanical weakening of the insulating material; vibrations of the whole machine, or torque pulsations related to the internal torque production, mechanically stress the insulation, as it is mechanically weaker than the iron core and copper windings; overvoltages due to the supply or fast switching transients in the electronic controller are yet a separate cause of insulation breakdown, albeit through a different physical mechanism, partial discharges (PDs) [7].

In low- and medium-voltage electrical machines, the turn-to-turn insulation is usually a thin layer of enamel, which often reveals an organic chemical composition (polyamide, polyimide, polyester-imide, etc.). Hence, PDs can be incepted in air spaces within the insulation (i.e., voids), leading to polymeric bonds breaking and resulting in conductor short circuits [8]. As degradation progresses, these short circuits result in the changes of the insulating material makeup and create pockets or voids in it, and discharges in these voids are initiated, a precursor of arcing and breakdown. Detection of these discharges is therefore useful although difficult to conduct online. The high values of the rate of voltage rise, dv/dt, resulting from high-frequency PWM both increase the PD occurrence and make it more difficult to distinguish the high frequencies of the signal and of the discharge [9].

Many parameters have been used to detect insulation degradation other than PDs. As insulation ages, the resistance of the insulation decreases, and hence a measurement of that resistance can lead to an understanding of the fault presence and severity estimation. The capacitance between turns and the grounded wall or other turns also depends on the insulation quality and either directly or indirectly is a measure of this fault [10].

Open circuit windings are important to detect in that they may disrupt the control algorithm, causing torque pulsations and overcurrents. They are caused most of the time by a fault in a power electronic switch, which stays open, and to a lesser extent by the breakage of a connection at the terminals or in internal connections of the windings due to corrosion, mechanical stress, or initial weakness. Nevertheless, they have to be detected, identified, and compensated for through the utilization of a modified control algorithm.

A special case is the open circuit of rotor bars or the end ring in a squirrel cage of induction machines. This occurs due to increased temperature and uneven temperature distribution and expansion between copper (or aluminum) bars and the steel core, and mechanical stresses, and in the case of a well-designed machine operating from a variable speed drive, during heavy loads. The asymmetries resulting from these breakages cause high-frequency stator currents and voltages, decreased developed torque, and increased torque pulsations. This fault propagates, as other rotor bars become increasingly loaded due to the failure of one AC [11].

4.2.2.3 Demagnetization of permanent magnets

Permanent magnet machines are increasingly being adopted. Unlike induction machines, the magnetizing field can be very high, more than $2T$, due to the high energy density of modern magnets, resulting in high torque density and efficiency. Their design has evolved into many variants and dramatically improved their already attractive characteristics.

For this discussion, we will consider two basic permanent magnet materials: rare earth and ferrites. There are a number of concerns in the use of permanent magnets: rare earth magnets, neodymium iron boron (NdFeB), and SmCo alloys, typically Sm2Co17, have very high energy density but are sensitive to demagnetization due to higher temperature and demagnetizing fields. NdFeB materials are high energy content and magnetization characteristics preferable to those of SmCo, but the latter can operate at higher temperatures. Demagnetization happens at high temperatures and under a demagnetizing field, which is often due to intended field weakening at high speeds. Fig. 4.5 shows the demagnetization of Permanent magnets due to high temperature and/or demagnetizing field.

An important concern is that certain materials, such as dysprosium, which in very small quantities is needed for high temperature operation, is not easily available, and hence its use results in high costs. Efforts to alleviate these problems have led to the use of lower cost, lower strength ferrite magnets that are not sensitive to high temperatures but are sensitive to demagnetization under a strong negative field. The problems associated with the use of magnets are as follows:

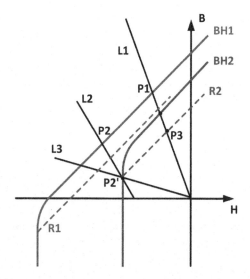

FIGURE 4.5   Demagnetization characteristics of NdFeB magnetic material. BH1: The original magnetization curve at low temperature; BH2: the same at higher temperature shifted; L1: a load curve without demagnetizing current; L2: a load curve with demagnetization current; P1, P2: operating points with and without demagnetizing current, of the "cold" magnet; P2': operating point below the knee; R2: the new recoil line after operating at P2'; P3: the operating point P1 shifted on the new recoil line.

- Avoidance of demagnetization through material selection, design, and operation, and
- Detection of demagnetization and continuous operation after partial demagnetization.

4.2.3 Capacitors

A very common and critical component in electrical drives is the DC link capacitor, used to filter the ripple of the voltage provided to the inverter. The inverter supplies AC current to the load of frequencies that depend on the switching frequency, and hence the current it draws from the source is not pure DC. This causes the voltage of the DC link to vary, and a capacitor is used to smooth it. Three types of capacitors are generally available for DC-link applications: aluminum electrolytic capacitors (Al-Caps), metallized polypropylene film capacitors (MPPF-Caps), and high-capacitance multilayer ceramic capacitors (MLC-Caps).

Al-Caps could achieve the highest energy density and lowest cost per joule but have the disadvantages of relatively high equivalent series resistance (ESR), low ripple current ratings, and wear out issue due to evaporation of electrolyte. MLC-Caps have smaller size, wider frequency range, and higher operating temperatures up to $200^{o}C$. However, they suffer from higher cost and mechanical sensitivity. MPPF-Caps provide a balanced performance for high-voltage applications (e.g., above 500 V) in terms of cost and ESR, capacitance, ripple current, and reliability. Nevertheless, they have large volume and moderate upper operating temperature.

The DC-link applications can have a high or low ripple current. The ripple current capability of the three types of capacitors is approximately proportional to their capacitance values. C1 is defined as the minimum required capacitance value to fulfill the voltage ripple specification. For low ripple current applications, capacitors with a total capacitance no less than C1 are to be selected by both the Al-Caps solution and the MPPF-Caps solution. For high ripple current applications, the Al-Caps with capacitance of C1 could not sustain the high ripple current stress due to low value of capacitance $(A/\mu F.)$ Therefore, the required capacitance is increased by the Al-Caps solution, whereas the one by the MPPF-Caps solution is C1. In terms of ripple current (i.e., $/A), the cost of MPPF-Caps is about one-third that of Al-Caps. This implies the possibility to achieve a lower cost, higher power density DC-link design with MPPF-Caps in high ripple current applications, like the case in electric vehicles.

Electrolyte vaporization is the major wear-out mechanism of small size Al-Caps due to their relatively high ESR and limited heat dissipation surface. For large-size Al-Caps, the wear-out lifetime is primarily determined by the increase of leakage current. An important reliability feature of MPPF-Caps is their self-healing capability. Initial dielectric breakdowns (e.g., due to overvoltage) at local weak points of a MPPF-Cap will be cleared and the capacitor regains its full ability except for a negligible capacitance reduction. With the increase of these isolated weak points, the capacitance of the capacitor is gradually reduced to reach end of life.

The metallized layers in MPPF-Caps are less than 100 nm in thickness and are susceptible to corrosion due to absorption of moisture. Severe corrosion occurs at the outer layers, resulting in the separation of metal film and the reduction of capacitance. Unlike the dielectric materials of Al-Caps and MPPF-Caps, the dielectric materials of MLC-Caps are expected to last for thousands of years at use-level conditions without significant degradation. An MLC-Cap could degrade much more quickly due to the "amplifying" effect from the large number of dielectric layers. A modern MLC-Cap could wear out faster through the increase of the number of layers. The failure of MLC-Caps may induce severe consequences to power converters due to the short circuit failure mode.

The dominant failure causes of MLC-Caps are insulation degradation and flex cracking. Insulation degradation results in increased leakage currents. Under high voltage and high temperature conditions. either with an abrupt burst of current leading to an immediate breakdown, or a more gradual increase of leakage current [12].

Failure modes, failure mechanisms, and critical stressors.

DC-link capacitors could fail due to intrinsic and extrinsic factors, such as design defect, material wear out, operating temperature, voltage, current, moisture and mechanical stress, and so on. Generally, the failure can be divided into catastrophic failure due to single-event overstress and wear-out failure due to the long-time degradation of capacitors. Based on these prior art research results, Table 4.1 gives a systematic summary of the failure modes, failure mechanisms, and corresponding critical stressors of the three types of capacitors.

4.2.4 Batteries

Although not always a component of a drive, batteries are quite commonly used, especially in transportation applications. Their health is directly related to the health of the drive, and often their health monitoring, fault diagnosis, and failure prognosis are integrated in the drive [13, 14]. Lithium-ion (Li-ion) batteries are the dominant general type used today, and this section discusses only this type.

Li-ion batteries consist of two electrodes as the anode and the cathode, which are separated by an electrolyte, where lithium ions move from the cathode to the anode during charging and back during discharging. Compared to nonrechargeable batteries containing metallic lithium, Li-ion batteries utilize a compound lithium electrode material.

4.2.4.1 Overview of the battery

- **Characteristics of Li-ion batteries.** The important characteristics of Li-ion batteries include their size (physical and energy density), longevity (capacity

TABLE 4.1 Overview of failure modes, critical failure mechanisms, and critical failure mechanisms of capacitors (Wang and Blaabjerg [12]).

Cap. type	Failure modes	Critical failure mechanisms	Critical stressors
Al-Caps	Open circuit	Self-healing dielectric breakdown	V_C, T_a, i_C
		Disconnection of terminals	Vibration
	Short circuit	Dielectric breakdown of oxide layer	V_C, T_a, i_C
	Wear out: electrical parameter drift (C, ESR, tanδ, I_{LC}, R_p)	Electrolyte vaporization	T_a, i_C
		Electrochemical reaction (e.g. degradation of oxide layer, anode foil capacitance drop)	V_C
MPPF-Caps	Open circuit (typical)	Self-healing dielectric breakdown	$V_C, T_a, dV_C/dt$
		Connection instability by heat contraction of a dielectric film	T_a, i_C
		Reduction in electrode area caused by oxidation of evaporated metal due to moisture absorption	Humidity
	Short circuit (with resistance)	Dielectric film breakdown	$V_C, dV_C/dt$
		Self-healing due to overcurrent	T_a, i_C
		Moisture absorption by film	Humidity
	Wear out: electrical parameter drift (C, ESR, tanδ, I_{LC}, R_p)	Dielectric loss	V_C, T_a, i_C, humidity
MLC-Caps	Short circuit (typical)	Dielectric breakdown	V_C, T_a, i_C
		Cracking; damage to capacitor body	Vibration
	Wear out: electrical parameter drift (C, ESR, tanδ, I_{LC}, R_p)	Oxide vacancy migration; dielectric puncture; insulation degradation; micro-crack within ceramic	V_C, T_a, i_C, vibration

V_C-capacitor voltage stress, i_C-capacitor ripple current stress, i_{LC}-leakage current, T_a-ambient temperature.

and life cycles), charge and discharge characteristics, cost, performance in a wider temperature range, self-discharge profile and leakage, gassing, and toxicity impact. On the positive side, they have high specific energy (230 Wh/kg) and power density (12 kW/kg) good energy density, excellent cycle life and long life, and good charging and discharging efficiency. On the negative side are the cost, the electronic protection system that is mandatory during charging and discharging, and the emissions during manufacturing and disposal.

The Li-ion battery has good charging and discharging electrical characteristics, as shown in Fig. 4.6. While charging, the charging capacity increases gradually with the charge voltage while maintaining a constant current. When the voltage reaches a maximum, the current decreases exponentially. However, the capacity discharge maintains an almost constant voltage and current to the load, with a small decrease and increase in the voltage and current values, respectively, until the cell capacity reaches the minimum acceptable level.

- *Li-ion components.* The Li-ion battery is composed of four primary components: the cathode, anode, electrolyte, and separator. The cathode is a lithium-metal-oxide powder. The lithium ions enter the cathode when the battery discharges and leave when the battery charges. The lithium ions leave the anode when the battery discharges and enter the anode when the

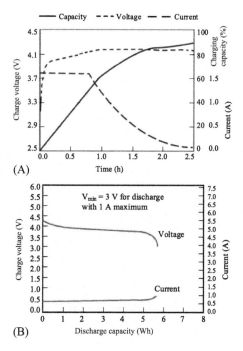

FIGURE 4.6 Typical characteristics of the Li-ion battery charging (A) and discharging (B). From Hannan *et al.* [13].

battery charges. The cathode and anode materials are made of lithium metal oxide and lithiated graphite in Li-ion batteries, where both structures are organized in layer of aluminium and copper current collectors, respectively. The electrolyte is composed of lithium salts and organic solvents; it allows for the transport of the lithium ions between the cathode and anode rather than electrons.

• *Li-ion battery formations.* Li-ion batteries can be constructed and packed in either metal cans in cylindrical or prismatic shapes, or laminate films (stacked cells) that are familiarized as Li-ion polymer batteries. They can be shaped as the cylindrical structure of rolled and plastered layers in metal cans with electrolytes. In the stacked form, the three layers are confined in laminate film and where their edges are heat-sealed aluminized plastic.

In general, the main sources of the active lithium ions in a battery are the positive electrode material or the cathode. Hence, to achieve high capacity, a huge amount of lithium is included in this material. Additionally, cathode materials follow a reversible process to exchange the lithium with slight structural modifications to its properties; in the electrolyte, the materials are prepared from high lithium ions that have diffusivity, good conductivity, and high efficiency. Those types of cathode materials involve lithium cobalt oxide (LiCoO2), lithium manganese oxide (LiMn2O4), lithium iron phosphate (LiFePO4), lithium nickel manganese cobalt oxide (LiNiMnCoO2), lithium nickel cobalt aluminium oxide (LiNiCoAlO2), and lithium titanate l (Li4Ti5O12):

- *Lithium cobalt oxide (LiCoO2).* This has high specific energy, is expensive because of the of cobalt, and has short life and restricted load capacity. It requires protection against overheating and excessive stress while charging quickly, and the charge and discharge rate need to be limited to a secure level.
- *Lithium manganese oxide (LiMn2O4).* Due to its structure, this has high thermal stability and safety but limited life span. LiMn2O4 has more specific energy than cobalt. This type of battery provides approximately 50% more energy than nickel-based batteries. Good design enhances the longevity and high current handling of this battery.
- *Lithium iron phosphate (LiFePO4).* This material is steady in the over-charged condition and can tolerate high temperatures without breaking down; the cathode material in this battery is more dependable and more secure than other cathode materials. Phosphates have a cell operating temperature range of -30^oC to $+60^oC$ and a cell packing temperature range of -50^oC to $+60^oC$ that deteriorates thermal runaway and prevents from burning out. The LiFePO4 battery has low resistance, long life span, high-load handling capability, improved security and thermic consistency, no toxic effects, and less expense.
- *Lithium nickel manganese cobalt oxide (LiNiMnCoO2).* The cathode blend of nickel-manganese-cobalt (NMC) can cause either high specific energy or power with high density. For the silicon-based anode, the capacity and life cycle compromise each other. The cathode mix of 33% nickel, 33% manganese, and 34% cobalt brings lower raw material costs because of the decreased cobalt content. Presently, this battery is in great demand for EV applications due to its high specific energy and minimum self-heating rate.
- *Lithium nickel cobalt aluminium oxide (LiNiCoAlO2).* The lithium nickel cobalt aluminium oxide battery (NCA) has a small amount of the world market share . Now automobile industries are emphasizing NCA battery production because of its high profile, such as high specific energy and power densities and long life span considering cost and safety [57].
- *Lithium titanate (Li4Ti5O12).* Lithium-titanate anodes have been commonly used for batteries since the 1980s. It has a spinel architecture and a high life span compared to that of a typical Li-ion battery. Moreover, Li4Ti5O12 batteries can be operated safely and have phenomenal features at cold temperatures. Because its specific energy is not high, unlike other Li-ions, the developments and research are focused on enhancing the specific energy and reducing the price.

4.3 Diagnosis methodology and tools

In general, effective fault diagnosis requires accurate measurements and tracking of signals in a motor drive. In this section, we discuss the types, characteristics, and selection. Fig. 4.4 show the basic methodology for fault diagnosis and classification.

4.3.1 Signal selection

It is seldom that one can perform a fault diagnosis of a drive without some prior knowledge of the drive operation under healthy conditions, and an expectation of the manifestation of all expected faults under different operating conditions and under different fault severities. This knowledge has to be developed from models—analytical, statistical, or based on artificial intelligence (AI)—that satisfy some basic criteria:

• Occupy minimal storage space,
• Are used in real or quasi-real time,
• Identify and discriminate between fault types and fault severities.

Knowledge of the behavior of the drive under a fault based on the physical understanding of the fault and an analytical model is the preferred technique, but it is possible only in a few scenarios. An obvious example is the monitoring of a stator current, which will indicate an open circuit; more complex but also possible is the monitoring of the harmonics of the currents or voltages of an induction machine drive during a rotor bar breakage. A number of faults either in the power electronics or in the machine may cause vibrations in the housing of the drive and a qualitative change in the stray magnetic field. Some raw data from these indicators are enough to identify the fault and its severity, but in most cases further analysis is needed.

The signals that are used can be the raw currents, voltages, or fluxes, or transformed to a rotating frame of reference, typically a field-oriented one, at least in a PMAC machine. The field orientation, although useful in an induction machine under healthy, normal operation, becomes inaccurate during some faults, as its operation and accuracy depends on the signals and parameters that are disrupted.

In an AC drive system, the stator currents and/or voltages are continuously available in the controller, either from measurements or from internal calculations. However, they may not be of the bandwidth required, and rectifying this may introduce additional costs. Other signals that are used are rotor position or speed, internal or stray magnetic field and vibrations, and temperature. Other signals may be specific to power electronics, such as voltages between gate, collector or emitter, and gate current.

The signals collected are processed first by filtering to remove unwanted frequency components and noise, and then by more complex operations. For example, in the case of a current-controlled, voltage source inverter, harmonics can be extracted, real and reactive power calculated, or the voltage command outputs of the current controller to the inverter can be extracted and from those the terminal voltages estimated. In many faults in electrical machines, the variable most closely connected to the fault is the magnetic flux, which is disrupted by the fault itself. It is therefore advantageous to attempt to measure the flux itself, a cumbersome and expensive proposal, or one of two variables: the voltage induced due to the magnetizing flux in the main power windings, or additional sensing windings, or the stray flux leaking from the stator frame. Among these faults are eccentricity, shorts, and demagnetization. They lead to changes of both the leakage flux and in the back EMF induced in that stator windings. These changes are not only in the amplitude but in the saturation level of the steel, and hence the relationship between stator magnetizing current and back EMF (e.g., see [15, 16]).

4.3.2 Signal features

Features of the signals used for the diagnosis are selected in one of three possible domains: time domain, frequency domain, and time-frequency domain. Of course, combinations of these and hybrid versions abound. Here we will limit ourselves to the basic concepts and applications.

In the time domain, the statistical parameters of signals that are mostly used to detect damage are for a signal of N sample points x_i, $i = 1 \cdots N$.

Use of time domain signals, when effective, offers significant reduction of effort compared to more complex techniques [17].

The use of harmonics of signals, voltages, currents, vibrations stray field, and so forth computed using fast Fourier transform requires operation in steady state—a condition not very common in electrical drives. Among other frequency analysis tools, bispectrum analysis [18], a higher-order spectrum (third-order spectrum), has unique advantages compared with the power spectrum, such as identification of nonlinear systems, retention of phase information, and elimination of Gaussian noise. It is usually used to detect quadratic phase coupling in nonlinear signals.

In stationary conditions, the frequency content of the signals is obtained through the Fourier transform. The time-frequency characteristics are obtained from like wavelet analysis and the general Cohen class transformations. These signal features are extracted from a relatively large collection of prior observations of healthy and faulty operation of similar drives. Signal processing techniques are used to detect the health condition based on statistical or AI models.

Most prevalent is time-frequency domain analysis, using wavelets and Cohen-class transformations. To perform monitoring of dynamic systems, an approach to retain key temporal information contained in the signatures is to use wavelet transforms (WTs). A WT is a powerful tool for time-frequency

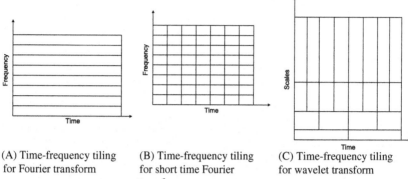

(A) Time-frequency tiling for Fourier transform

(B) Time-frequency tiling for short time Fourier transform

(C) Time-frequency tiling for wavelet transform

FIGURE 4.7 Time-frequency tiling for Fourier transform, Short time Fourier transform and Wavelet transform.

analysis in a dynamic system. Wavelet analysis is suitable for nonstationary signals. The discrete WT (DWT) greater flexibility than short time Fourier transform. Different basis functions, or mother wavelets, are used in wavelet analysis, whereas the basis function for Fourier analysis is always the sinusoid. Unlike sinusoids, wavelets have finite energy concentrated around a point. One can choose or design a wavelet to achieve the best results for a specific application.

Time-frequency tiling for the DWT is shown in Fig. 4.7(C). Unlike Fourier methods (Fig. 4.7(B), tiling for the DWT is variable, allowing for both good time resolution of high-frequency components and good frequency resolution of low-frequency components in the same analysis. However, it requires to predefine a specific band of frequency of interest to perform the analysis. In the case when the area of frequency interest is dynamically changing, the application of WT becomes challenging. Computational complexity has also been challenging to implement in a commercial DSP until now, or field-programmable gate array has been seldom utilized. A drawback of using wavelets is that once selected, the same transformation must be applied to all possible fault scenarios. The Cohen distributions make up a generalized class of time-frequency distributions that includes several types popular in electric machine diagnostics, such as Wigner-Ville, Choi-Williams, and Zhao-Atlas- Marks distributions.

4.3.3 Classification

4.3.3.1 Nearest neighbor rule

In k-nearest neighbors (kNN) classification, the output is a class membership. An object is classified by a plurality vote of its neighbors, a set of previously classified points, with the object being assigned to the class most common among its kNN (k is a positive integer, typically small). If $k = 1$, then the object is simply assigned to the class of that single nearest neighbor. To categorize a sample point in d-dimensional space, it is assumed that observations which are close to each other (in some appropriate metric) will have the same classification. This categorization problem can be approached in two different ways: first by assuming that some statistical distribution is given for the data, and second by assuming no knowledge of a distribution except for what can be concluded from the samples.

In calculating the minimum distance, some appropriate measure needs to be used. Any dissimilarity measure (4.4) would be applicable; however, the most commonly used dissimilarity measures are the Minkowski p metrics (4.5), where d is the dimensionality of the vectors $\mathbf{X}_m$ and $\mathbf{X}_n$.

$$d(\mathbf{X}_m, \mathbf{X}_n) = g\left[\sum_{i=1}^{d} f_i(X_{im}, X_{in})\right] \tag{4.4}$$

$$d(\mathbf{X}_m, \mathbf{X}_n) = \left[\sum_{i=1}^{d} |X_{im} - X_{in}|^p\right]^{\frac{1}{p}} \quad (p \geq 1) \tag{4.5}$$

The three most often used Minkowski metrics are the taxi-cab distance (4.6) where $p = 1$, the Euclidean metric (4.7) for which $p = 2$, and the maximum coordinate distance (4.8) where $p = \infty$.

$$d(\mathbf{X}_m, \mathbf{X}_n) = \sum_{i=1}^{d} |X_{im} - X_{in}| \tag{4.6}$$

$$d(\mathbf{X}_m, \mathbf{X}_n) = \left[\sum_{i=1}^{d}(X_{im} - X_{in})^2\right]^{\frac{1}{2}} \tag{4.7}$$

$$d(\mathbf{X}_m, \mathbf{X}_n) = \max_{1 \leq i \leq d}\{|X_{im} - X_{in}|\} \tag{4.8}$$

4.3.3.2 Linear discriminant analysis

A second approach to categorizing points in a d-dimensional space relies on the use of discriminant functions. In the implementation of discriminant functions, no prior knowledge of a probability distribution among the sample points is assumed. The space is divided into K regions, each having its own weighting coefficients. In this work, linear discriminant functions (4.9) are used,

$$D_k(\mathbf{x}) = x_1\alpha_{1k} + x_2\alpha_{2k} +, \ldots, +x_N\alpha_{Nk} + \alpha_{N+1,k}$$
$$k = 1, 2, \ldots, K, \tag{4.9}$$

where x is the N-dimensional sample vector and α are the normalized weighting coefficients for the k-th class. Linear discriminant functions were chosen for the algorithm since they are the most computationally efficient form. A sample vector belongs to a particular class if its discriminant function is greater for that

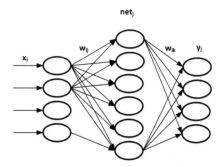

FIGURE 4.8 A typical neural network, x is the input layer, y is the output layer and one hidden layer.

class than for any other class—that is, $\mathbf{x}_i$ belongs to class C_j if

$$D_j(\mathbf{x}) > D_k(\mathbf{x}) \quad \text{for every } k \neq j.$$

The weighting coefficients are adjusted from their initial guess through a training procedure using sample vectors of which the proper classification is known. The algorithm for this procedure makes adjustments to the weighting coefficients until each sample vector is correctly classified.

Once a sample vector is correctly classified, no adjustment to the weighting coefficients is made. When a sample vector is incorrectly classified, or

$$D_j(\mathbf{x}) \leq D_l(\mathbf{x}),$$

where

$$D_l(\mathbf{x}) = \max_{l \neq j} [D_1(\mathbf{x}), \dots, D_K(\mathbf{x})],$$

adjustments are made to α_j (4.10) and α_l (4.11) only:

$$\alpha_j(i+1) = \alpha_j(i) + a\mathbf{x}_i \tag{4.10}$$

$$\alpha_l(i+1) = \alpha_l(i) - a\mathbf{x}_i, \tag{4.11}$$

where a is a gain constant. Fig. 4.8 shows a typical neural network with four inputs and outputs and one hidden layer.

4.3.3.3 Support vector machines

For classification problems, the support vector machine (SVM) attempts to find a hyperplane, the optimal one, to separate the data points according to their classes such that the separation between the classes is maximum [19]. Consider a two-class training set consisting of data points in which x_i is the i-th real valued–dimensional input vector and y_i is the corresponding class of x_i. The optimal hyperplane, which successfully separates the points according to their classes, can be given by the equation $\mathbf{w}^T x_i + b = 0$.

In this equation, $\mathbf{w}$ and b denote a weight vector and a bias term, respectively, and the goal of SVM is to find the values of $\mathbf{w}$ and b such that the separation between the classes is maximum. This problem is solvable in linear cases, but nonlinear versions have been developed to address real-life problems. They utilize a transformation of the training data into a higher (extended) dimension space. The original nonseparable data may become separable in the expanded space. The transformation is accomplished by using kernel functions (linear, polynomial, or sigmoid)

4.3.3.4 AI-based tools

AI-based tools extract inherent characteristics of data to classify faults from healthy conditions. These tools can perform classification tasks on a mixture of electrical and mechanical signals. Neural networks and fuzzy and neuro-fuzzy system–based artificial techniques are mainly used for motor fault diagnosis.

The flexibility of the neural networks makes them an immensely popular choice in machine fault diagnosis. Artificial neural networks (ANNs) provide an approach that makes the use of small processing elements called *neurons* that are interconnected in a manner similar to the brain. The amount of influence that one neuron exerts on another is determined by the weight corresponding to the interconnection between them and is a tunable parameter of the network. Numerous variations of neural networks have been proposed, as well as many techniques to update the weights. We discuss the simplest possible version here. The input to the network x and the vector of all weights w determine its output.

Forward propagation. In the input layer, the output of neuron i equals the value of x_i. In the hidden layer, with the neuron j of the hidden layer, its input value net_j is each unit's weighted sum of the prior layer. The output layer is the weighted sum of its inputs.

$$net_j = \sum_i w_{ij} x_i \tag{4.12}$$

$$a_j = f(net_j) \quad f \text{ is the sigmoid function} \tag{4.13}$$

$$y_k = \sum_k w_{jk} a_j \tag{4.14}$$

Back propagation. For targeted output d_k, the error function is defined as

$$E = (1/2) \sum_k (d_k - y_y)^2 \tag{4.15}$$

and it is used to correct the weights:

$$\Delta w_{jk} = -\eta \frac{\partial E}{\partial w_{jk}}. \tag{4.16}$$

4.4 Faults, their manifestation, and diagnosis

We first discuss the commonalities between various drive types and the diagnosis tools used there. We then proceed to discuss some of the specific methods that have been used in the more common types of drives.

The diagnosis process is similar to the diagnosis of most faults in nonelectrical systems and starts with the measurement of variables of interest. What is different in electrical drives is that the signals used, as well as the source of the signals, are primarily electrical or magnetic. In all cases, the first step is measurement of available signals. What happens next depends on the technique used.

An electrical drive is a complex system consisting minimally of a power electronics converter, a motor, a controller, and sensors. Each of these components consists of parts that, as discussed earlier, can fail in more than one way and have multiple manifestations of the fault, each depending on its own particular characteristics as well as the conditions they are operating. Faults in general can be detected and identified, their severity determined, and their progress monitored and predicted based on signals from sensors that are either in place as part of the drive controller or installed for this purpose. It is preferable that the entire fault detection, identification, and management process is conducted when the drive is operating normally, although some testing could be done during off times. It is also important that extensive testing be done at drive commissioning, both to detect inherent existing weaknesses and faults and to establish a baseline of healthy operation.

The controllers in electrical drives have as integral parts the speed and position, current, and possibly voltage sensors. It is then reasonable to expect that these sensors are the first to be considered for use in diagnosis. In addition, sensors that may be used are those that can measure vibrations, temperature, and a stray magnetic field. If an analytical model is established for a healthy drive and for all anticipated fault types and severities, then this model can be used to compare its output to that of the drive. When signal- and data-based models are used, features again are extracted in the time, frequency, and/or time-frequency domain. These are processed and compared to those resulting from extensive training tests to determine fault identification and severity. Faults can be detected based on different techniques. A more or less accurate model of the drive operation already exists in the controller and can be used to detect deviations of the expected operation. Using an analytical model of the machine, one can estimate the magnetic flux, torque, and terminal voltage waveforms of the healthy stage and compare them to the ones commanded or measured. The deviation can be a source of information about the health condition.

Some of the signals can be used in their raw form; to take a very simplistic example, a zero current is an indication of an open circuit. The closer the measured variable characteristics are to those of predetermined fault, the higher its significance and usefulness. A change in the flux, whether measured or estimated through the induced voltage, is closely related to a demagnetization, eccentricity, or short circuit. However, bearing degradation may result in low-amplitude high-frequency torque pulsations, which are distantly related to torque and speed pulsations and current harmonics, which may also be caused by other faults. Detection and identification of faults and estimation of their severity then is not often possible based on the raw signals.

4.4.1 Winding faults in AC machines

An open circuit can be detected from the current measurement in the corresponding phase. The open circuit may be located in the winding itself or caused by the malfunction of one of the power electronic switches feeding that winding. This is more difficult to detect when only one switch is damaged.

A short circuit in the windings, between turns, between phases, or between a winding and the ground is a more complex fault to detect. There are a number of variables that affect the characteristics and severity: the number of shorted turns, speed, and load current. The design of the machine makes a big difference in the effects. Induction machines are usually constructed with distributed windings, where the current in one phase and a shorted turn affects the induced voltages and current in the others. In PMAC machines, windings can be concentrated or distributed; in concentrated windings, coupling between phases is minimized, thus effectively isolating magnetically and thermally the damaged phase from the others.

In induction machines, the magnetic field created by the rotor currents is controlled by the stator current through a time constant of typically a few hundred milliseconds. In PMAC machines, this is not the case, as the magnetic field is not controlled, except though field weakening, which is usually encountered at higher speeds. This means that the magnets, due to inertia, keep on rotating and continue to feed a short circuit until the speed is decreased or the demagnetizing stator current is applied.

Beyond the signals available either directly or calculated in the controller, additional signals include those measured in the stray field or by imbedded sensors. For measuring the health of bearings, the most common and reliable variable is vibrations, although they can also be caused by the load or even short circuits. In addition, measurements of the stator currents have been attempted with the results strongly dependent on the machine and application. For the detection of the health of power electronics, other than load voltages and currents, the condition of the gate voltage, and of the collector-emitter, V_{CE} is detected.

A higher possible level of processing to obtain the condition of the drive, using these signals and analytical models, will stay close to the source of the fault, such as flux estimated through the back EMF, real and reactive power, currents transformed from natural to a synchronously rotating frame of reference, and harmonics related to the slip of an induction machine, among others. A third level, which makes little use of an analytical model, is based on the signal

TABLE 4.2 Time domain features.

Peak Value	$\frac{1}{2}[max(x_i) - min(x_i)]$
RMS Value	$\sqrt{\frac{1}{N} \sum_{i=1}^{N} (x_i)^2}$
Standard Deviation	$\sqrt{\frac{1}{N} \sum_{i=1}^{N} (x_i - \bar{x})^2}$
Kurtosis	$\sum_{i=1}^{N} (x_i - \bar{x})^4$/RMS value
Skewness	$\frac{1}{N} \sum_{i=1}^{N} (x_i - \bar{x})^3 / \frac{1}{N} (\sum_{i=1}^{N} (x_i - \bar{x})^2)^{3/2}$
Crest Factor	(Peak value)/RMS
Impulse Factor	Peak value/$(\frac{1}{N} \sum_{i=1}^{N} |x_i|)$

characteristics. These are often in the time domain: mean, variance, skewness, and kurtosis as shown in Table 4.2.

4.4.2 Bearing faults

As discussed in Section 4.2.2.1, faults may be caused by a variety of operating conditions that are not always easy to pinpoint and to associate with the bearing health condition. This observation renders the use of variables available in the controller to be of lesser value. The general idea is that at least at the early stage of a fault, where there are distinct craters, the mechanical resistance will appear whenever a rolling element tries to pass the defect. This in turn produces a train of impulses, which can be detected. The frequencies with the anomaly appear depending on the location of the fault:

Outer raceway $\quad f_o = \frac{N_b}{1} f_r \left(1 - \frac{D_b}{D_c} \cos \beta \right)$

Inner raceway $\quad f_i = \frac{N_b}{1} f_r \left(1 + \frac{D_b}{D_c} \cos \beta \right)$

Rolling element $\quad f_b = \frac{D_c}{D_b} f_r \left(1 - \frac{D_b^2}{D_c^2} \cos^2 \beta \right)$

where D_b, D_c are dimensions of the bearing, N_b is the number of the rolling elements, and f_r is the mechanical frequency of rotation [20]. The detection can then be based on the resulting harmonics of the torque-producing stator current.

Much more acceptable is the use of direct measurement of vibrations and extracting the signature using time-frequency analysis and various classification methods such as neural networks with Fourier transform [21]. In this case, the algorithm followed is shown in Fig. 4.9.

This algorithm time segments the input signal. The time-segmented vibration signals are used to obtain spectral contents. The signal then is filtered and further enhanced. In Fig. 4.10 the original spectrum of the signal is shown, while in

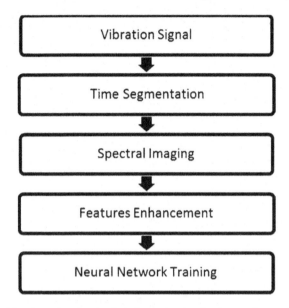

FIGURE 4.9 Flowchart of the bearing fault diagnosis using Fourier transform and ANN [21].

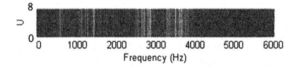

FIGURE 4.10 Exemplary motor vibration spectral image [21].

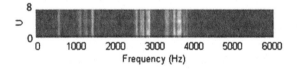

FIGURE 4.11 Enhanced image of an exemplary motor vibration signal. From Amar et al. [21].

Fig. 4.11, the enhanced version, resulting from filtering is shown, with coherent patterns enhanced and incoherent or noise spectra depreciated.

Supervised learning of the neural networks was used, and the steepest descent method was used to adjust the weights and biases. For the four classes considered, from the confusion matrix (see Fig. 27), the hit and false alarm probability pairs are (1, 0), (0.87, 0), (1, 0), and (1, 0.04), respectively. The overall hit and false alarm probabilities of the classifier are 0.96 and 0.01, respectively. High hit probability (0.96) and low false alarm probability (0.01) indicate that VSI with ANN is capable to distinguish different classes accurately, showing a very

TABLE 4.3 Confusion matrix indicating hit and false alarm probability pairs [21].

		Confusion matrix				
Output class	**1**	320	0	0	0	100%
		25%	0.0%	0.0%	0.0%	0.0%
	2	0	280	0	0	100%
		0.0%	21.9%	0.0%	0.0%	0.0%
	3	0	0	320	0	100%
		0.0%	0.0%	25%	0.0%	0.0%
	4	0	40	0	320	88.9%
		0.0%	3.1%	0.0%	25%	11.1%
		100%	87.5%	100%	100%	96.9%
		0.0%	12.5%	0.0%	0.0%	3.1%
		1	**2**	**3**	**4**	
		Target class				

high correct detection percentage with very low false alarms even under a poor signal-to-noise ratio (Table 4.3).

A new multispeed fault diagnostic approach was proposed by Hao et al. [22]. They used self-adaptive WT components generated from bearing vibration signals. The proposed approach was capable of discriminating signatures from four conditions of rolling bearing: normal bearing and three different types of defected bearings on outer race, inner race, and roller separately. Particle swarm optimization and Broyden-Fletche-Goldfarb-Shanno-based quasi-Newton minimization algorithms are applied to seek optimal parameters of the impulse modeling–based continuous WT model.

The impulse modeling–based continuous WT model was introduced for decomposing vibration signals obtained from roller bearings with WT. After that, three-dimensional statistical parameters are applied to extract fault characteristics. The nearest neighbor classifier using Mahalanobis distance is adopted to map samples into corresponding categories. The method provides very accurate results, as shown in Table 4.4.

4.4.3 Insulation

As we shall see later, incipient short circuits in machine windings are difficult to detect and the insulation breakdown can develop precipitously. In only a few cases can a short circuit fault be detected in time to be managed using the power frequency signals and related harmonics. It is therefore preferable to monitor the health of the insulation.

TABLE 4.4 Multispeed resulting trust rate of fault detection and identification on roller bearing [22].

The number of speed	Speeds	The number of training data sets	The number of testing data sets	The number of misclassified data sets	Trust rate	Testing Accuracies(%)
1	1000	400	400	0	98.94	100
2	1000,1500	800	800	0	98.38	100
3	1000,1500,2000	1200	1200	0	94.45	100
4	1000,1500,2000,2500	1600	1600	0	91.77	100
5	1000,1500,2000,2500,3000	2000	2000	0	91.10	100

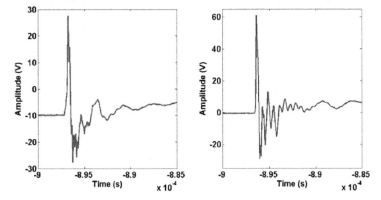

FIGURE 4.12   Turn-to-turn voltages for IGBT (red) and SiC (blue) inverters between the first and second turns with PWM-like signal ($U_{bus} = 700V$). From Acheen et al. [24].

The two phenomena, PDs and general insulation breakdown, have been studied extensively. For the first, there has been an IEEE guide published [23], although it is applicable to relatively high voltage machines. Detection is complex, as these discharges are caused by switching of power electronics, and the frequencies in the current resulting from PDs are close to the harmonics of switching frequency. Fig. 4.12 shows a typical case of different turn-turn stress in the same windings using IGBTs and SiC inverters. Fig. 4.13 shows the effect of PD on the current on the affected phase

When performing measurement using a nonintrusive sensor, it is usual to connect a high-pass analog filter to remove noise coming from the inverter drive switches or power amplifier. Typical cut-off frequency randing from 200 to 500 MHz been used in the following experiments. The nonintrusive sensor used to detect PD is taking advantage of a capacitive effect between the sensor fitting and the cable core.

It is necessary then to use more complex signal analysis and processing. There has been better success in detecting PDs and insulation health in offline measurement of noise, light, or radiated electromagnetic signals [25]. For the second,

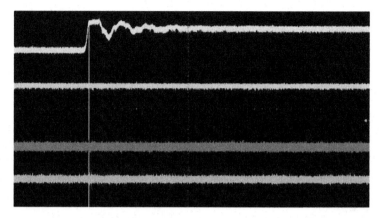

FIGURE 4.13 Phase-to-phase voltage (yellow), and PDs on phase 1, 2, and 3, respectively (blue/purple/green). From Acheen et al. [24].

general insulation breakdown, high-frequency signals and measurements are required [26]. It is interesting to notice that the sharp rise times of the power electronics switches that can be the cause of the insulation deterioration can also produce the current signals that can be used to detect the health of the insulation (e.g., [27, 28]). There an insulation state indicator (ISI) is introduced for the assessment of the insulation condition in one phase. It is based on quantifying the change in the machine's high-frequency behavior by comparison of the amplitude spectra recorded for a healthy machine condition (reference) and that recorded during later condition assessment. The root-mean-square deviation (RMSD) was chosen as a comparative value and serves as the ISI for the respective phase:

$$ISI_{p,k} = RMSD_{p,k}(x_1, x_2) \tag{4.17}$$

$$= \frac{\sqrt{\sum_{g=n_{\text{low}}}^{n_{\text{high}}} \left(Y_{\text{ref},p}(g) - \left|Y_{\text{con},p,k}(g)\right|\right)^2}}{n_{\text{high}} - n_{\text{low}}}$$

Here, Y_{ref} and Y_{con} are the Fourier transformed signals for a healthy machine condition (reference) and a later condition assessment, respectively. The index p defines the investigated phase (U,V,W). The variables n_{high} and n_{low} define the compared frequency range and depend on the sampling rate and investigated window length.

The performance and applicability of the method has been proven for different machine ratings and winding systems like random wound winding and preformed coils. One must notice that the sampling rate is significant, required to be in 2.6 MS/s for an induction machine of 1.4 MW. Jensen et al. [27] proposed a method to lower the necessary sampling rate.

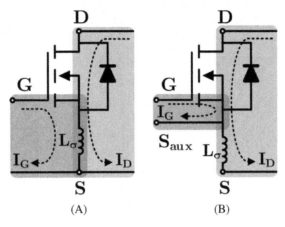

(A)   (B)

FIGURE 4.14 Circuit implementation of the monitoring system.

4.4.4 Power electronics

We have discussed some of the more common failure mechanisms of power electronics, particularly IGBTs. Oh et al. [29] presented a comprehensive review of condition monitoring and prognostics of IGBTs. A sensitive signal for the IGBT failure prognosis, using as precursor of the failure of wire binding, is V_{CE}, but its measurement cannot be reliable. Using more than one online measurement of damage indicators, such as the on-state voltage of the semiconductor and the voltage drop in the bond wires, a more accurate diagnosis can result (e.g., [30, 31]).

In the work of Gonzalez-Hernando [31], a monitoring system for both IGBTs and SiC MOSFETS is proposed and shown in Fig. 4.14.

The measurements are synchronized with the PWM signals generated in the digital controller. For V_{DSmax} , the measurement circuit is based on the online V_{CE} measurement. The circuit complexity limits the measurement speed of the system, thus defining the minimum settling time and maximum switching frequency. The diagnosis of the inverter circuits, especially when redundancies are used, is complex. The limited availability of sensors and measured signals lead most of the time to the use of simpler time domain methods.

Incipient IGBT faults may cause a spurious junction resistance increase at the junction point associated with junction degradation [32]. A normalized feature called *mean current vector* (MCV) gets arranged with respect to different types of spurious resistance faults. The normalized features thus obtained are ranked and an optimized set of effective features is computed using the SVM-based recursive feature elimination (SVM-RFE) algorithm. The method uses rather limited computations, a welcome development compared to most of the techniques available.

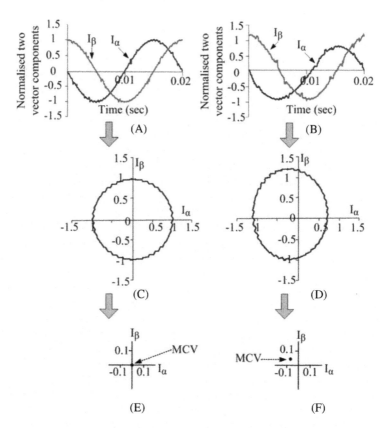

FIGURE 4.15 Normalized I_α and I_β current from a healthy inverter (A) and a faulty inverter (B). Concordia pattern on the $\alpha - \beta$-plane for a healthy inverter (C) and a faulty inverter (D). MCV on the $\alpha - \beta$-plane for a healthy inverter (E) and a faulty inverter (F).

When a plot of this normalized I_α versus normalized I_β currents is traced on a two-dimensional $\alpha - \beta$ reference frame, the normalized Concordia pattern is formed. This pattern is a perfect circle with unit radius for a healthy system and a distorted shape for a faulty system, as shown in Fig. 4.15. This is converted to a normalized Concordia pattern into a single point that is called *MCV*. The MCV for a healthy system is at the center of the circle, whereas for a faulty system it lies at a particular position on the $\alpha - \beta$ plane, as shown in Fig. 4.15(E) and (F), respectively).

Methods based on time domain analysis of the output current of the inverter, such as normalized current vector, slope methods, current profile shape, reference current error, and derivative of the absolute current Park's vector, have been in use for locating faults in the inverter portion of the drive. The whole array of statistical modes of the current can be used as features, and a variety of classifiers, ANNs, nearest neighbor, and SVM have been used. The problems of creating

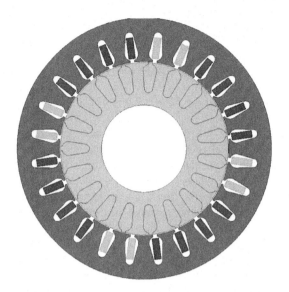

FIGURE 4.16 Cross section of a squirrel cage induction machine.

a database to use for classification and the selection of the most appropriate features limit the applicability of such methods.

4.4.5 Induction motor drives

Induction motor drives suffer often from open and short circuit faults, with the majority of these faults in the power electronics rather than in the motor. The single open-circuited phase winding is the most common form of winding fault, arising from either a fault in the winding itself or in the converter leg that it is connected to. The greatest impact is seen in the two phases that are physically located adjacent to the faulted phase in the stator winding, which exhibit an increase in input current in an attempt by the controller to provide the air-gap MMF that is missing in their locality. Fig. 4.16 shows a cross section of a typical rotor cage induction machine, with the stator windings and rotor bars shown.

In cage rotor induction machines, an additional problem may arise, in which a rotor bar is damaged to the point that its resistance becomes high or infinite, and similarly a section of the end ring may become damaged,

$$f_{SH} = (1 \pm 2s)f_s, \tag{4.18}$$

where f_s is the supply frequency and s is the slip. This open circuit fault causes increased stator currents, decreased torque, and increased torque pulsations. The stator current has frequency components related to the slip frequency, and these can be detected both in the case of operation off the grid and during transients under controlled operation. Motor current and motor voltage signature analysis has been shown to work well in most of the cases of broken rotor bars, although they are more accurate under extreme operating cases, such as heavy loading, corresponding to high slip.

Generally, the frequencies produced by interturn shorts are

$$f_{SHc} = f_s[m(1-s)/p \pm k], \tag{4.19}$$

with $m = 0, 2, 1$, $k = (0, 1, 3, 5)$. These frequency components already exist in the current since they already exist in the drive due to imbalances and so forth. Furthermore, under the short-circuit condition, a significant rise in rotor-slot harmonic components occurs.

Since many induction motor drives are using a voltage source current-controlled inverter, incipient short circuits are difficult to detect using motor current signature analysis, as it is close to the commanded current. However, the phase voltages are seldom directly measured. This leaves two options: either use the estimation of the stator voltages from the commanded ones, or use additional variables calculated from these voltages and currents. They include the instantaneous active and reactive power to extract signatures [16, 33], the negative sequence component of the currents or voltages, and the corresponding reactances, as well as the Park transformation of these variables. In these methods, the fault signature is a function of slip, which requires a lookup table with large data memory and complex interpolation for compensation. As most techniques cannot estimate fault severity, AI-based techniques have been proposed.

Monitoring the stray field of the machines has been a recently proposed tool, either alone or in conjunction with other signals [34, 35].

The average current methods detect the fault from the mean value of currents over one fundamental period. They either use the phase currents or the current space vector in the stator frame, applying Clarke's transformation. The magnitude of the average current space vector over one period is compared to a threshold value. The angle of this space vector indicates the faulty transistor. Since this method is dependent on the load, the current is normalized. In the slope method, the slope of the trajectory of the current space vector in the stator frame is observed. It is assumed that the slope is rather constant for a quarter of a period during the fault. The slope depends on which inverter leg experiences the fault. To determine the faulty transistor, Schmitt triggers are used for the detection of the polarity of the phase currents during a fault [36].

Having more than three phases improves the ability to continue operation after a fault, as well as windings, where each phase may have multiple coils. Parallel paths ameliorate the effects of open circuit faults by allowing currents to flow in some of the coils in the faulted phase, thereby providing a measure of fault tolerance.

Methods have been proposed based on neural networks, using every available variable as input: currents, speed, and voltages in the time or frequency domain.

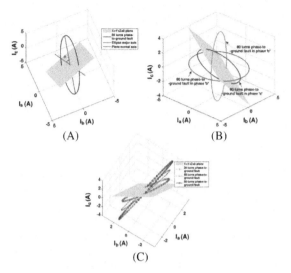

FIGURE 4.17 (A) Phase-to-ground fault locus in phase "a." (B) Phase-to-ground fault locus in different phases. (C) Phase-to-ground fault in different fault severities in phase "a.". From Eftekhari et al. [37].

A fuzzy decision system has been proposed, which has been shown to be accurate, although it requires extensive data for training.

In addition, most of the proposed methods deal with the detection procedure and not with the location of the fault, unless the neutral point is accessible or intrusive sensors are installed inside the machine or even extra sensors (e.g., voltage or flux sensors) are used. Meanwhile, a turn-to-turn fault in two phases is eventually possible to simultaneously occur, and it may even lead to more catastrophic consequences than the case of a fault in only one stator winding; however, there has not been much research effort in this area [37]. Fig. 4.17 shows the stator current locus for the case of phase-ground faults in different phases. This work is based on developing the best fit of an ellipse describing the currents in a three-dimensional current locus with the circular pattern of the healthy motor.

The short-circuit fault has to be recognized safely and quickly, in a few milliseconds, since, especially in IM distributed windings, the fault can propagate rapidly between phases that have turns in the same slots, resulting in a catastrophic fault that cannot be mitigated or compensated for.

4.4.6 PMAC drives

This general category, including interior and surface permanent magnet machines, with rare earth or ferrite magnets, concentrated or distributed windings, or axial or radial flux, is the fastest growing segment of electrical drives at

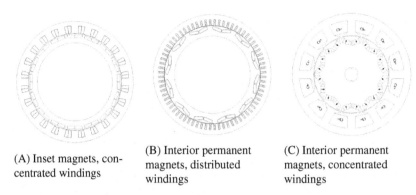

| (A) Inset magnets, con-centrated windings | (B) Interior permanent magnets, distributed windings | (C) Interior permanent magnets, concentrated windings |

FIGURE 4.18   Typical cross sections of radial flux PMAC machines.

present. Fig. 4.18 shows the cross sections of some typical Permanent Magnet AC machines. Unlike induction machines, PMSMs require power electronics converters to operate and form a drive, and the faults of these drives can be either internal to the machine or the converter. Some of these faults have common characteristics, making it difficult to separate and discriminate, and one fault can precipitate the other. They have to be isolated, since any mitigation or other action will depend on the fault type. All of these faults have been the subject of a plethora of detailed individual studies, with a comprehensive review published by Choi et al. [38].

PMAC drives operate under one of two control principles: either field orientation (discussed in Section 4.1.1) that requires the continuous updating of the position of the rotor and online transformation of currents and voltages, measured, estimated, or commanded, to a frame of reference related to this rotor position, or a DTC method (discussed in Section 4.1.2).

Eccentricity fault in these machines is often caused or initiated by manufacturing flaws, which can be detected before commissioning, at the end of the production line, by mechanical radial loads causing bearing wear, and by partial demagnetization. They result in imbalance of radial forces and magnetic pull, which will worsen the wear and increase the fault severity. They lead to vibrations, rotor-stator interference, and rubbing.

Detecting eccentricity is based on one of two principles. The first is detecting sideband components of the stator current [39] affected by the machine design (slots and saturation) and analyzed using WT or simply spectrum analysis. The second is based on changes in the flux [40], which can be estimated through the back EMF and identified through the changes in the flux saturation. Changes in the stray flux, measured outside the machine, are a promising indicator. Similar are the tools for detecting partial demagnetization of magnets.

An open circuit in a PMAC drive can be in the windings, the terminals, or the switches. The first two are relatively straightforward to detect, given that typically there are current sensors in the drive controller. More complex is the

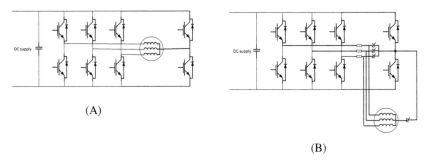

(A)

(B)

FIGURE 4.19 Typical connections to mitigate inverter fault.

case of a fault in the inverter supplying the machine, as it is possible that only one switch may be open, discussed by Eickhoff et al. [41].

Again, it is the currents that can be used for the detection of the fault, along with the rotor position. In this case, it is important to detect which switch is malfunctioning, the number of shorted turns, and the resistance of the open. The indicators that have been proposed include the negative sequence component of the stator current and voltage and the real and reactive power estimation in the controller. In most cases of open circuit fault, mitigation is possible, especially if the drive has been designed with that in mind, such as multiple phases or access to the neutral access to the neutral and an additional inverter leg that can be connected to it. Two such options are shown in Fig. 4.19. To accomplish this, it will be necessary to provide access to the neutral and utilize additional power electronics components [38].

A short circuit in PMAC machine windings can lead very quickly to a subsequent catastrophic fault. This is because even if the fault is detected rapidly and the machine is disconnected from the supply, its rotational speed will not drop as quickly, and thus the magnetic field will continue feeding the short circuit. The short circuit can be detected through the measurement of the harmonics of the current and commanded voltages, and the estimation of real and reactive power fed to the motor. Then the current in the short-circuited portion of the winding can possibly be limited through design of high inductance, and through the injection on a demagnetizing current. A disadvantage of this scheme is that very fast sampling may be needed.

Since the short-circuit current increases rapidly and a detection may be completed too late for the system to react, a reasonable approach is instead, or in parallel, to monitor the health of insulation rather than the operating variables of the drive, and predict its failure through the calculation of the RUL. This is a rapidly developing and promising area. Separating the various faults has been a daunting experience. One of the many proposals was by Haddad et al. [42, 43], with results shown in Fig. 4.20 and mitigating a short circuit is not always possible. To limit damage, in the case of concentrated windings with

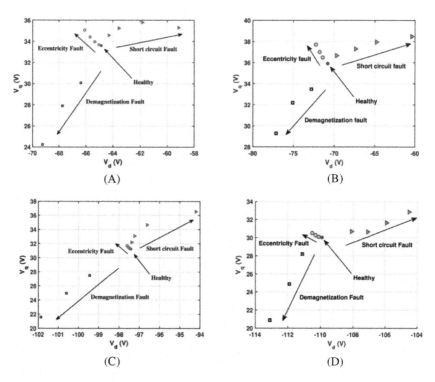

FIGURE 4.20 Experimental and simulation results of the effects of three fault types on PMAC machines. From Haddad et al. [43].

electrical, mechanical, and thermal isolation, the only possible solution is to inject demagnetizing current in the stator (e.g., see [44]).

4.4.7 Switched reluctance machines

The operation of thee machines is based entirely on developing torque using the reluctance of the magnetic circuit that varies with the rotor position. A simplistic cross section of the cross section of such a machine is shown in Fig. 4.21(A), and one of the many possible switch configurations is shown in Fig. 4.21(B). The machine operates by consequently energizing the coils in windings around the stator teeth A, B, and C. It is clear that these coils are not in physical proximity and hence have little thermal or magnetic coupling.

It is generally accepted that SRM drives are fault tolerant by their nature but not completely fault free [45].

An exhaustive methodology to identify possible faults, open and short of diodes and switches in the inverter, has been presented by Gopalakrishnan et al. [46]. Certain faults can be only handled by disabling the complete drive. In most cases, disabling a phase can lead to the continued operation of the drive.

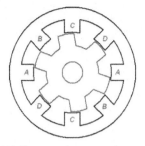

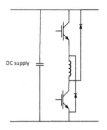

(A) Conceptual cross section
of a switched reluctance machine

(B) Components of a switched
reluctance drive

FIGURE 4.21 Cross section of a 6/8 switched reluctance machine, and diagram of a one-phase inverter leg.

Phase-to-phase shorts are generally of little interest here because of the physical separation of the phases.

4.5 Failure prognosis, fault mitigation, and reliability

4.5.1 From diagnosis to prognosis

The obvious goal of failure prognosis is to estimate the RUL of a drive, component, or part of a drive. This knowledge is used to schedule maintenance, employ redundancies, or, in general, avoid unexpected failures and manage the health of the drive. To do so in a profitable manner, the estimation of RUL has to be accurate, and hence a degree of certainty has to be associated with this prediction, related to the width of the prediction interval. The prediction interval is defined as an estimate of the time interval in which a future observation (in our case a failure) will fall, with a certain probability, given what has already been observed. The RUL estimate, accounting for this interval, has to be longer than the time it takes to schedule and perform maintenance or mitigate an incipient fault. It should be pointed out that decisions based on diagnosis alone implicitly include failure prognosis, although they do not provide an estimate of time to fail.

Two extreme examples illustrate the concept, its usefulness, and limitations. Bearings degrade almost always at a slow rate, but it is very seldom that a fault of a bearing can be mitigated. Extensive research has shown that the bearing RUL can be estimated accurately and well in advance of a failure. This allows to schedule maintenance with a low chance of catastrophic failure. The opposite extreme is the short-circuit fault in the windings of a PMAC. In a limited number of cases, the fault can be mitigated by a quick modification of the control algorithm, but in most cases, the short-circuit current is fed by the rotating magnets and increases faster than any action can be taken. If it is the short circuit that is being monitored and predicted, there is not always enough time to manage

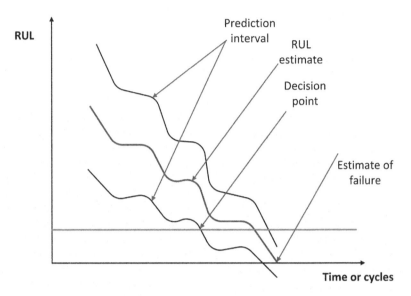

FIGURE 4.22   The RUL decision interval and the threshold determine the point of action.

the fault, and hence in that case, the RUL estimate is essentially useless. But if it is insulation health that is being determined, there may be adequate time to plan maintenance.

Prognosis of drive failure can increase the reliability of the system and decrease cost of operation and chances of unexpected failure. As the state of the health approaches failure, a well-designed prediction system will provide a RUL estimate that approaches zero, and the estimate of the confidence in it increases. Action is needed when the threshold for decision (mitigation, shutdown, etc.) falls within this interval, and the decision should be made well ahead of that point so that the time to act is adequate. If the decision threshold is well defined in relation to the prediction interval, early decision for action, followed by the necessary time to act, will be effective. If instead this threshold is set too low, then the decision will be too late, leaving insufficient time for action. If the threshold is set too high, the decision will cause early interruption of service, as shown in Fig. 4.22.

4.5.2 Prognosis tools

To predict the RUL with high precision, it is necessary to have the diagnostic tools in place first, from which the state of health of the device can be estimated. Beyond this, methods have been developed to identify trends of the features used for diagnosis, based on Baysian statistics and AI. To identify trends in the degradation of a component or subsystem, it is necessary to have stored histories of similar components, and at least part of the history of the one that is being monitored. The establishment of the relationship between physical degradation and its manifestation gives validity to a prognostic technique. Although physics-based methods offer a direct connection between operation and degradation, they often become too complicated and require more observed variables.

4.5.2.1 Kalman filter

The Kalman filter is the optimal linear estimator for linear system models (e.g., see [47]). The extended Kalman filter (EKF) [48] and the unscented Kalman filter linearize the model but are not optimal [49].

A Kalman filter is a model-based state estimator that estimates the state utilizing a linear model, given inaccurate inputs and inaccurate measurements. The continuous state-space model is first discretized with a timestep t_s, using the Euler backward method, resulting in a discrete state-space model with a timestep.

An EKF can be used to calculate the RUL of the data that are measured. It uses an estimate of the expected trajectory in the state variables to predict future values of those variables when the input measurements are noisy. The EKF will predict the next values of the state variables, receive a measurement from the system, and then update the prediction of the next value along with updating the parameters of the expected trend that the variables will follow. During this process, a certain level of white noise is expected in the system model and in the input measurements. The nonlinear system model used for the EKF is shown in Eqs. 4.20 through 4.25. Here, x represents the state variables, F is the state transition matrix, w is the process noise covariance, v is the measurement noise covariance, H is the output matrix, and z is the measured output. The uncertainty matrices, M and P, are used to update the Kalman gain, K, and the predicted value of the state variables as shown in Eqs. 4.20 through 4.25. Matrix R represents the measurement noise covariance, and matrix Q represents the process noise covariance.

$$\hat{x}_k = F_{k-1}x_{k-1} + e_{k-1} \tag{4.20}$$

$$z_k = H_k\hat{x}_k + u_k \tag{4.21}$$

$$M_k = F_{k-1}P_{k-1}F_{k-1}^T + Q_{k-1} \tag{4.22}$$

$$K_k = M_kH_k^T\left(H_kM_kH_k^T + R_k\right)^{-1} \tag{4.23}$$

$$x_k = \hat{x}_k + K_k(Z_k - H_kx_k) \tag{4.24}$$

$$P_k = (1 - K_kH_k)M_kx \tag{4.25}$$

4.5.2.2 Particle filters

Particle filters are an attractive alternative, discussed elsewhere (e.g., [50–52]). The Monte Carlo method is a Baysian model–based estimation of internal states in dynamical systems when partial observations are made. The solution of the filtering problem is computed by recursive estimation. The filtering problem is to estimate the first two moments of the state vector that is governed by the dynamic state-space model having noisy observation. A discrete time-controlled process can be expressed in state-space form by the stochastic difference equation of the form:

$$x_k = f_k(x_{k-1}, w_{k-1}) \qquad (4.26)$$

and a measurement equation $y \in \mathfrak{R}^k$ given by

$$y_k = h_k(x_k, v_k). \qquad (4.27)$$

Eq. (4.26) is called the *state transition (dynamic) equation*, whereas Eq. (4.27) is called the *correction, update, or output equation*. At time t_k, x_k is the state vector, w_k is the dynamic noise, y_k is the real observation vector, and v_k is the observation noise vector. The function f_k gives the relationship between the previous state and the current state, and the function h_k links the current state to the output.

In Bayesian form, instead of the future state vector, the probability density (pdf) of the future state vector is estimated. Following the pattern of update and measurement equations, the prior pdf is calculated using the update equation and the posterior pdf using the measurement equation. Eq. (4.26) gives the predictive conditional transition density, $p(x_k|x_{k-1}, y_{k-1})$, of the current state, given the *previous* states and *previous* observations. The observation or measurement equation, Eq. (4.27), gives the likelihood function of the current measurement given the current state, $p(y_k|x_k)$. If $p(x_{k-1}|y_{k-1})$ is defined as the previous posterior density, then the prior pdf $p(x_k|y_{1:k-1})$ is defined using the Baye's rule as

$$p(x_k|y_{1:k-1}) = \int p(x_k|x_{k-1})p(x_{k-1}|y_{k-1})dx_{k-1} \qquad (4.28)$$

The correction step generates the posterior probability density function from

$$p(x_k|y_{1:k}) = c * p(y_k|x_k)p(x_k|y_{1:k-1}). \qquad (4.29)$$

The filtering problem is to recursively estimate the first two moments of x_k given y_k. For a general distribution, p_x, this consists of recursive estimation of the expected value of any function of x, say $\langle g(x) \rangle_{p(x)}$, using Eqs. 4.28 and 4.29.

$$\langle g(x) \rangle_{p(x)} = \int g(x)p(x)dx \qquad (4.30)$$

4.5.2.3 Hidden Markov model

The hidden Markov model (HMM) is a stochastic technique for modeling signals that evolve through a finite number of states. The states are assumed hidden and responsible for producing observations. A HMM assumes that the system is

Markovian (i.e., the behavior depends only on the current state). The objective is to characterize the states given the observations. S_k is the hidden state at time k and O_k is the observation sequence, assuming that there are C possible states. The main objective is to determine hidden parameters (states) from the observable parameters. The problem to be solved in our case is as follows. Given the observation sequence $y = \{y_1 y_2 ... y_k\}$ and set of model parameters $\theta = \{\pi, A, B\}$, how do we choose the corresponding state sequence $x = \{x_1, x_2 ..., x_k\}$, which is optimal to generate the observation sequence. The optimal measure can be the maximum likelihood. The model developed has three elements:

π: C x 1 initial state distribution vector where the i_{th} element is the probability of being in state i at time $k = 0$, $p(S_0 = i)$.
A: C x C state-transition matrix where the $(i, j)^{th}$ element is the probability of being in state j at time $k + 1$, given that it is in state i at time k, $p(S_{k+1} = j|S_k = i)$.
B: State-dependent observation density B. Its jth element is the probability of observing O_k at time k given the system is in state j, $b_j(O_k) = p(O_k|S_k = j)$.

The model parameters are collectively denoted by $\lambda = \{\pi, A, B\}$. To implement the HMM-based prognosis algorithm, the model parameters need to be trained. The state transition probabilities (**A**) and state dependent observation densities (**B**) are generally obtained from the historical data collected from a large number of observations, and the initial state probability distributions (π) depend on the implementation area and the nature of operation of the system being studied.

4.5.3 Applications and new developments

The prognosis techniques that have been described, as well as similar ones, have been applied to a level that the end users have found acceptable and useful and have incorporated in industrial systems. The main applications include the following.

Bearings . This problem has been addressed in multiple publications; an experimental platform, PRONOSTIA [53], provided data for a large number of research efforts and a competition, which advanced the state of the art. The tools used extend the whole gamut of feature extraction, both time and time-frequency domain, and both data and AI methods for diagnosis and prognosis [3].

In the work of Kim et al. [54], intelligent fault diagnosis and health states estimation of discrete failure degradation was performed using a range of classification algorithms, such as ANNs, SVM, classification and regression trees and random forests, and linear regression. Among the available classifiers, SVM showed outstanding performance in the classification process compared with the other classifiers. The health state probability estimations were conducted using the classification ability of SVM and with subsequent machine prognostics being conducted based on the probability distributions of each health state. The method

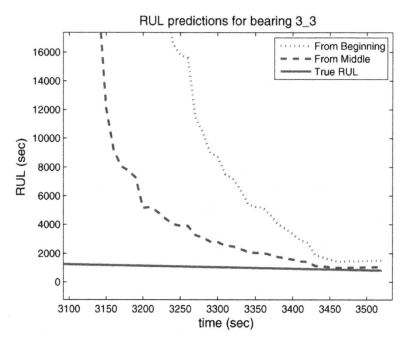

FIGURE 4.23 RUL estimation with different EKF tracking start times. From Singleton et al. [57].

shows the usefulness of identifying degradation states, in that case five, to be used as an estimation tool for machine remnant life prediction in real-life industrial applications.

Using more than one set of features allows an in-depth understanding of the degradation of bearings [55]. There the authors used primarily two features (skewness of 236–256 Hz band and entropy of 160–200 Hz band) of six selected features to detect change points in signals with high volatility, such as bearing vibration data. The more features used, the more the change-point detection algorithm becomes robust to the noise and can identify more distinct change points. The same authors used the time-frequency domain as well as frequency-domain features and Kalman filtering [56]. In the last case, in Fig. 4.23 it becomes clear that the duration of the estimation process plays a role and that RUL is more accurately estimated close to failure.

Soualhi et al. [58] combined data-driven (time domain features of the vibration signal) as health indicators and experience-based approaches (artificial ant clustering) for classification. The imminence of the next degradation state in bearings is given by HMMs, and the estimation of the remaining time before the next degradation state is given by the multistep time series prediction and the adaptive neuro-fuzzy inference system. Fig. 4.24 gives the predication of the RUL based on this method.

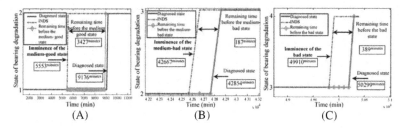

FIGURE 4.24 Prediction of the imminence and the remaining time of a bearing: the medium-good state (A), the medium-bad state (B), and the bad state (C). From Soualhi et al. [58].

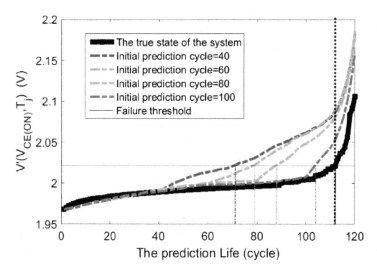

FIGURE 4.25 IGBT RUL prediction results of V' ($V_{CE(on)}$, T_j) using particle filters under different initial prediction periods. From Rao et al. [59].

Power electronic switches . Extensive efforts discussed earlier have led to advanced and affordable techniques to identify failure precursors and utilize modifications, either of the control system of built-in redundancies. Rao et al. [59] fused information of the junction temperature and collector emitter voltage to establish a more accurate precursor, using a Particle Filter algorithm Fig. 4.25.

An extensive review of topologies and their uses is given by Zhang et al. [60]. These topologies, some shown in , require only partial redundancies and modification of the control algorithm. Alternatives [61, 62] include keeping the same controller but modifying the PWM scheme to decrease the switching losses and temperatures.

Winding insulation . Methods to determine the development of incipient faults and predict RUL well ahead of failure have been published both by industry and academia (see [26]). Nussbaumer et al. [28] utilized the response

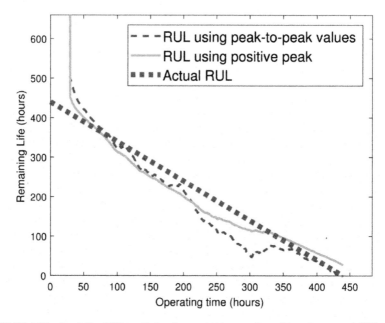

FIGURE 4.26 Insulation RUL prediction from switching transients. From Jensen et al. [27].

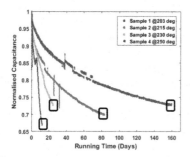

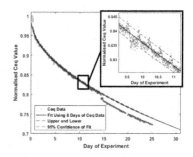

(A) Normalized capacitance change for different temperatures indicating effect of temperature on aging.

(B) Output space vector diagram after one switch open.

FIGURE 4.27 Changes in output space vector with one switch open. From Tsyokhla et al. [63].

to high-frequency testing and the switching impulses provided by the inverter. EKF has been used for prognosis, and innovative schemes have been proposed to avoid high-frequency sensors [27]. There the authors utilized the peaks the current transients at the PWM edges resulting from the use of fast switches (Fig. 4.26).

Tsyokhla et al. [63] used the capacitance data as a prognostic tool. In Fig. 4.27, the change of normalized capacitance is shown for different tem-

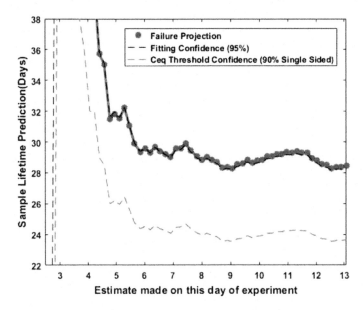

FIGURE 4.28   Insulation failure prediction and confidence along with C_{eq} versus time. From Tsyohkla et al. [63].

peratures and the level of filtering used. In Fig. 4.28, the RUL estimate and its threshold confidence are plotted versus the operating time.

Batteries . The concerns related to battery health as being vital to the health of electric transportation has been increasing. Fault mitigation methods have been proposed, using analytical models, Monte Carlo, and particle filters (e.g., see [64]).

4.5.4   Decisions based on prognosis and mitigation

Prognosis tools can give adequate warning of an impending failure but not the ability to recover or mitigate the fault. The resulting decision may include a complete shutdown at a convenient moment before the anticipated failure for bearings, gears, couplings, and decaying insulation. If the estimation of RUL leaves limited time for such action, redundancies such as the (1) use of a different motor or inverter or (2) operation with reduced phases or with a neutral inverter leg should be utilized. Such failures can be at the switches, windings, and so forth, and the control algorithm can be changed, for instance, to inject negative d-axis current to offset the effects of a short circuit fed by the rotating permanent magnet.

Health monitoring and the possible detection of a fault and its severity are steps toward the decision to some action. This action may be

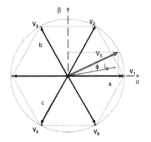 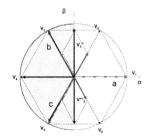

(A) Output space vector diagram before fault

(B) Output space vector diagram after one switch open

FIGURE 4.29 Changes in output space vector with one switch open. From Ginart et al. [67].

Continue operation considering the drive to be healthy,
Plan maintenance without disruption of present operation,
Take mitigating action including deployment of redundancies,
Continue operation recognizing the possibility of a failure, or
Emergency stop.

Both in cases of prognosis of an expected fault or of outright fault identification, some of the appropriate action may allow the continuation of operation. These are based primarily on the use of redundancies, modifications of the control algorithm, or both.

Several techniques have been proposed to handle phase loss in an induction machine, as well as the case where an open fault is internal to the inverter. A detailed survey was published by Mirafzal [65]. The case where the switch remains open but the antiparallel diode remains intact is discussed by Eickhoff et al. [66]. A disadvantage of these mitigation schemes is that they cannot produce the complete range of voltages [67], thus limiting the operational ability of the drive (Fig. 4.29). A few considerations are needed: the inverter fault has to be detected very quickly, in the order of microseconds; the drive has to have a more complex topology, including a fourth inverter leg; and possibly a split DC link capacitor, thyristors, and fuses.

A case of a shorted switch can also be dealt with by injecting demagnetizing current before the fault expands, thus limiting the extend of the damage but also decreasing produced torque and introducing torque pulsations.

Batteries can be managed by prognosis and topology changes for stress mitigation. A converter equipped with combinations of battery cells and capacitors forms a unit with increased power density and lowered electromagnetic interference. Electrolytic capacitors, as well as supercapacitors, placed in addition to batteries are used to reduce stress and losses that medium-frequency current pulsations cause in battery packs (see [68]).

In a recent article, Lee et al. [69] discuss emerging challenges in this area and the possible directions of research. For failure prognosis to become more widely applied, open questions such as how to decrease the amount of data used to train the algorithm and how to improve the confidence, giving adequate time for reaction have to be further addressed. Methods being investigated include the use of translational models, such as adapting past results from similar systems without extensive new tests, and hybrid methods combining statistical methods with neural networks. Another important line of research is in applying advanced algorithms to fault classification and prognostics of electric drives.

As indicated earlier, prognosis is a natural and necessary step after most diagnosis events. Since it is based on and requires methods to identify and utilize trends, it results in a further level of complexity. Advanced data-based methods are evolving, tested, and proposed, and they offer accurate predictions, albeit often based on a long history of observations. Fusion of sensed data and hybrid physical/data-based systems also offer a promise of reducing the testing effort and data storage [70].

References

[1] F. Niu, B. Wang, A.S. Babel, K. Li, E.G. Strangas, Comparative evaluation of direct torque control strategies for permanent magnet synchronous machines, *IEEE Transactions on Power Electronics* 31 (2) (2016) 1408–1424.

[2] R. Wu, F. Blaabjerg, H. Wang, M. Liserre, F. Iannuzzo, Catastrophic failure and fault-tolerant design of IGBT power electronic converters—An overview, Proceedings of the 39th Annual Conference of the IEEE Industrial Electronics Society (IECON), 2013, pp. 507–513.

[3] A. Muetze, E.G. Strangas, The useful life of inverter-based drive bearings: Methods and research directions from localized maintenance to prognosis, *IEEE Industry Applications Magazine* 22 (4) (2016) 63–73.

[4] A. Aggarwal, E.G. Strangas, J. Agapiou, Analysis of unbalanced magnetic pull in PMSM due to static eccentricity, Proceedings of the 2019 IEEE Energy Conversion Congress and Exposition (ICCE), 2019, pp. 4507–4514.

[5] T. Plazenet, T. Boileau, C. Caironi, B. Nahid-Mobarakeh, A comprehensive study on shaft voltages and bearing currents in rotating machines, *IEEE Transactions on Industry Applications* 54 (4) (2018) 3749–3759.

[6] G.C. Stone, I. Culbert, E.A. Boulter, H. Dhiran, *Electrical Insulation for Rotating Machines: Design, Evaluation, Aging, Testing, and Repair*, IEEE Press Series on Power Engineering, 2nd, Wiley–IEEE Press, Hoboken, NJ, 2014.

[7] S. Ul Haq, M.K.W. Stranges, B. Wood, A proposed method for establishing partial discharge acceptance limits on API 541 and 546 sacrificial test coils, *IEEE Transactions on Industry Applications* 53 (1) (2017) 718–722.

[8] V. Madonna, P. Giangrande, W. Zhao, G. Buticchi, H. Zhang, C. Gerada, M. Galea, Reliability vs. performances of electrical machines: Partial discharges issue, Proceedings of the 2019 IEEE Workshop on Electrical Machines Design, Control, and Diagnosis (WEMDCD), 1, 2019, pp. 77–82.

[9] T.J. Å. Hammarström, Partial discharge characteristics within motor insulation exposed to multi-level PWM waveforms, *IEEE Transactions on Dielectrics & Electrical Insulation* 25 (2) (2018) 559–567.

[10] P. Maussion, A. Picot, M. Chabert, D. Malec, Lifespan and aging modeling methods for insulation systems in electrical machines: A survey, Proceedings of the 2015 IEEE Workshop on Electrical Machines Design, Control, and Diagnosis (WEMDCD), 2015, pp. 279–288.

[11] V. Climente-Alarcon, J.A. Antonino-Daviu, E.G. Strangas, M. Riera-Guasp, Rotor-bar breakage mechanism and prognosis in an induction motor, *IEEE Transactions on Industrial Electronics* 62 (3) (2015) 1814–1825.

[12] H. Wang, F. Blaabjerg, Reliability of capacitors for DC-link applications in power electronic converters—An overview, *IEEE Transactions on Industry Applications* 50 (5) (2014) 3569–3578.

[13] M.A. Hannan, M.M. Hoque, A. Hussain, Y. Yusof, P.J. Ker, State-of-the-art and energy management system of lithium-ion batteries in electric vehicle applications: Issues and recommendations, *IEEE Access* 6 (2018) 19362–19378.

[14] K. Smith, Y. Shi, S. Santhanagopalan, Degradation mechanisms and lifetime prediction for lithium-ion batteries—A control perspective, Proceedings of the 2015 American Control Conference, 2015.

[15] O. Vitek, M. Janda, V. Hajek, P. Bauer, Detection of eccentricity and bearings fault using stray flux monitoring, Proceedings of the 8th IEEE Symposium on Diagnostics for Electrical Machines, Power Electronics, and Drives, 2011, pp. 456–461.

[16] K.N. Gyftakis, A.J. Marques Cardoso, Reliable detection of stator interturn faults of very low severity level in induction motors, *IEEE Transactions on Industrial Electronics* 68 (4) (2021) 3475–3484.

[17] A. Bellini, Quad demodulation: A time domain diagnostic method for induction machines, Proceedings of the 2007 IEEE Industry Applications Annual Meeting, 2007, pp. 2249–2253.

[18] Y. Jiang, C. Tang, X. Zhang, W. Jiao, G. Li, T. Huang, A novel rolling bearing defect detection method based on bispectrum analysis and cloud model-improved EEMD, *IEEE Access* 8 (2020) 24323–24333.

[19] B.-K. Yeo, Y. Lu, Expeditious diagnosis of linear array failure using support vector machine with low-degree polynomial kernel, *IET Microwaves, Antennas & Propagation* 6 (13) (2012) 1473–1480.

[20] M. Blodt, P. Granjon, B. Raison, G. Rostaing, Models for bearing damage detection in induction motors using stator current monitoring, *IEEE Transactions on Industrial Electronics* 55 (4) (2008) 1813–1822.

[21] M. Amar, I. Gondal, C. Wilson, Vibration spectrum imaging: A novel bearing fault classification approach, *IEEE Transactions on Industrial Electronics* 62 (1) (2012) 494–502.

[22] Z. Huo, Y. Zhang, P. Francq, L. Shu, J. Huang, Incipient fault diagnosis of roller bearing using optimized wavelet transform based multi-speed vibration signatures, *IEEE Access* 5 (2017) 19442–19456.

[23] IEEE. P1434/D14, Jul 2014—IEEE Approved Draft Guide for the Measurement of Partial Discharges in AC Electric Machinery. IEEE, Los Alamitos, CA

[24] R. Acheen, C. Abadie, T. Billard, T. Lebey, S. Duchesne, Study of partial discharge detection in motors fed by SiC MOSFET and Si IGBT inverters, Proceedings of the 2019 IEEE Electrical Insulation Conference (EIC), 2019, pp. 497–500.

[25] A. Bhure, E.G. Strangas, J. Agapiou, R.M. Lesperance, Partial discharge detection in medium voltage stators using an antenna, Proceedings of the IEEE 11th International Symposium on Diagnostics for Electrical Machines, Power Electronics, and Drives (SDEMPED), 2017, pp. 480–485.

[26] K. Younsi, P. Neti, M. Shah, J.Y. Zhou, J. Krahn, K. Weeber, C.D. Whitefield, On-line

capacitance and dissipation factor monitoring of AC stator insulation, *IEEE Transactions on Dielectrics & Electrical Insulation* 17 (5) (2010) 1441–1452.

[27] W.R. Jensen, E.G. Strangas, S.N. Foster, A method for online stator insulation prognosis for inverter-driven machines, *IEEE Transactions on Industry Applications* 54 (6) (2018) 5897–5906.

[28] P. Nussbaumer, M.A. Vogelsberger, T.M. Wolbank, Induction machine insulation health state monitoring based on online switching transient exploitation, *IEEE Transactions on Industrial Electronics* 62 (3) (2015) 1835–1845.

[29] H. Oh, B. Han, P. McCluskey, C. Han, B.D. Youn, Physics-of-failure, condition monitoring, and prognostics of insulated gate bipolar transistor modules: A review, *IEEE Transactions on Power Electronics* 30 (5) (2015) 2413–2426.

[30] U. Choi, F. Blaabjerg, S. Jorgensen, S. Munk-Nielsen, B. Rannestad, Reliability improvement of power converters by means of condition monitoring of IGBT modules, *IEEE Transactions on Power Electronics* 32 (10) (2017) 7990–7997.

[31] F. Gonzalez-Hernando, J. San-Sebastian, A. Garcia-Bediaga, M. Arias, F. Iannuzzo, F. Blaabjerg, Wear-out condition monitoring of IGBT and MOSFET power modules in inverter operation, *IEEE Transactions on Industry Applications* 55 (6) (2019) 6184–6192.

[32] I. Bandyopadhyay, P. Purkait, C. Koley, Performance of a classifier based on time-domain features for incipient fault detection in inverter drives, *IEEE Transactions on Industrial Informatics* 15 (1) (2019) 3–14.

[33] M. Drif, A.J.M. Cardoso, Stator fault diagnostics in squirrel cage three-phase induction motor drives using the instantaneous active and reactive power signature analyses, *IEEE Transactions on Industrial Informatics* 10 (2) (2014) 1348–1360.

[34] C. Yang, T. Kang, D. Hyun, S.B. Lee, J.A. Antonino-Daviu, J. Pons-Llinares, Reliable detection of induction motor rotor faults under the rotor axial air duct influence, *IEEE Transactions on Industry Applications* 50 (4) (2014) 2493–2502.

[35] F.E. Prahesti, D.A. Asfani, I.M. Yulistya Negara, B.Y. Dewantara, Three-phase induction motor short circuit stator detection using an external flux sensor, Proceedings of the 2020 International Seminar on Intelligent Technology and Its Applications (ISITIA), 2020, pp. 375–380.

[36] H.T. Eickhoff, R. Seebacher, A. Muetze, E.G. Strangas, Enhanced and fast detection of open-switch faults in inverters for electric drives, *IEEE Transactions on Industry Applications* 53 (6) (2017) 5415–5425.

[37] M. Eftekhari, M. Moallem, S. Sadri, M. Hsieh, Online detection of induction motor's stator winding short-circuit faults, *IEEE Systems Journal* 8 (4) (2014) 1272–1282.

[38] S. Choi, M.S. Haque, M.T.B. Tarek, V. Mulpuri, Y. Duan, S. Das, V. Garg, Fault diagnosis techniques for permanent magnet AC machine and drives—A review of current state of the art, *IEEE Transactions on Transportation Electrification* 4 (2) (2018) 444–463.

[39] B.M. Ebrahimi, J. Faiz, M.J. Roshtkhari, Static-, dynamic-, and mixed-eccentricity fault diagnoses in permanent-magnet synchronous motors, *IEEE Transactions on Industrial Electronics* 56 (11) (2009) 4727–4739.

[40] A. Aggarwal, E.G. Strangas, Review of detection methods of static eccentricity for interior permanent magnet synchronous machine, *Energies* 21 (12) (2019) 4105.

[41] H.T. Eickhoff, R. Seebacher, A. Muetze, E.G. Strangas, Enhanced and fast detection of open-switch faults in inverters for electric drives, *IEEE Transactions on Industry Applications* 53 (6) (2017) 5415–5425.

[42] R.Z. Haddad, E.G. Strangas, On the accuracy of fault detection and separation in permanent

magnet synchronous machines using MCSA/MVSA and LDA, *IEEE Transactions on Energy Conversion* 31 (3) (2016) 924–934.

[43] R.Z. Haddad, C.A. Lopez, S.N. Foster, E.G. Strangas, A voltage-based approach for fault detection and separation in permanent magnet synchronous machines, *IEEE Transactions on Industry Applications* 53 (6) (2017) 5305–5314.

[44] J.G. Cintron-Rivera, S.N. Foster, E.G. Strangas, Mitigation of turn-to-turn faults in fault tolerant permanent magnet synchronous motors, *IEEE Transactions on Energy Conversion* 30 (2) (2015) 465–475.

[45] C. Gan, J. Wu, S. Yang, Y. Hu, W. Cao, Wavelet packet decomposition-based fault diagnosis scheme for SRM drives with a single current sensor, *IEEE Transactions on Energy Conversion* 31 (1) (2016) 303–313.

[46] S. Gopalakrishnan, A.M. Omekanda, B. Lequesne, Classification and remediation of electrical faults in the switched reluctance drive, *IEEE Transactions on Industry Applications* 42 (2) (2006) 479–486.

[47] K. Reif, R. Unbehauen, The extended Kalman filter as an exponential observer for nonlinear systems, *IEEE Transactions on Signal Processing* 47 (8) (1999) 2324–2328.

[48] R. Dhaouadi, N. Mohan, L. Norum, Design and implementation of an extended Kalman filter for the state estimation of a permanent magnet synchronous motor, *IEEE Transactions on Power Electronics* 6 (3) (1991) 491–497.

[49] M. Mosallaei, K. Salahshoor, Comparison of centralized multi-sensor measurement and state fusion methods with an adaptive unscented Kalman filter for process fault diagnosis, Proceedings of the 4th International Conference on Information and Automation for Sustainability, 2008, pp. 514–524.

[50] N. Patila, D. Dasa, M. Pecht, A prognostic approach for non-punch through and field stop IGBTs, *Microelectronics Reliability* 52 (3) (2012) 482–488.

[51] M.S. Haque, S. Choi, J. Baek, Auxiliary particle filtering-based estimation of remaining useful life of IGBT, *IEEE Transactions on Industrial Electronics* 65 (3) (2018) 2693–2703.

[52] M.S. Arulampalam, S. Maskell, N. Gordon, T. Clapp, A tutorial on particle filters for online nonlinear/non-Gaussian Bayesian tracking, *IEEE Transactions on Signal Processing* 50 (2) (2002) 174–188.

[53] P. Nectoux, R. Gouriveau, K. Medjaher, E. Ramasso, B. Chebel-Morello, N. Zerhouni, C. Varnier, PRONOSTIA: An experimental platform for bearings accelerated degradation tests, Proceedings of the 2012 IEEE International Conference on Prognostics, 2012.

[54] H.-E. Kim, A.C. Tan, J. Mathew, B.-K. Choi, Bearing fault prognosis based on health state probability estimation, *Expert Systems with Applications* 39 (2012) 5200–5213.

[55] R.K. Singleton, E.G. Strangas, S. Aviyente, Discovering the hidden health states in bearing vibration signals for fault prognosis, Proceedings of the 40th Annual Conference of the IEEE Industrial Electronics Society (IECON), 2014, pp. 3438–3444.

[56] R.K. Singleton, E.G. Strangas, S. Aviyente, Extended Kalman filtering for remaining-useful-life estimation of bearings, *IEEE Transactions on Industrial Electronics* 62 (3) (2015) 1781–1790.

[57] R.K. Singleton, E.G. Strangas, S. Aviyente, Extended Kalman filtering for remaining-useful-life estimation of bearings, *IEEE Transactions on Industrial Electronics* 62 (3) (2015) 1781–1790.

[58] A. Soualhi, H. Razik, G. Clerc, D.D. Doan, Prognosis of bearing failures using hidden Markov models and the adaptive neuro-fuzzy inference system, *IEEE Transactions on Industrial Electronics* 61 (6) (2014) 2864–2874.

[59] Z. Rao, M. Huang, X. Zha, IGBT remaining useful life prediction based on particle filter with fusing precursor, *IEEE Access* 8 (2020) 154281–154289.

[60] W. Zhang, D. Xu, P.N. Enjeti, H. Li, J.T. Hawke, H.S. Krishnamoorthy, Survey on fault-tolerant techniques for power electronic converters, *IEEE Transactions on Power Electronics* 29 (12) (2014) 6319–6331.

[61] Y. Song, B. Wang, Evaluation methodology and control strategies for improving reliability of HEV power electronic system, *IEEE Transactions on Vehicular Technology* 63 (8) (2014) 3661–3676.

[62] E. Ugur, S. Dusmez, B. Akin, An investigation on diagnosis-based power switch lifetime extension strategies for three-phase inverters, *IEEE Transactions on Industry Applications* 55 (2) (2019) 2064–2075.

[63] I. Tsyokhla, A. Griffo, J. Wang, Online condition monitoring for diagnosis and prognosis of insulation degradation of inverter-fed machines, *IEEE Transactions on Industrial Electronics* 66 (10) (2019) 8126–8135.

[64] R. Xiong, Y. Zhang, J. Wang, H. He, S. Peng, M. Pecht, Lithium-ion battery health prognosis based on a real battery management system used in electric vehicles, *IEEE Transactions on Vehicular Technology* 68 (5) (2019) 4110–4121.

[65] B. Mirafzal, Survey of fault-tolerance techniques for three-phase voltage source inverters, *IEEE Transactions on Industrial Electronics* 61 (10) (2014) 5192–5202.

[66] H.T. Eickhoff, R. Seebacher, A. Muetze, E.G. Strangas, Post-fault operation strategy for single switch open-circuit faults in electric drives, *IEEE Transactions on Industry Applications* 54 (3) (2018) 2381–2391.

[67] A.E. Ginart, P.W. Kalgren, M.J. Roemer, D.W. Brown, M. Abbas, Transistor diagnostic strategies and extended operation under one-transistor trigger suppression in inverter power drives, *IEEE Transactions on Power Electronics* 25 (2) (2010) 499–506.

[68] A. Kersten, O. Theliander, E.A. Grunditz, T. Thiringer, M. Bongiorno, Battery loss and stress mitigation in a cascaded H-Bridge multilevel inverter for vehicle traction applications by filter capacitors, *IEEE Transactions on Transportation Electrification* 5 (3) (2019) 659–671.

[69] S. Lee, G. Stone, J. Antonino-Daviu, K. Gyftakis, E. Strangas, P. Maussion, C. Platero, Recent challenges in condition monitoring of industrial electric machines, IEEE Industrial Electronics Magazine (2020). Early access, June 8

[70] A. Chehade, Z. Shi, Sensor fusion via statistical hypothesis testing for prognosis and degradation analysis, *IEEE Transactions on Automation Science & Engineering* 16 (4) (2019) 1774–1787.

Chapter 5

Intelligent fault diagnosis for dynamic systems via extended state observer and soft computing

Paul P. Lin[#]

Fellow of the American Society of Mechanical Engineers (ASME); Professor Emeritus, Mechanical Engineering Department, Cleveland State University, United States; Visiting Scholar, Kaohsiung University of Science and Technology, Taiwan

Overview. There have been many studies on observer-based fault detection and isolation (FDI), such as using an unknown input observer and a generalized observer. Most of them require a nominal mathematical model of the system. Unlike sensor faults, actuator faults and process faults greatly affect the system dynamics. The main function of an observer, also known as estimator, is to extract information of the otherwise immeasurable variables for a vast number of applications that include feedback controls and system health monitoring or fault diagnosis. Over the past few decades, two classes of observer design have emerged. One relies on mathematical plant models to produce state estimates; the other uses available plant knowledge to estimate not only the state but also the part of the physical process that is not described in the plant model (i.e., disturbances). For the first class, however, it requires an accurate mathematical model of the plant that is often unavailable in practice. In contrast, the second class provides practical state and disturbance estimation when significant non-linearity and uncertainty are present in a dynamic system.

This chapter presents a new process fault diagnosis technique without exact knowledge of the plant model via extended state observer (ESO) and soft computing. The ESO's augmented or extended state is used to compute the system dynamics in real time and thereby provides a foundation for real-time process fault detection. Based on the input and output data, the ESO identifies the

---

[#]**Contributors:** Zhiqiang Gao, Cleveland State University, United States; Qing Zheng, formerly worked at Gannon University, United States; Jimmy Zhu, Facebook Inc., United States

Fault Diagnosis and Prognosis Techniques for Complex Engineering Systems. DOI: 10.1016/B978-0-12-822473-1.00009-4
181

unmodeled or incorrectly modeled dynamics combined with unknown external disturbances in real time and provides vital information for detecting faults with only partial information of the plant, which cannot be easily accomplished with any existing methods. Another advantage of the ESO is its simplicity in tuning only a single parameter. Without the knowledge of the exact plant model, fuzzy inference was developed to isolate faults. A strongly coupled three-tank nonlinear dynamic system was chosen as a case study. In a typical dynamic system, a process fault such as pipe blockage is likely incipient requires the degree of fault determination at all time. Neural networks were trained to identify faults and also instantly determine the degree of fault. The simulation results indicate that the proposed FDI technique effectively detected and isolated faults and also accurately determined the degree of fault. Soft computing (i.e., fuzzy logic and neural networks) makes fault diagnosis intelligent and fast because it provides intuitive logic to the system and real-time input-output mapping.

For a typical MIMO (multiple-input, multiple-output) nonlinear dynamic system, FDI usually aims at process faults with an assumption that actuator faults and sensor faults do not occur at the same time, which is not always the case. Simultaneous faults of different types turns out to be quite complex, which may explain why there have been very few studies on this topic. This study investigates the coupling relationship among process faults, actuator faults, and sensor faults, and presents how a combination of different types of faults could lead to no-fault detection or false FDI. Finally, a method to isolate actuator faults from process faults is presented.

5.1 Introduction

The term *fault diagnosis* generally refers to FDI. The fault diagnosis for nonlinear dynamic systems using model-free or model-based approaches has received much attention lately [1–3]. The model-free approach relies on a rich data collection to train neural networks in conjunction with the use of a fuzzy inference system (FIS). Such an approach might prove to be impractical, if not impossible, to collect rich experimental data. The model-based approach uses a linear or linearized model of the supervised system to generate a series of fault-indicating signals. In particular, the observer-based FDI methodologies have been developed along with the observer theory, and some of them have been successfully applied to industrial processes [4–6]. To deal with the nonlinearity and uncertainty of a dynamic system, nonlinear fault diagnosis has recently become an active research topic. There have been many observer-based residual generation methods for fault diagnosis in nonlinear dynamic system. Frank [7] first proposed a nonlinear identity observer approach for fault diagnosis, followed by a survey on diagnostic observers [8] and a survey on robust residual generation and evaluation methods used in observer-based fault detection [9]. Later, Isermann [10] presented the status and applications of model-based fault detection and diagnosis. Observer-based fault diagnosis was applied to robot manipulators using a mathematical technique called *algebra of functions* to design the nonlinear diagnostic observer [11]. Adaptive observers [12] and nonlinear robust-based observer schemes [13, 14] both developed an algorithm to adjust the gain matrix of the observer to track the fault parameters of the system online and have been applied to practical processes successfully. Additionally, a new concept of practical optimality using disturbance estimation for health monitoring has been proposed [15]. However, the common drawback of these observer-based fault diagnosis methods is the dependency on detailed knowledge of the process represented by its mathematical model.

This study begins with discussion on diagnosing process faults that affect the plant of a nonlinear dynamic system. The sensors and actuators are assumed healthy when process faults occur. More specifically, the presented fault diagnosis technique aims at a nonlinear dynamic system with an uncertain system model and unmodeled or incorrectly modeled dynamics combined with unknown external disturbances. The complexity of fault isolation due to simultaneous faults of different types will be discussed later.

Based on the parameterized ESO, a new FDI technique is proposed in this chapter, which is organized as follows. Section 5.2 describes the design of the improved ESO and its estimation error convergence. Section 5.3 presents a case study on a MIMO nonlinear dynamic system. Section 5.4 describes fault detection by means of the ESO, whereas Section 5.5 describes fault isolation, fault identification, and degree-of-fault determination. Section 5.6 discusses simultaneous faults of different types, followed by isolation of process faults and actuator faults in Section 5.7. Finally, our conclusion and future work are presented in Section 5.8.

5.2 Extended state observer

To extend FDI to the processes beyond the scope of existing methods, consider a nonlinear dynamic system that can be described by

$$y^{(n)} = f\left(t, y, \dot{y}, \cdots, y^{n-1}, d\right) = bu \qquad (5.1)$$

where $y^{(n)}$ denotes the n-th time derivative of y, f, short for $f(t, y, \dot{y}, \cdots, y^{(n-1)}, d)$, is a lumped nonlinear time-varying function of the plant dynamics and the unknown external disturbance d, u is the system's input and b is a constant. In all physical systems, f and b are both bounded. From the fault diagnosis point of view, the f can be thought of as lumped unknown unmodeled or incorrectly modeled dynamics combined with the unknown external disturbances. Instead of separating unmodeled dynamics from the disturbance, the term f in its totality is to be estimated as an extended state of the system, together with the states of the system. Normally, an observer only provides the state estimation; however, with what is known as ESO [16–19], the term f is treated as another state and estimated in real time.

Such additional information proves to be crucial for FDI purposes, as will be shown in this chapter. The ESO technique first developed by Han [16, 17], however, is rather complex, and its implementation requires the adjustments or tuning of several parameters, which can be difficult and time consuming. Later, Gao [18] improved the ESO technique and made it more practical by using a particular parameterization method that reduces the number of tuning parameters to 1. Such parameterized ESO has been successfully applied in many applications, particularly in the context of the active disturbance rejection control [19].

In this section, the design of the improved ESO is described, followed by the proof of the observer's estimation error convergence.

5.2.1 ESO design

The main idea of ESO is to use an augmented state space model of Eq. (5.1) that includes f as an additional state. Thus, Eq. (5.1) can be represented in state space form as

$$\begin{cases} \dot{x}_1 = x_2 + bu = f + bu \\ \dot{x}_2 = \dot{f} = \eta(x, u, d, \dot{d}) \end{cases} \tag{5.2}$$

where both f and η are assumed unknown.

Alternatively, in the case of single output (i.e., $y = x_1$), Eq. (5.2) can be written in matrix form as

$$\begin{cases} \dot{x} = Ax + Bu + E\eta \\ y = Cx \end{cases} \tag{5.3}$$

where

$$A = \begin{bmatrix} 0 & 1 \\ 0 & 0 \end{bmatrix}; \quad B = \begin{bmatrix} b \\ 0 \end{bmatrix}; \quad C = [1 \ 0]; \quad E = \begin{bmatrix} 0 \\ 1 \end{bmatrix}$$

The ESO can be expressed in matrix form as

$$\begin{cases} \dot{z} = Az + Bu + L(y - \hat{y}) \\ \hat{y} = Cz \end{cases} \tag{5.4}$$

or

$$\begin{cases} \dot{z}_1 = z_2 + l_1(x_1 - z_1) + bu \\ \dot{z}_2 = l_2(x_1 - z_1) \end{cases} \tag{5.5}$$

where $L = [l_1 \ l_2]^T$ is the observer gain vector that can be obtained using any known method, such as the pole placement technique. When properly selected, the ESO provides an estimate of the state in Eq. (5.3) (i.e., z_i estimates x_i, where $i = 1, 2$), where $\hat{y}$ is the estimate of system output y. More specifically, z_1 tracks the system output, whereas z_2 tracks f that includes system internal dynamics and external disturbance. The choice of the observer gain vector L, originally consisting of a set of nonlinear gains [16, 17], was simplified with linear gains so that it can be parameterized by solving the characteristic equation of the observer [18]. For instance, if gains are chosen as $L = [2\omega_o \ \omega_o^2]^T$, then the characteristic polynomial of Eq. (5.4) becomes

$$\lambda_0(s) = (s + \omega_o)^2 \tag{5.6}$$

where ω_o is the observer bandwidth, which needs to be tuned in practice to ensure that the ESO operates effectively, and this is a complex argument (Laplace's variable). In comparison with the original ESO, this is regarded as the improved ESO since the observer bandwidth is the only parameter that needs to be tuned. The analysis of ESO was briefly given in the work of Gao [18]; a more elaborate account is given in the work of Zheng et al. [19]. For practitioners, however, perhaps it is just as interesting to see the various applications of ESO and their success in providing a practical solution in dealing with uncertainties [18, 20]. The estimation error of the ESO is described in the next section.

5.2.2 Estimation error convergence

In this section, we will mathematically prove that, with plant dynamics largely unknown, the ESO can accurately estimate the unknown dynamics and disturbances with upper-bounded estimation error. Let

$$\tilde{\xi}_i(t) = x_i(t) - z_i(t), \ i = 1, \ 2 \tag{5.7}$$

From Eq. (5.2) and Eq. (5.4), the observer estimation error for states x_1 and x_2 can be described as

$$\begin{aligned}\dot{\tilde{\xi}}_1 &= \tilde{\xi}_2 - l_1\tilde{\xi}_1 \\ \dot{\tilde{\xi}}_2 &= \eta - l_2\tilde{\xi}_1\end{aligned} \tag{5.8}$$

Now let us scale down the observer estimation error $\tilde{\xi}_i(t)$ by ω_o^{i-1}—that is, let

$$\varepsilon_i(t) = \frac{\tilde{\xi}_i(t)}{\omega_o^{i-1}}, \ i = 1, \ 2$$

Then Eq. (5.8) can be written as

$$\dot{\varepsilon} = \omega_o A_\varepsilon \varepsilon + B_\varepsilon \frac{\eta(x, u, d, \dot{d})}{\omega_o} \tag{5.9}$$

where

$$A_\varepsilon = \begin{bmatrix} -2 & 1 \\ -1 & 0 \end{bmatrix}, \ B_\varepsilon = \begin{bmatrix} 0 \\ 1 \end{bmatrix}$$

Here, A is Hurwitz for $L = [l_1 \ l_2]^T = [2\omega_o \ \omega_o^2]^T$.

Theorem 1. Assuming $\eta(x, u, d, \dot{d})$ is bounded, there exists a constant $\sigma_i > 0$ and a finite time $T_1 > 0$ such that $|\tilde{\xi}_i(t)| \leq \sigma_i, i = 1, 2, \forall t \geq T_1 > 0$, and $\omega_o > 0$.

Note that

$$\sigma_i = O\left(\frac{1}{\omega_o^k}\right) \tag{5.10}$$

where O is a function representing the order of the reciprocal of bandwidth to the order of a positive integer k. The boundedness of $\eta(x, u, d, \dot{d})$ (i.e., $\dot{f}$) means that the rate of change of the combined effect of internal dynamics and external disturbances is finite, which leads to an assumption that the combined effect and the control input are continuous. Here, η is essentially the derivative of acceleration. In a typical motion system, η being bounded means that the force applied to the body does not change infinitely within a very short period of time. In other words, the jerk (i.e., time derivative of acceleration) is finite. This is a reasonable assumption for a typical motion.

Proof. Solving Eq. (5.9) gives

$$\varepsilon(t) = e^{\omega_o A_s t}\varepsilon(0) + \int_0^t e^{\omega_o A_s(t-\tau)}B_\varepsilon \frac{\eta(x, u, d, \dot{d})}{\omega_o}d\tau \tag{5.11}$$

Let

$$p(t) = \int_0^t e^{\omega_o A_s(t-\tau)}B_\varepsilon \frac{\eta(x(\tau), u, d, \dot{d})}{\omega_o}d\tau \tag{5.12}$$

Since $\eta(x(\tau), u, d, \dot{d})$ is bounded—that is, $\eta(x(\tau), u, d, \dot{d}) \leq \delta$, where δ is a positive constant, for $i = 1, 2$—then

$$|p_i(t)| \leq \frac{\delta}{\omega_o^2}\left[|(A_\varepsilon^{-1}B_\varepsilon)_i| + |(A_\varepsilon^{-1}e^{\omega_o A_s t}B_\varepsilon)_i|\right] \tag{5.13}$$

With

$$A_\varepsilon^{-1} = \begin{bmatrix} 0 & -1 \\ 1 & -2 \end{bmatrix}, B_\varepsilon = \begin{bmatrix} 0 \\ 1 \end{bmatrix}$$

the following can be written:

$$|(A_\varepsilon^{-1}B_\varepsilon)| = \begin{cases} 1|_{i=1} \\ 2|_{i=2} \end{cases} \tag{5.14}$$

Since A_ε is Hurwitz, there exists a finite time $T_1 > 0$ such that

$$|[e^{\omega_o A_\varepsilon t}]_{ij}| \leq \frac{1}{\omega_o^2} \tag{5.15}$$

for all $t \geq T_1, i, j = 1, 2$. Hence,

$$|[e^{\omega_o A_\varepsilon t}B_\varepsilon]_i| \leq \frac{1}{\omega_o^2} \tag{5.16}$$

for all $t \geq T_1$, $i = 1, 2$. Note that T_1 depends on $\omega_o A_\varepsilon$. Combining

$$A_\varepsilon^{-1} = \begin{bmatrix} 0 & -1 \\ 1 & -2 \end{bmatrix} = \begin{bmatrix} S_{11} S_{12} \\ S_{21} S_{22} \end{bmatrix}$$

and Eq. (5.16), which means

$$\begin{cases} \left| \left[e^{\omega_o A_\varepsilon^t} B_\varepsilon \right]_1 \right| = d_1 \leq \frac{1}{\omega_o^2} \\ \left| \left[e^{\omega_o A_\varepsilon^t} B_\varepsilon \right]_2 \right| = d_2 \leq \frac{1}{\omega_o^2} \end{cases}$$

gives the expression

$$\left| \left(A_\varepsilon^{-1} e^{\omega_o A_\varepsilon^t} B_\varepsilon \right)_i \right| = |s_{i1} d_1 + s_{i2} d_2| \leq |s_{i1} d_1| + |s_{i2} d_2| \leq \begin{cases} \frac{1}{\omega_o^2} |_{i=1} \\ \frac{3}{\omega_o^2} |_{i=2} \end{cases} \quad (5.17)$$

for all $t \geq T_1$. Eq. (5.13) can be expressed in terms of Eq. (5.14) and Eq. (5.17) as follows:

$$|p_i(t)| \leq \frac{2\delta}{\omega_o^2} + \frac{3\delta}{\omega_o^4} \quad (5.18)$$

for all $t \geq T_1$, $i = 1, 2$. Let $\varepsilon_{sum}(0) = |\varepsilon_1(0)| + |\varepsilon_2(0)|$; it follows that

$$\left| \left[e^{\omega_o A_\varepsilon^t} \varepsilon(0) \right]_i \right| \leq \frac{\varepsilon_{sum}(0)}{\omega_o^2} \quad (5.19)$$

for all $t \geq T_1$, $i = 1, 2$. Eq. (5.11) yields the following constraint:

$$|\varepsilon_i(t)| \leq \left| \left[e^{\omega_o A_\varepsilon^t} \varepsilon(0) \right]_i \right| + |p_i(t)| \quad (5.20)$$

Substituting

$$\varepsilon_i(t) = \frac{\tilde{\xi}_i(t)}{\omega_o^{i-1}}$$

and Eq. (5.18) into Eq. (5.20) leads to a conclusion that the absolute estimation error is, indeed, upper bounded.

$$|\tilde{\xi}_i(t)| \leq \frac{1}{\omega_o^{3-i}} \left[|\tilde{\xi}_1(0)| + \left| \frac{1}{\omega_o} \tilde{\xi}_2(0) \right| \right] + \frac{2\delta}{\omega_o^{3-i}} + \frac{3\delta}{\omega_o^{5-i}} = \sigma_i$$

for all

$$t \geq T_1, i = 1, 2 \quad (5.21)$$

Theorem 1 has been mathematically proved that, in the absence of the plant model, the estimation error of the ESO as described in Eq. (5.4) is bounded and its upper bound monotonously decreases with the observer bandwidth. As long as the bandwidth is sufficiently large, the ESO can be used to estimate the state and the extended state f, which includes system internal dynamics and external disturbance. The ESO's ability to estimate and track the system's output state, y, and the extended state, f, provides a foundation for the proposed FDI schemes.

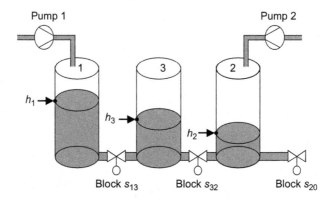

FIGURE 5.1 Schematic diagram of the three-tank system.

Since the extended state f, which includes system internal dynamics and external disturbances, is estimated by the ESO in real time and canceled in the control law in real time, the ESO achieves high disturbance rejection performance and strong robustness performance.

5.3 Case study: three-tank dynamic system

To illustrate how the presented ESO can be used to track a nonlinear dynamic system, a three-tank nonlinear dynamic system [3] as shown in Fig. 5.1 was chosen for a case study. The system consists of three tanks (T_1, T_2, and T_3) that are connected by three pipes. The system has two controlled inputs (pump flow rates), three measurable outputs (h_1, h_2, and h_3; water levels), and three possible faults (pipe blockages). It is, indeed, a strongly coupled MIMO system.

Using Torricelli's law, the following three dynamic system equations can be obtained:

$$\begin{cases} A_T \frac{dh_1}{dt} = -s_{13}a_1 sign(h_1 - h_3)\sqrt{2g|h_1 - h_3|} + Q_1 \\ A_T \frac{dh_2}{dt} = s_{32}a_3 sign(h_3 - h_2)\sqrt{2g|h_3 - h_2|} - s_{20}a_2\sqrt{2gh_2} + Q_2 \\ A_T \frac{dh_3}{dt} = s_{13}a_1 sign(h_1 - h_3)\sqrt{2g|h_1 - h_3|} - s_{32}a_3 sign(h_3 - h_2)\sqrt{2g|h_3 - h_2|} \end{cases},$$

$$(5.22)$$

where

A_T is the circular cross-sectional area of each tank (assumed the same for all);
a_1, a_2, and a_3 denote the circular cross-section area of each pipe;
s_{13}, s_{32}, and s_{20} denote pipe blockage;
Q_1 and Q_2 denote the pump's flow rate; and
h_1, h_2, and h_3 denote the water level of tanks T_1, T_2, and T_3, respectively.

The blockage is in terms of degree of fault between 0 and 1, where 0 and 1 correspond to complete blockage and no blockage, respectively. Eq. (5.22) can be rewritten as

$$\begin{cases} \dot{h}_1 = f_1 + \frac{1}{A_T}Q_1 \\ \dot{h}_2 = f_2 + \frac{1}{A_T}Q_2, \\ \dot{h}_3 = f_3 \end{cases} \tag{5.23}$$

where

$$f_1 = -\frac{1}{A_T}\left[s_{13}a_1 sign(h_1 - h_3)\sqrt{2g|h_1 - h_3|}\right]$$

$$f_2 = \frac{1}{A_T}\left[s_{32}a_3 sign(h_3 - h_2)\sqrt{2g|h_3 - h_2|} - s_{20}a_2\sqrt{2gh_2}\right]$$

$$f_3 = \frac{1}{A_T}\left[s_{13}a_1 sign(h_1 - h_3)\sqrt{2g|h_1 - h_3|} - s_{32}a_3 sign(h_3 - h_2)\sqrt{2g|h_3 - h_2|}\right].$$

Let $y(t)$ and $u(t)$ be the system's output and input vector, respectively,

$$y(t) = [h_1\,h_2\,h_3]^T;\; u(t) = [Q_1\,Q_2\,0]^T, \tag{5.24}$$

where h_1, h_2, and h_3 denote the water level of tanks T_1, T_2, and T_3, respectively, and Q_1 and Q_2 denote the flow rate of pumps 1 and 2, respectively. Essentially, the water levels are the system output variables and the flow rates are the system input variables. Combining Eq. (5.23) and Eq. (5.24) gives

$$\dot{y}(t) = f + b_o u(t) \tag{5.25}$$

where

$$b_o = \frac{1}{A_T}\begin{bmatrix} 1 & 0 & 0 \\ 0 & 1 & 0 \\ 0 & 0 & 0 \end{bmatrix} \quad f = \begin{bmatrix} f_1 \\ f_2 \\ f_3 \end{bmatrix}.$$

The f_1, f_2, and f_3 are called the *generalized system dynamics* of tank T_1, T_2, and T_3, respectively, and $u(t)$ is the system's inputs. Note that the constant b_o can be determined by the system, which in this case is simply the reciprocal of the tank's area.

Eq. (5.25) can be represented in state space form as

$$\begin{cases} \dot{x}_1 = x_2 + b_o u \\ \dot{x}_2 = v \\ y = x_1 \end{cases} \tag{5.26}$$

where $u(t) = [Q_1\,Q_2\,0]^T$ is the system input, $y = x_1 = [h_1\,h_2\,h_3]^T$ is the system output, $x_2 = f = [f_1\,f_2\,f_3]^T$ is an augmented state, and v is the time derivative of f. Rewriting Eq. (5.26) in matrix form gives

$$\begin{cases} \dot{x} = Ax + Bu + Dv \\ y = Cx \end{cases}, \tag{5.27}$$

where

$$x = \begin{bmatrix} x_1 \\ x_2 \end{bmatrix}_{6 \times 1}, \quad A = \begin{bmatrix} 0 & I \\ 0 & 0 \end{bmatrix}_{6 \times 6}, \quad B = \begin{bmatrix} b_o \\ 0 \end{bmatrix}_{6 \times 3}, \quad D = \begin{bmatrix} 0 \\ I \end{bmatrix}_{6 \times 3}$$
$$C = [1 \ 1 \ 1 \ 0 \ 0 \ 0]$$

and I is a 3×3 identity matrix. Note that the expression for C in Eq. (5.27) is for three outputs, whereas that for C in Eq. (5.3) is for a single output.

Employing the ESO design (Eqs. 5.2–5.6), denoting y as the measured or actual output, $\hat{y} = \begin{bmatrix} \hat{h}_1 & \hat{h}_2 & \hat{h}_3 \end{bmatrix}^T$ as the estimated output, and incorporating the difference between the two outputs, the ESO of Eq. (5.26) can be rewritten as

$$\begin{cases} \dot{z}_1 = z_2 + l_1(x_1 - z_1) + b_0 u \\ \dot{z}_2 = l_2(x_1 - z_1) \end{cases}. \tag{5.28}$$

The state space observer can be constructed as

$$\begin{cases} \dot{z} = Az + Bu + L(x_1 - z_1) \\ \hat{y} = Cz \end{cases}, \tag{5.29}$$

where

$$z = [z_1 \ z_2]^T \ \left(\text{i.e., } z_1 = [z_{11} \ z_{12} \ z_{13}]^T ; \ z_2 = [z_{21} \ z_{22} \ z_{23}]^T \right).$$

Eq. (5.22) shows that three-tank system consists of three simultaneous first-order differential equations. Thus, the observer gain matrix, L, can be expressed as

$$L = \begin{bmatrix} A\omega_o & 0 & 0 \\ 0 & 2\omega_o & 0 \\ 0 & 0 & 2\omega_o \\ \omega_o^2 & 0 & 0 \\ 0 & \omega_o^2 & 0 \\ 0 & 0 & \omega_o^2 \end{bmatrix}. \tag{5.30}$$

With a chosen bandwidth ω_o, the z vector can be used to estimate the system outputs and the system dynamics in real time. As proved in Section 5.2, the ESO's estimation error is upper bounded and monotonously decreases with the bandwidth. With a sufficiently large bandwidth and as time proceeds, z_1 quickly approaches y (i.e., h_1, h_2, and h_3) and z_2 approaches f (i.e., f_1, f_2, and f_3). In other words, z_1 tracks the system's outputs, and z_2 tracks the unmodeled system dynamics combined with external disturbance. More specifically, as stated in Eq. (5.29), $z_1 = [z_{11} \ z_{12} \ z_{13}]^T$ estimates the state variables x_1 (i.e., the water levels h_1, h_2, and h_3), and $z_2 = [z_{21} \ z_{22} \ z_{23}]^T$ estimates the extended state f (i.e., f_1, f_2, and f_3).

$$\begin{cases} z_{11} = \hat{h}_1 \to h_1; \quad z_{12} = \hat{h}_2 \to h_2; \quad z_{13} = \hat{h}_3 \to h_3 \\ z_{21} \to f_1; \quad z_{22} \to f_2; \quad z_{23} \to f_3 \end{cases}. \tag{5.31}$$

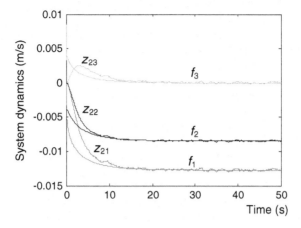

FIGURE 5.2   System dynamics tracking with $\omega_o = 1$ and 5% noise.

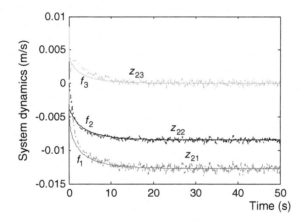

FIGURE 5.3   System dynamics tracking with $\omega_o = 5$ and 5% noise.

The value of the bandwidth ω_o affects the system's tracking speed and the state estimation's sensitivity to measurement noise. Figs. 5.2 and 5.3 show the simulation results on the sensitivity of the ω_o value to the measurement noise (with sampling time, $\Delta t = 0.01$ s).

The simulation results demonstrate the effectiveness of the ESO in tracking the outputs and the dynamics of the system. The smaller the ω_o, the slower the ESO tracks the system. As the ω_o increases, the ESO tracks the system more quickly, but it also becomes more sensitive to the measurement noise. Choosing the appropriate ω_o is a trade-off between the tracking speed and sensitivity to noise.

5.4 Fault detection by means of ESO

This section presents how faults can be detected by means of the ESOs based on real-time estimation of the system dynamics.

5.4.1 Fault detection scheme

As mentioned earlier, the faults to be detected are neither the sensor faults nor the actuator faults. Rather, they are the process faults possibly caused by structural deterioration. The process faults, in this case, are the pipe blockage faults s_{13}, s_{32}, and s_{20} as shown in Fig. 5.1. Traditionally, faults are considered detected when the outputs exceed the expected values by a preset tolerance. This approach, however, has some drawbacks in open-loop and closed-loop controls. When using the ESO for closed-loop control, observing the system's output does not provide useful information about the health of the system because the controller tries to augment the inputs in an effort to stabilize the system. Thus, the health does not surface until the system finally collapses. Using the ESO for open-loop control also encounters a problem before the system reaches its steady states. In other words, an abrupt change on the system output does not necessarily mean the system is becoming faulty. Thus, solely relying on monitoring the system output could trigger a false alarm or miss detection of possible faults.

It is worthwhile to note that the ESO's unique feature is its ability to estimate the general system dynamics (i.e., the unmodeled system dynamics and unknown external disturbance) in real time, which provides crucial information for the presented fault detection technique. Our study found that the system outputs and the general system dynamics both exhibit abrupt changes as soon as a fault occurs. However, the rate of change on the general system dynamics is more profound. Furthermore, the system outputs potentially contain the process faults (i.e., the pipe blockage faults) and the actuator faults (i.e., the actuating faults in the pumps), whereas the general system dynamics solely contain the process faults. Considering that the goal of this study was to diagnose the process faults, our proposed fault detection scheme is based on the general system dynamics, f. More specifically, a fault is considered detected when the rate of change of general system dynamics, $\Delta f / f$, exceeds the predetermined threshold value.

5.4.2 Fault detection without exact knowledge of the plant model

As mentioned earlier, the ESO estimates the states of z_{21}, z_{22}, and z_{23} that track the system dynamics f_1, f_2, and f_3. The only information needed for fault detection is to estimate the value of b_o. Our study found that the value of b_o is, indeed, not critical to fault detection. Fig. 5.4 shows the simulation result of successfully detecting two sequential faults using the exact b_o values of 127. Fig. 5.5 further indicates that the same faults can be detected even with a b_o value of 635, which is five times as much as the exact one. The simulation assumes

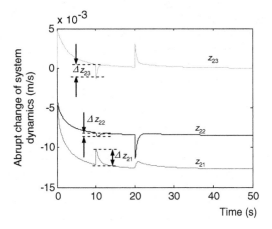

FIGURE 5.4 Detection of multiple faults ($s_{13} = 0.8$ at $t = 10$ s and $s_{32} = 0.6$ at $t = 20$ s) with $b_o = 127$ (the exact value).

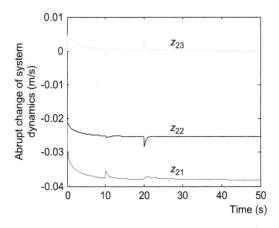

FIGURE 5.5 Detection of multiple faults ($s_{13} = 0.8$ at $t = 10$ s and $s_{32} = 0.6$ at $t = 20$ s) with $b_o = 635$ (rough estimated value).

that the first blockage fault $s_{13} = 0.8$ (i.e., 80% blocked) in the pipe connecting tanks 1 and 3 occurs at $t = 10$ s, followed by the second blockage fault $s_{32} = 0.6$ (i.e., 60% blocked) in the pipe connecting tanks 3 and 2 occurring at $t = 20$ s. The first fault affects the dynamics of tanks 1 and 3 (f_1 and f_3), which reflects the abrupt changes in the estimated states z_{21} and z_{23}. The second fault affects the dynamics of tanks 3 and 2 (f_3 and f_2), which reflects abrupt changes in the estimated states z_{23} and z_{22}.

Note that ESO's estimated z_{21}, z_{22}, and z_{23} closely track the system dynamics f_1, f_2, and f_3, respectively. The b_o value is associated with the physical system, which is the cross-sectional area of the pipes connecting tanks. Figs. 5.4 and

5.5 clearly demonstrate that the b_o value is not critical to fault detection, which suggests that knowledge of the exact system model is not required.

The presented ESO-based fault detection technique suggests that the accuracy of b_o is not critical to fault detection. It should be noted that although faults can be detected without exact knowledge of the plant model, some knowledge about the model, such as the order of the system, is needed.

The changes of these three extended states are worth observing. For instance, as shown in Fig. 5.4, when the first fault just occurred, Δz_{23} was negative, Δz_{22} was close to 0, and Δz_{21} was positive. However, when the second fault was added 10 s later, Δz_{23} became positive and Δz_{22} became negative, but Δz_{21} remained positive but smaller. The changing signs of the states and the levels of the state values (i.e., low, medium, and high) provide useful information for fault isolation.

5.5 FAULT isolation and fault identification

The fault isolation to be presented here is based on the assumption that the exact system model is unknown. However, to verify the effectiveness of the presented technique, the referenced system outputs need to be generated first.

5.5.1 Generation of reference values

The outputs, in the case of the three-tank system, can be obtained by using such as piezo-resistive pressure sensors with resolution of 0.1 mm to measure the water levels. With sufficient input-output correspondence, a back-propagation neural network can be trained. The trained network can then be used to predict the outputs with reasonably good accuracy.

Alternatively, the system outputs can be estimated in real time using the ESO based on the assumption that the exact plant model is known. With this alternative approach, the first step for identifying faults is to associate all faults with the system dynamics. First of all, Eq. (5.22) containing the pipe dynamics (the dynamics between two outputs) can be extracted as follows:

$$\begin{cases} P_{13} = a_1 sign(h_1 - h_3)\sqrt{2g|h_1 - h_3|} \approx a_1 sign(z_{11} - z_{13})\sqrt{2g|z_{11} - z_{13}|} \\ P_{32} = a_3 sign(h_3 - h_2)\sqrt{2g|h_3 - h_2|} \approx a_3 sign(z_{13} - z_{12})\sqrt{2g|z_{13} - z_{12}|}, \\ P_{20} = a_2\sqrt{2gh_2} \approx a_2\sqrt{2gz_{12}} \end{cases}$$
$$(5.32)$$

where z_{11}, z_{12}, and z_{13} are the ESO's system outputs, the water level in each tank as shown in Eq. (5.29). Substituting Eq. (5.32) into Eq. (5.23) gives the expressions for the general system dynamics f as follows:

$$\begin{cases} f_1 = -\frac{1}{A_T}s_{13}P_{13} \\ f_2 = \frac{1}{A_T}(s_{32}P_{32} - s_{20}P_{20}), \\ f_3 = \frac{1}{A_T}(s_{13}P_{13} - s_{32}P_{32}) \end{cases}$$
$$(5.33)$$

where A_T is the circular cross-sectional area of each tank (assumed the same for all). Note that b_o is reciprocal of the A_T.

Furthermore, if the exact plant model were known, the degree of each fault for the three-tank system could be easily determined by

$$
\begin{cases}
\hat{s}_{13} = -\dfrac{z_{21}A_T}{P_{13}} \\[2mm]
\hat{s}_{20} = -\dfrac{(z_{21} + z_{22} + z_{23})A_T}{P_{20}} \\[2mm]
\hat{s}_{32} = -\dfrac{(z_{21} + z_{23})A_T}{P_{32}}
\end{cases}
\tag{5.34}
$$

In the case of an uncertain plant model, not only does fault isolation becomes more difficult but also degree-of-fault determination becomes a major task. These will be addressed in the following two sections.

5.5.2 Fault isolation by means of fuzzy inference and ESO

In addition to monitoring the system outputs, the system dynamics, f, used for fault detection can be used for fault isolation. Referring to Fig. 5.4 when the first fault occurs at $t = 10$ s, if Δz_{21} (the ESO's estimated Δf_1) is positive, Δz_{22} (the ESO's estimated Δf_2) is negative, and Δz_{23} (the ESO's estimated Δf_3) is negative, then a blockage fault between tanks 1 and 3 (i.e., s_{13}) likely has occurred. When the second fault occurs at $t = 20$ s, if Δz_{21} is positive, Δz_{22} is negative, and Δz_{23} is positive, then a blockage fault between tanks 3 and 2 (i.e., s_{32}) likely has occurred. The observations suggest some intuitive logic, better known as fuzzy logic, can be employed to classify the faults.

An FIS consists of input membership functions, output membership functions, and the if-then fuzzy logic rules. Among them, constructing the proper input membership functions is critical and can be most difficult if there is no prior knowledge about how input data are distributed. The best way to determine data distribution is through the use of histograms. The FIS's input variables are Δz_{21}, Δz_{22}, and Δz_{23}, which are normalized to the range of $[-1, 1]$. The output variables are the degree of fault for s_{13}, s_{32}, and s_{20}, which are normalized to the range of $[0, 1]$, where "0" represents no fault and "1" represents complete fault.

The input membership functions for Δz_{21}, Δz_{22}, and Δz_{23} are the same, which are LNG (large negative), SNG (small negative), and POS (positive). The output membership functions for faults s_{13}, s_{32}, and s_{20} are also the same, which are normal and faulty. The crisp input variables are first fuzzified and then processed by the fuzzy logic rules. Afterward, they are defuzzified into the range between 0 and 1, which indicates the fault occurrence confidence between 0% and 100%. The six if-then fuzzy rules for a single fault are as follows:

Rule 1: If (Δz_{21} is POS) and (Δz_{22} is SNG) and (Δz_{23} is LNG), then (s_{13} is faulty) and (s_{32} is normal) and (s_{20} is normal).

TABLE 5.1 Result of fault isolation and identification

Pump 1 flow rate, Q_1 (L/min)	Pump 2 flow rate, Q_2 (L/min)	Assumed degree of fault	Fault occurrence confidence	NN predicted degree of fault	Degree-of-fault error
6.5	8.25	$s_{13} = 0.17$	s_{13} with 96%	0.1700	0%
7	9	$s_{13} = 0.38$	s_{13} with 96%	0.3802	0.05%
10	6.75	$s_{13} = 0.63$	s_{13} with 96%	0.6306	0.10%
6.5	8.25	$s_{32} = 0.12$	s_{32} with 96%	0.1200	0%
7	9	$s_{32} = 0.44$	s_{32} with 96%	0.4403	0.07%
10	6.75	$s_{32} = 0.57$	s_{32} with 96%	0.5704	0.07%
6.5	8.25	$s_{20} = 0.23$	s_{20} with 96%	0.2301	0.04%
7	9	$s_{20} = 0.45$	s_{20} with 96%	0.4503	0.06%
10	6.75	$s_{20} = 0.70$	s_{20} with 96%	0.7009	0.13%

NN: Neural network.

Rule 2: If (Δz_{21} is POS) and (Δz_{22} is LNG) and (Δz_{23} is LNG), then (s_{13} is faulty) and (s_{32} is normal) and (s_{20} is normal).

Rule 3: If (Δz_{21} is POS) and (Δz_{22} is LNG) and (Δz_{23} is SNG), then (s_{13} is faulty) and (s_{32} is normal) and (s_{20} is normal).

Rule 4: If (Δz_{21} is POS) and (Δz_{22} is LNG) and (Δz_{23} is POS), then (s_{32} is faulty) and (s_{13} is normal) and (s_{20} is normal).

Rule 5: If (Δz_{21} is POS) and (Δz_{22} is SNG) and (Δz_{23} is POS), then (s_{32} is faulty) and (s_{13} is normal) and (s_{20} is normal).

Rule 6: If (Δz_{21} is POS) and (Δz_{22} is POS) and (Δz_{23} is POS), then (s_{20} is faulty) and (s_{13} is normal) and (s_{32} is normal).

The FIS essentially gives the confidence in a fault occurrence. A component is considered faulty when the confidence exceeds or is equal to 80%.

5.5.3 Fault identification via neural networks

With the given three-tank system, incipient faults are likely to occur, which will require monitoring and determining the degree of fault at all time. However, the degree of fault, in theory, cannot be determined unless the exact plant model is known. The only alternative is to use experimental data. In the absence of experimental data, simulation data using Eq. (5.2) were generated.

Table 5.1 shows examples of single fault in which the FIS was able to isolate all faults with 96% confidence, which was the maximum output value by design. The error of each predicted degree of fault was extremely small. In this simulation, the system input variables are the pump rates: $Q_1 = 6$ liter/min and $Q_2 = 4$ liter/min. To demonstrate the ESO's effectiveness in filtering noise, 5% white noise was added to each input variable. More studies on model-free fault diagnosis can be found in other works [21–23].

With the given three-tank system, incipient faults are likely to occur, which will require monitoring and determining the degree of fault at all time. However, the degree of fault, in theory, cannot be determined unless the exact plant model is known. The only alternative is to use experimental data. In the absence of experimental data, simulation data using Eq. (5.2) were generated.

To do so, a back-propagation neural network for each fault using randomly selected inputs and their corresponding outputs was trained via the Matlab Neural Network Toolbox. The input variables of each neural network are the pump flow rates, whereas the output variable is the degree of fault between 0 and 1. As soon as the fault is isolated, the respective neural network is fired to instantly predict the degree of fault.

5.6 Simultaneous faults of different types

In 2007, Lin and Singh [21] developed an intelligent model-free diagnosis technique for multiple faults in a nonlinear dynamic system; however, the technique was limited to the same type of process faults. Later, Zhang et al. [22] investigated the issue of isolation of process faults and sensor faults for a class of nonlinear uncertain systems but did not address the more difficult issue on isolation of actuator faults and process faults. In fact, our literature search found no studies conducted on simultaneous actuator faults and process faults. This study investigates the complexity due to simultaneous occurrence of process faults, actuator faults, and sensor faults, and proposes a method to isolate actuator faults from process faults. To better explain how a system behaves with simultaneous faults of different types, a strongly coupled MIMO three-tank dynamic system is used in this study.

5.6.1 Isolation of process faults

In the given three-talk dynamic system, there exist three possible types of faults: process faults, sensor faults, and actuator faults. The three process faults in this case refer to the pipe blockage for each tank, which was investigated and presented earlier in this chapter. The sensor faults refer to faults in sensing the water level of each tank, whereas the actuator faults refer to failure in pumping the right water to tanks 1 and 2. Isolation of simultaneous process faults, in general, is not a difficult task. However, it can become complex for a strongly coupled dynamic system such as the three-tank system. This is when hard computing alone cannot accomplish the isolation task. The isolation can be more reliable with the aid of soft computing.

5.6.2 Isolation of sensor faults

The occurrence of a sensor fault typically causes a bias to occur in the measurements of the affected sensor. The sensor faults investigated here were introduced via an instantaneous numerical offset at a specific time after reaching the system's steady state.

The unique behavior of the sensor fault that distinguishes it from the actuator and process faults can be mathematically explained via the matrix algebra of the ESO as described in Eq. 5.5. This equation is repeated here.

$$\dot{z}_1 = z_2 + l_1(x_1 - z_1) + bu$$

$$\dot{z}_2 + l_2(x_1 - z_1)$$

For multiple output systems with n variables, the variables x_1, z_1, and z_2 are column vectors with n elements. Correspondingly, the observer gains l_1 and l_2 are n by n diagonal matrices where each nonzero element represents the gain corresponding to a specific element of $(x_1 - z_1)$.

In the event of a sensor fault, the immediate change in one of the measured variables causes one of the elements of $(x_1 - z_1)$ to become nonzero (as the ESO values previously matched the measurements, all elements in $(x_1 - z_1)$ were zero). After multiplication by the diagonal gain matrix, the resulting vector also contains only one nonzero element, with the same index.

Sensor faults demonstrate distinct patterns in both the observable state variables and extended state variables, which makes them relatively easy to detect and separate from process and actuator faults. Specifically, if the i-th sensor associated with the i-th observable state variable has a fault, the associated i-th extended state variable will have a spike corresponding to the step change in the observed state variable. The changes in other observable and extended state variables are negligible.

Using the same three-tank system, the following data were used to perform simulation. Figs. 5.6 and 5.7 exhibit abrupt changes corresponding to the occurrence of three consecutive sensor faults:

Initial heights $h_o = [6\ 3\ 4]$ m
Bandwidth $\omega_o = 10$
Tank area, $A_t = 3$ m^2
Simulation period $= 90$ s
Acceleration of gravity $g = 9.81$ m/s^2
Pipe areas $a_1 = a_2 = a_3 = 1$ m^2
Flux coefficients or blockage $s_{13} = s_{32} = s_{20} = 1$ unless specified otherwise
Pump inputs $Q_1 = Q_2 = 5$ m^3/s unless specified otherwise.

Fig. 5.7 shows how the Z_2 values that track extended state variables responded to the three sensor faults occurring sequentially at three different times. Each time a fault occurred, only the corresponding observable state and extended state

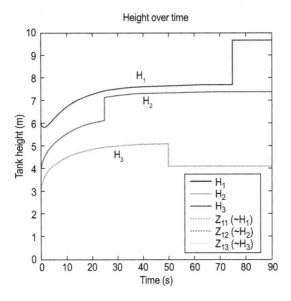

FIGURE 5.6 Water level spikes due to consecutive sensor faults at 25, 50, and 75 s.

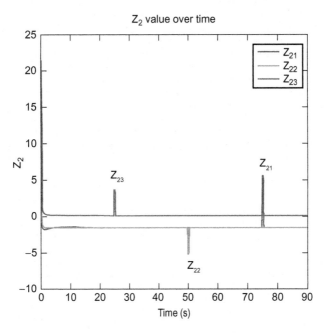

FIGURE 5.7 Z_2 value spikes responding to the sensor faults at 25, 50, and 75 s.

variable exhibit changes (a step and a spike), which clearly demonstrates the isolated nature of sensor faults. Despite having multiple faults occur (each in a different tank), each one is isolated to its respective variables. Thus, when a sensor fault occurs alone, a remark can be made as follows.

Remark 1. If the i-th sensor experiences a fault, only the i-th elements of the observable state variable and the extended state variables are affected despite multiple faults.

5.6.3 Isolation of actuator faults

It should be noted here that the given three-tank system is a MIMO system consisting of two inputs and three outputs. The two inputs are the pump rates, and the three outputs are the water level of each tank. Thus, sensors are only available to measure the output variables (i.e., the water levels).

When an actuator fault alone occurs, the calculated values of the associated extended state variable from the ESO do not converge to the expected value. This is because ESO uses system input u to calculate the extended state variables. Thus, when an actuator fault occurs, the corresponding value in u deviates from the designed theoretical value, which causes the extended state variable to change at the equilibrium. Likewise, occurrence of a process fault alone can cause changes in state variables, which leads to changes in extended state variables. This, indeed, further complicates the isolation of simultaneous process faults and actuator faults.

5.7 Isolation of simultaneous process faults and actuator faults

A process or actuator fault usually affects values of more than one observable variables and extended state variables as calculated by the ESO. Isolation of process faults from actuator faults exhibits complex coupling effects for a typical MIMO system. Isolation of process faults and actuator faults are closely examined in the following.

5.7.1 Characteristics of process faults and actuator faults

To distinguish actuator faults and process faults, the way they affect the convergence of extended state variables calculated by the ESO must be examined. Process faults alter the dynamics of the system, and thus the final steady state values of the state variables (in this case the tank heights) are different. Actuator faults cause unexpected discrepancies in the system input and thus also result in different steady state variable values. However, due to the calculation method of the ESO, the steady state values of the extended state variables of z_2 are also affected. At steady state, all time derivatives become zero and the ESO's estimated values match the measured values, which cause Eq. (5.5) to reduce to the following:

$$0 = z_2 + l_1(0) + bu(i.e.\ z_2 = -bu)$$

In the case of an actuator fault, values in vector u will be affected by the fault and deviate from the designed or expected theoretical values. However, the ESO will not know this and will still use the theoretical values. This will cause the elements of z_2 to converge to incorrect values, which do not match the steady state values of corresponding functions $f_1, f_2, \ldots f_n$. Without the ability to compare z_2 to f (as in most situations, f that represents plant dynamics must be assumed unknown), this discrepancy cannot be observed and thus the only distinguishing characteristic of actuator faults is unobservable.

Recall that a general nonlinear system that can be modeled by the ESO is expressed in the following:

$$y^{(n\cdot)} = f + bu,$$

where $y^{(n)}$ denotes the n-th time derivative of y, f is essentially the extended state representing the lumped nonlinear time-varying function of the plant dynamics and external disturbances, u is the system input, and b is a constant that is related to the physical model.

The occurrence of a process fault causes a change in f, whereas an actuator fault causes a change in u. However, either type of fault results in a net change to the same variable, $y^{(n)}$, which updates the system state. This makes it possible for both types of faults to produce quite similar behavior in the system state and extended state variables in the case of a single fault, or cause the appearance of an undisturbed system with simultaneous actuator faults and process faults. Mathematically, this fault ambiguity can be represented by ϵ:

$$y^{(n)} = f + bu + \epsilon \tag{5.35}$$

For actuator faults or process faults, ϵ represents a change in either the input $b(u + \Delta u)$ or the system dynamics $f + \Delta f$. The fact that ϵ could be either $b\Delta u$ or Δf allows for either type of a single fault to produce similar states if $b\Delta u = \Delta f$. Furthermore, simultaneous process faults and actuator faults can result in a seemingly undisturbed state if $b\Delta u = -\Delta f$. The computer simulation supports this mathematical finding. Figs. 5.8 and 5.9 show how false isolation could be concluded when an actuator fault and process faults occur at the same time ($t = 25$ s).

Thus, the following remark can be made.

Remark 2. For a typical MIMO nonlinear system, actuator faults, in general, cannot be isolated from process faults unless one or more additional sensor measurements are made.

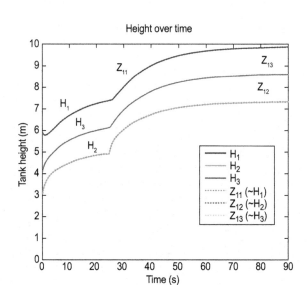

FIGURE 5.8 Water levels changed when an actuator fault and a process fault are introduced at 25 s.

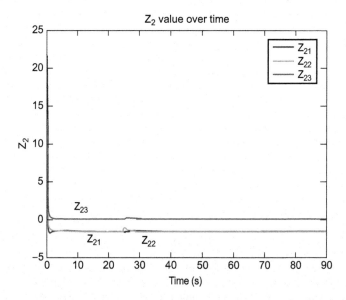

FIGURE 5.9 Z_2 values that tracked ESO show virtually no changes due to combined faults.

5.7.2 Utilizing an outflow sensor to isolate actuator faults

As discussed earlier, at least a sensor measurement must be added to resolve the ambiguity between process faults and actuator faults. Taking the presented three-tank system as an example, one can add an outflow sensor at the very end of the pipe (right outside the right-hand side of tank 2) to measure the system's net outflow. At steady state, conservation of mass dictates that the outflow of the three-tank system must be equal to the sum of the inputs $Q_1 + Q_2$ (i.e., pump flow rates). The theoretical steady state value of this quantity is only dependent on system specifications, and the actual outflow must always converge to the theoretical value except in the event of an actuator fault. As no nonactuator fault can affect Q_1 or Q_2, measuring the net outflow allows for a means to isolate actuator faults. This concept will work only if the added outflow sensor itself is not faulty. In fact, there is a simple way to determine if the outflow sensor itself is faulty. After the system reaches the steady state, if the net outflow does not equal $Q_1 + Q_2$ but there is no noticeable disturbance in the observable state and the extended state variables, then it is likely the outflow sensor is faulty. However, if the net outflow does not equal $Q_1 + Q_2$ and there is noticeable disturbance in those variables, then there exists a fault in actuator 1 or 2. The only means to isolate one actuator fault from the other is to directly measure each actuator's output.

5.8 Conclusion and future work

This study mathematically proved that the ESO's estimation error is upper bounded and its upper bound monotonously decreases with the observer bandwidth. This important proof allows for applying the improved ESO to be an effective means for FDI. The main advantage of the presented FDI technique is its robustness against uncertainty in the plant dynamics as well as disturbances. The parameterized ESO that requires tuning of only a single parameter (the observer bandwidth) makes it easy to be implemented in fault diagnosis. The bandwidth affecting the system's tracking speed and sensitivity to measurement noise can be easily tuned to meet the individual need for diagnosis.

From the model-based FDI point of view, the issue of how much knowledge about a nonlinear dynamic system is needed has been of great interest to researchers for years. This study concludes that ESO-based fault detection requires little knowledge about the plant model, not much beyond the order of the system.

The ESO-based fuzzy inference proved to be an effective mean for fault isolation. The fuzzy inference is particularly good at handling uncertainty in the plant model. Furthermore, this study went beyond the traditional FDI by adding the capability of determining the degree of fault via neural networks. Such capability is particularly important for diagnosis of incipient faults.

The detection and isolation of process faults by means of ESO and fuzzy inference have been presented. This study was conducted via computer simulation in which the equations for the three-tank system were used to calculate the theoretical values. To simulate unmodeled dynamics, 5% to 10% of external disturbance was introduced. The ESO was found capable of filtering system noise and correctly detecting process faults even when the system was not correctly modeled (when using $b_o = 635$ as supposed to the exact value of 127). However, in reality, process faults could be accompanied by sensor faults and/or actuator faults. This study investigated the complexity of fault isolation for simultaneous process faults, actuator faults, and sensor faults. Among them, sensor faults can be easily detected and isolated. However, for a strongly coupled MIMO nonlinear system, the combination of process faults and actuator faults exhibits complex coupling effects because these two types of faults affect the values of both observable state and extended state variables. For the presented three-tank system, a remedy was proposed to isolate actuator faults from process faults. Future work will include developing a general methodology to isolate actuator faults from process faults for a strongly coupled dynamic system.

References

[1] P. M. Frank, S. X. Ding, T. Marcu. Model-based fault diagnosis in technical processes. Transactions of the Institute of Measurement and Control 2000;22(1):57–101.

[2] V. Venkatasubramanian, R. Rengaswamy, K. Yin, S. N. Kavuri. A review of process fault detection and diagnosis: Part I. Quantitative model-based methods. Computer & Chemical Engineering 2003;27(3):293–311.

[3] P. P. Lin, X. Li. Fault diagnosis, prognosis and self-reconfiguration for nonlinear dynamic system using soft computing techniques. In *Proceedings of the 2006 IEEE Conference on Systems, Man, and Cybernetics.*

[4] R. Tarantino, F. E. Szigeti, Colina-Morles. Generalized Luenberger observer-based fault-detection filter design: An industrial application, Control Engineering Practice 8 (2000) 665–671.

[5] S. K. Dash, R. Rengaswamy, V. Venkatasubramanian. Fault diagnosis in a nonlinear CSTR using observers. In Proceedings of the 2001 Annual AIChE Meeting. Paper 282i.

[6] A. Z. D. Odloak. Sotomayor, Observer-based fault diagnosis in chemical plants, Chemical Engineering Journal 112 (2005) 93–108.

[7] P.M. Frank, Advanced fault detection and isolation schemes using nonlinear and robust observers, Proceedings of the 10th IFAC World Congress (1987).

[8] P. M. Frank. Online fault-detection in uncertain nonlinear-systems using diagnostic observers—A survey. International Journal of Systems Science 1994;25(12):2129–2135.

[9] P. M. Frank, X. Ding. Survey of robust residual generation and evaluation methods in observer-based fault detection systems. Journal of Process Control 1997;7(6):403–424.

[10] R. Isermann. Model-based fault-detection and diagnosis—Status and applications. Annual Reviews in Control 2005;29: 71–85.

[11] V. F. Filareretov, M. K. Vukobratovic, Observer-based fault diagnosis in manipulation robots, Mechatronics 9 (1999) 929–939.

[12] A. Xu, Q. Zhang. Nonlinear system fault diagnosis based on adaptive estimation. Automatica 2004;40: 1181–1193.

[13] M. Fang, Y. Tian, L. Guo. Fault diagnosis of nonlinear system based on generalized observer. Applied Mathematics & Computation 2007;185: 1131–1137.

[14] H. B. Wang, J. L. Wang, Robust fault detection observer design: Iterative LMI approaches, ASME Journal of Dynamic Systems, Measurement & Control 129 (2007) 77–82.

[15] A. Radke, On Disturbance Estimation and Its Application on Health Monitoring, Ph.D. Dissertation, Cleveland State University, 2006.

[16] J. Han. A class of extended state observers for uncertain systems. Control & Decision 1995;10(1):85−88 (in Chinese).

[17] J. Han, Nonlinear design methods for control systems, Proceedings of the 14th IFAC World Congress (1999).

[18] Z. Gao. Scaling and parameterization based controller tuning. In Proceedings of the 2003 American Control Conference. 4989−4996.

[19] Q. Zheng, L. Q. Gao, Z. Gao. On estimation of plant dynamics and disturbance from input-output data in real time. In Proceedings of the 2007 IEEE Multi-Conference on Systems and Control. 1167−1172.

[20] Z. Gao. Active disturbance rejection control: A paradigm shift in feedback control system design. In Proceedings of the 2006 American Control Conference. 4989−4996.

[21] P. P. Lin, H. Singh. Intelligent model-free diagnosis for multiple faults in a nonlinear dynamic system. In *Proceedings of the 2007 IEEE/ASME Conference on Advanced Intelligent Mechatronics (AIM)*.

[22] X. Zhang, M. M. Polycarpou, T. Parisini. Isolation of process and sensor faults for a class of nonlinear uncertain system. In *Proceedings of the 2008 American Control Conference*.

[23] P. Zhang, A model-free approach to fault diagnosis of continuous-time systems based on time domain data, International Journal of Automation & Computing 4 (2) (2007) 189–194.

[24] M.J. Korbicz. Kowal, Fault detection under fuzzy model uncertainty, International Journal of Automation & Computing 4 (2) (2007) 117–124.

Chapter 6

Fault diagnosis and failure prognosis in hydraulic systems

Jie Liu [a], Yanhe Xu [a], Kaibo Zhou [b] and Ming-Feng Ge [c]

[a] School of Civil and Hydraulic Engineering, Huazhong University of Science and Technology, Wuhan, China. [b] School of Artificial Intelligence and Automation, Huazhong University of Science and Technology, Wuhan, China. [c] School of Mechanical Engineering and Electronic Information, China University of Geosciences, Wuhan, China

6.1 Application status of sensor detection technology

Based on the energy conversion relationship between water and machinery, and the hydraulic characteristics, dynamic characteristics and structural characteristics of hydraulic machinery are the main research objects, and the main task is to ensure high efficiency and safe and stable operation of various hydraulic machines. The main contents of the research are energy characteristics of hydraulic machinery, cavitation characteristics, and operational stability [1]. Sensor detection technology is an important method for researchers to grasp the operating status information of hydraulic units. This chapter introduces the relevant standards for hydraulic machinery sensor detection, two typical application scenarios of model test rigs and field prototypes, and the development and application status of sensor detection technology.

6.1.1 Relevant standards of hydraulic machinery sensor detection technology

The sensor detection system can obtain online status information of equipment in real time, record the operation data of the equipment comprehensively, reduce the downtime of the unit, and find the fault symptoms in advance. This has always been a research hotspot in the industry. The extensive application of sensor detection technology can improve the accuracy of test experiments, reduce errors, extend the preliminary test period of equipment, and provide a decision-making basis for the power plant to carry out state maintenance and optimized maintenance. Industrial organizations such as the China Electrical Equipment Industry Association, the China Water Turbine Standardization

Fault Diagnosis and Prognosis Techniques for Complex Engineering Systems. DOI: 10.1016/B978-0-12-822473-1.00011-2
207

TABLE 6.1 Relevant standards of hydraulic machinery sensor detection technology.

ID	Name
GB/T15613	Model acceptance test of hydraulic turbine, energy storage pump, and pump turbine
GB/T10969	Technical conditions for flow components of hydraulic turbines
GB/T15613	Regulations on site acceptance test for hydraulic performance of hydraulic turbines, energy storage pumps, and pump turbines
SL142-2008	Test rules turbine model muddy water acceptance
IEC60041	Field acceptance tests to determine the hydraulic performance of hydraulic turbines, storage pumps, and pump turbines
IEC60193	Hydraulic turbines, storage pumps, and pump-turbines: model acceptance tests
IEC60308	Hydraulic turbines–testing of control systems
IEC994	Guide for field measurements of vibrations and pulsations in hydraulic machines

Technical Committee, the International Electrotechnical Commission (IEC), and other industry organizations pay close attention to hydraulic mechanical sensor detection systems and have launched a series of related standards successively, among which the important ones are shown in Table 6.1.

6.1.2 Instrumentation for the hydraulic turbine prototype

Hydraulic machinery is a subject mainly based on experimental science. High-precision hydraulic machinery test equipment and sensor testing technology are the necessary conditions for the design and verification of new hydraulic machinery products, and also important means to study and solve abnormal failures in actual field operations [2].

(1) *Brief introduction to the hydraulic mechanical model test bench.* The high-precision hydraulic machinery model test bench is an indispensable tool to advance the technological progress of hydraulic machinery disciplines. The main universal hydraulic machine model test beds worldwide include VOITH, ALSTOM, EPFL, Rainpower, ANDRIZ, Chinese Institute of Water Resources and Hydropower Research (IWHR), Harbin Institute of Electrical Machinery (HEC), and Dongfang Electric Co., Ltd. (DEC). At present, the highest test head of the hydraulic mechanical model test stand with an international advanced level can reach 150 m, the maximum test flow can reach 2.2 m^3/s, and the comprehensive efficiency error of the efficiency test

TABLE 6.2 Comparison of parameters of the test bench for an advanced-level hydraulic machinery model.

Test bench	VOITHUHD2-2	ALSTOMT3	EPFLPF2	IWHR-TP1	HEC-H	DF-100
Highest test head/m	240(T) /250(P)	100(T) /150(P)	120	150	150	100
Test flow (m³/s)	1.5	0.9	1.4	2.2	2.0	1.5
Model runner diameter/mm	300–500	300–500	300–500	250–500	300–500	350–500
Power/kW	600	360	300	540	500	500
Rotate speed /rpm	2000	1000–2400	2500	2600	300–2500	3000
Water sup- ply pump	2	—	—	24SA-10	24SA -10B	2
Motor power /kW	1600	—	—	724*2	600*2	700
Motor rotate speed/rpm	1490	—	—	1200	—	1100
Integrated efficiency error/%	±0.2	±0.25	<±0.25	±0.2	±0.2	<±0.25

is less than 0.2%. The detailed parameters of each test bench are shown in Table 6.2.

(2) *Application of sensor detection technology in performance test model test.* Relying on the high-precision hydraulic machinery model test bench and its supporting sensor detection system, a series of hydraulic machinery performance test experiments can be carried out. Typical basic performance tests include energy, runaway characteristics, pressure pulsation, and sediment wear tests.

Model test of energy characteristics of hydraulic turbines: The model test of the energy characteristics of the turbine is used to measure the efficiency of the model turbine under various working conditions, and it is the main method to grasp the data of the energy characteristics of the turbine.

Model test of hydraulic turbine flight characteristics: When the turbine is thrown off suddenly and all of the generator output power is zero, the turbine vane cannot be closed for various reasons. When the input water flow energy is balanced with various mechanical losses and hydraulic losses, the stable maximum speed is reached. The turbine flight characteristics test is an

TABLE 6.3 Types and measurement quantities of sensors for an energy characteristic model test of a hydraulic turbine.

Sensor type	Measured quantity	Comments
Liquid pressure sensor	Head (H)	Object of test observation
High-frequency pressure pulsation sensor	Pressure pulsation	Object of test observation
Optical interference profile sensor	Cumulative surface wear depth	Object of test observation
Angular displacement sensor	Guide vane opening (α)	Determine test conditions
Rotational speed sensor	Rotational speed (n)	Determine test conditions
Flow sensor	Flow (Q)	Determine test conditions
Power sensor	Power (P)	Determine test conditions

important way to understand the performance characteristics of the turbine under sudden failure.

Model test of hydraulic pressure pulsation: The hydraulic pressure pulsation model test measures the pressure pulsation characteristics of the model unit at various key positions such as the volute inlet, double row cascade, and tail water outlet under various operating conditions, and is the key test to ensure the unit's operating stability.

Muddy water model test of turbine silt wear: Through the model test of muddy sand wear and muddy water of the turbine, the wear parts and relative abrasion strength of the hydraulic machinery can be observed, the wear laws of the hydraulic machinery in the sand-bearing water flow can be explored, and the relevant protective measures can be studied by the experiment [3]. It has important significance in the life and reliability analysis of hydraulic machinery, as well as wear protection. The monitoring sensors and their measurement quantities commonly used in basic performance test model tests are shown in Table 6.3.

The sensor detection equipment has undertaken important tasks, such as measurement data and determination of test conditions in the hydraulic turbine model test. It is an indispensable foundation for carrying out the hydraulic turbine model test.

(3) *Application of related advanced technology*. With the development of sensor detection technology, some cutting-edge technologies are gradually applied to hydraulic machinery model test stands, providing researchers with a new dimension to fully understand the performance characteristics of model machines.

Internal flow field measurement technology. The internal flow field measurement technology is a direct means to study the internal flow mechanism of fluid machinery. It can intuitively and accurately reflect the true state of internal flow of fluid machinery [4]. The testing of external characteristics of hydraulic machinery includes performance testing (head, flow, power, etc.) and pressure pulsation testing. The corresponding parameters can be directly or indirectly measured by pressure sensors, power sensors, flow meters, and so on. The internal flow field measurement is to measure the velocity field, temperature field, and other related statistics in real time. The internal flow field test methods currently include high-speed camera technology, the hot wire/film anemometer (HWFA) test method [5, 6], and the laser Doppler velocimetry test method, among others.

Internal fluid testing technology has been improving continuously. The HWFA test method has been invented for 100 years, and it has made great contributions to turbulence research, which is still in use. Although HWFA performs contact-type point measurement, which can only target a specific coordinate point in space and interfere with the flow field, it has always played an important role in the transient single-point velocity test of the flow field.

Since the development of laser Doppler velocimetry testing technology in the 1960s, it has achieved noncontact measurement of the flow field. It uses the Mie scattering of the particles in the flow field to the incident laser light. The scattered light will generate Doppler relative to the incident laser light. With the amount of frequency shift, measuring the Doppler frequency shift, you can get the velocity of the particles in the flow field [7]. This technique has good temporal resolution and spatial resolution, and can be used for three-dimensional velocity measurement. However, like HWFA, it is still a single-point measurement technology. With the advancement of technology, especially the rapid development of computer technology, microelectronics, and optical technology, many advanced measurement technologies have been derived, such as laser hologram technology, laser scattering technology, and particle image velocimetry (PIV) technology [8]. Among them, PIV test technology realizes the noncontact measurement of the flow field, and the measurement accuracy of the two- and three-dimensional flow fields is high, and the PIV technology can be adapted to a variety of flowing media, so it has attracted great attention from people and gained rapid development [9].

PIV high-speed photography technology [10]. PIV technology is a transient, multipoint, noncontact hydrodynamic velocity measurement method. It has been continuously improved and developed in recent decades, can record a large number of spatial points in the same transient velocity distribution information, and can provide a wealth of spatial structure and flow characteristics of the flow field, help to further study the internal turbulence mechanism of rotary hydraulic machinery, and provide direct clues for the optimal design of turbomachinery [11]. The use of PIV to measure the internal flow field of the turbine requires visual processing of the measurement location of the hydraulic mechanical

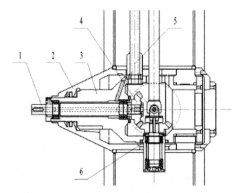

FIGURE 6.1 Schematic diagram of the typical structure of a hydrostatic bearing: 1) Main shaft, 2) Static sleeve, 3) wing sleeve, 4) Bevel gear, 5) Cantilever, 6) Mechanical bearing.

model test bench, such as replacing the stainless steel structure with organic glass. The PIV test experiment system mainly includes the pulse laser generator, charge coupled device (CDD) high-speed camera, synchronous controller, high-speed image data interface board, image data analysis and processing system, and acquisition computer. The laser provides an intensity sheet light source for flow field illumination. The CCD high-speed camera realizes the synchronous action of the CCD camera and the laser generator through the synchronous controller, and outputs the collected image data to the computer in real time. The original image data obtained through PIV is postprocessed by the acquisition computer. In addition, in the PIV system measurement, the laser surface is used to illuminate the measured area in the flow field, and to display the trajectory of the flow field, it is necessary to show the fluid flow phenomenon through the movement of the tracer particles [12].

High-sensitivity static pressure bearing. The hydrostatic bearing is a key device for studying the force of the runner of the hydromechanical model [13]. Its function is to support the runner and measure the energy lost due to the friction of the mechanical bearing. Because the hydrostatic bearing is installed in the flow channel, its design space is limited, but sufficient oil cavity width is required to provide excellent oil film stiffness to meet bearing load requirements; static pressure bearings work in water for a long time, which requires a good sealing performance, but if the sealing is too tight, it will bring greater friction. The larger friction force will reduce measurement accuracy. Similar conditions restrict each other, which greatly increases the difficulty of designing and manufacturing the static pressure bearing. The hydrostatic bearing is mainly composed of the main shaft, mechanical bearing, static sleeve, swing sleeve, bevel gear, and dynamometer system, among others (Fig. 6.1). Its main working method is as follows: the oil film is formed between the static pressure bearing swing sleeve and the static sleeve, and a pressure oil chamber is provided between the swing sleeve and the static sleeve. The external pump station is used to provide stable and durable oil supply pressure. When the load changes, it can use the oil pressure adjustment system to automatically adjust the pressure of each oil chamber, always keeping the swing sleeve suspended above the oil film. A hydraulic fulcrum is provided in the upstream and downstream directions of the bearing swing sleeve, which are located on the left and right sides of the force measuring arm to achieve stable support [14].

Measurement of dynamic stress of turbine runner blades. Due to the rotating state of the turbine runner during operation, there are great difficulties in the arrangement of strain gauges, the power supply of the sensors, and the signal output [15]. The traditional way is mainly to send signals to the collection equipment through the contact brush ring and the wire-collecting ring. Otherwise, a noncontact capacitive or inductive transmitter, electromagnetic wave transmitter, and so on are used to transmit signals. The poor signal to noise of these devices limits the accuracy of the entire test system and reduces the reliability of the measured data [16, 17]. To study the dynamic stress of hydraulic turbine runners, the first step is to solve the test technique of hydraulic turbine runner dynamic stress [18]. In recent years, with the rapid development of electronic application technology, computer technology, and wireless digital transmission technology, and the hydraulic turbine runner stress test, the technology has begun to make new progress. Internationally renowned hydropower group companies use a test device similar to the "aircraft black box" to complete the field test of the runner stress. The test results provide a strong technical basis for the research of the stress of the runner component and the analysis and treatment of the runner blade cracks [19, 20].

6.1.3 On-site detection for hydraulic turbines

Prototype testing technology of hydropower units is to carry out comprehensive signal measurement of the operating state of the unit through a large number of sensors. Combined with computer data storage technology, mathematical analysis methods are used to process the measured data to understand the operating state of the unit. It provides data foundation and basis for optimizing the operation of the unit, and eliminating the unit hidden dangers and state maintenance. On-site tests usually include the turbine efficiency test, unit stability test, and force characteristic test.

(1) *Efficiency test.* The efficiency of the turbine has a significant impact on the economic benefits of the hydropower plant. To make the turbine work better in the high-efficiency area, it is necessary to conduct an efficiency test on the turbine. Through the turbine efficiency test, not only can we understand the operation status of the unit, to provide technical information for maintenance and reconstruction, but also through the adjustment of the turbine, it can be operated in a higher-efficiency area, make full use of hydraulic resources, and improve the economic efficiency of the hydropower

plant to guide the economic operation of hydropower units and even the power grid. Through the analysis of the efficiency test data, the actual dynamic characteristics of the unit under the test head can be obtained, and the characteristic curves of the unit operation can be drawn, such as the unit efficiency characteristic curve, the turbine efficiency characteristic curve, the water consumption rate characteristic curve, the flow rate and the guide, the relationship curve of the blade, and the flow characteristic curve of the turbine, among others. By comparing the measured efficiency characteristic curve with the model efficiency characteristic curve, it can be verified whether the turbine meets the efficiency guarantee value provided by the manufacturer, and can be used as an important standard for the field acceptance test of the newly put into production unit [21].

Unit efficiency, flow measurement of large-diameter pipes, and on-site detection of water consumption rate are major scientific and technological research topics for hydropower production and construction [22]. In recent years, many scientific research institutes and colleges and universities have vigorously conducted theoretical research and device development, and achieved gratifying results. The ZFM-01 type microcomputer differential pressure flow monitoring device developed by the Tianjin Institute of Geological Exploration and Electrical Engineering was put into operation at the Panjiakou Hydropower Station in Hebei Province; the YLI-01 type microcomputer differential pressure flow monitoring device developed by the Wuhan University of Water Resources and Electric Power, Dongjiang Hydropower Station was put into operation in Hunan Province; and the SJ-gn ultrasonic flow monitoring device developed by Nanjing Automation Research Institute was installed and commissioned in Tuoxi Hydropower Station in Hunan Province. There are still some power stations that have built online unit efficiency monitoring and analysis systems, such as the PLC monitoring system of Gezhouba Hydropower Plant, the unit measurement and vibration monitoring system of Wanjiazhai Hydropower Station in Inner Mongolia, and the impact unit online efficiency of Yilihe Power Plant in Yunnan, on-line monitoring system for unit flow and efficiency of Zhejiang Shafan Hydropower Station, and so on. The completion and use of these systems has laid the foundation for the future development of comprehensive condition monitoring systems and accumulated valuable experience [23].

(2) *Unit stability test.* With the improvement of design and manufacturing capabilities, the capacity of the turbine is increasing, the structural size of each component of the unit has also increased accordingly, and the receiving member dynamic, static load, and random noise also increase. Inevitably, there will be various mechanical, hydraulic and electromagnetic vibration problems [24, 25]. Therefore, more and more attention has been paid to the stability of the turbine in operation. As we all know, when the Francis turbine deviates from the optimal working condition, there is a problem of unstable operation. Due to the improvement of manufacturing technology, the increase of single unit capacity, and people's general pursuit of high efficiency and high utilization of materials, the unstable zone of Francis turbines has a tendency to expand, and the intensity of pressure pulsation and hydraulic excitation has increased. The trend has led to many failures, such as hydraulic structure destruction, runner cracks, and so on. To reduce accidents, improve the stability of the operation of the unit, and extend the service life of the unit, conduct an on-site stability test of the already-operated unit. During operation of the unit, a special vibration test instrument is used to measure the relative vibration (swing) and mechanical. The absolute vibration of the frame and the top cover, comprehensive measurement of the vibration and swing of the unit, mastering the objective data of the unit's operating stability and vibration or unstable operation area are important for the evaluation of the unit's vibration performance and vibration isolation operation significance.

The turbine vibration level is an important criterion to judge the operation stability of the turbine. The assessment method for the stability of hydraulic turbines is usually measured by three indicators: vibration, swing, and hydraulic pressure fluctuation [26]. In the stability test of hydraulic turbines, the sensor detection system plays an important role. An eddy current sensor is generally used to measure the relative swing of the upper and water guides. Using a low-frequency velocity pickup, measure the absolute vibration velocity in the radial direction of the upper frame, the radial direction, and the vertical direction of the top cover. The absolute vibration displacement of the top cover in the radial and vertical directions was measured with a low-frequency velocity pickup and an integrating amplifier. Use an intelligent data acquisition instrument to record the swing and vibration data under each stable test condition, and make spectrum analysis.

(3) *Online monitoring of the operating status of the unit.* In recent years, the rapid development of the state monitoring industry of hydropower units has promoted the further advancement of state maintenance in the application of hydropower plants, and the technical level and functional application of state monitoring products are also quite different. From the current industry point of view, there are mainly three types of products for online monitoring of hydro-generator unit status: 1) Basic swing and vibration monitoring devices. This type of device mainly completes the real-time measurement and monitoring and alarm functions of the unit's vibration and swing, mainly for the purpose of replacing manual dial indicator measurement. Most of these products either do not provide or provide only simple analysis functions. Currently, only a few power plants are still in use. 2) The general condition monitoring system. It not only monitors the vibration and swing of the unit, but also completes the monitoring of various objects, such

as air gap, magnetic flux, generator partial discharge, runner cavitation cavitation, and so on. Different monitoring devices are connected, thus data integration, unified operation, and standardized functions are performed. However, this type of system is mainly oriented to experimental institutions and industry experts. Its complex fault mechanism is difficult to grasp by the maintenance personnel of the power plant. As a set of powerful real-time online monitoring systems, the power plant has not yet really provided guidance. The role of state maintenance can only be done with the help of experts for some targeted unit tests. This is a widespread phenomenon at present, and it is also the fundamental reason the state online monitoring system has not been popularized and recognized in power plants. 3) Remote analysis and intelligent comprehensive fault diagnosis system. To enable engineering and technical personnel at different levels to master the use of condition monitoring systems without a professional background, such systems not only realize the functions they have but also provide users with the characteristics of intelligence, automation, practicality, easy operation, and so on. Based on the quantitative evaluation of the indicators based on the characteristics of the faults that frequently occur in the unit, the general maintenance personnel can understand the operation status of the unit and judge the occurrence of faults on this basis [27].

At the same time, the development of Internet/Intranet networked remote communication technology has made the automation of power systems more and more comprehensive. The establishment of a remote analysis and diagnosis center is the latest stage of the development of a condition monitoring system that can be realized at present. The remote center can make full use of remote experts for the unit. The analysis and diagnosis service is both decentralized and centralized. At present, a remote analysis and diagnosis center model based on power group companies, power generation basin companies, research and development institutions, provincial pilot laboratories, and other institutions has been formed, and its use is not only at the technical level but also provides new ideas for the management model.

Today, the abroad monitoring and analysis systems of hydropower units are the Vibrocontrol 4000 system of Schenck in Germany, which is mainly used for monitoring and analysis of vibration of turbines; the ZOOM 2000 system and AGMS system of VibroSystM in Canada are used to monitor the generator air gap and vibration of the hydro-generator set; the VM600 system of VIBRO-METER company in Switzerland; the Hydro VU system of Bently Nevada Corporation in the United States; the Scard system of Siemens Company in Germany, and so on. Domestically, there are the PSTA system of Tsinghua University and Beijing Orient General Electrical Technology Institute, the HSJ system of Huazhong University of Science and Technology, and the EN-8000 of Envada. UK. Ltd. Beijing, and so on.

With the rapid growth of computer information technology and DSP technology, the data processing efficiency of data acquisition systems has been greatly improved, and with the advancement of IC manufacturing technology, CMOS, and iCMOS (industrial CMOS) processes, the development time and production cost of data acquisition systems are significantly reduced, and began to manufacture data acquisition systems with extremely fast acquisition speed, higher acquisition accuracy, and better stability. With the development of data collection technology and the guidance of growing market demand, the data collection system of hydraulic machinery will be developed in the direction of miniaturization, information, and intelligence.

The future sensor detection and diagnosis system will be an integrated, Internet-based remote diagnosis system. The system should rely on a complete sensor perception system to provide the data basis, and use different inference modes according to the characteristics of different subsystems and different problems. Several different reasoning models are used for mixed reasoning. In a system, the advantages of various inference models will be fully utilized to achieve the purpose of improving the speed of inference and the accuracy of the solution. At the same time, this system is based on a computer network, connecting scattered monitoring systems, and various service and management systems to achieve resource sharing and overall coordination.

6.2 Cavitation research

With the rapid development of hydropower exploitation and major cross-basin water conservancy projects, a series of large and medium-sized power stations and pumping stations have been put into operation or are under construction or planning. Meanwhile, the number of water turbines or pumps with abrasion has increased greatly. The operating experience of established pumping stations and power stations shows that pumping stations and hydropower stations operating in sandy rivers often suffer from wear, cavitation erosion, and the combined action of both. Therefore, the damage is very serious. Abrasion damage not only causes huge economic loss but also seriously affects the safe operation of the hydropower station and pump station. Due to the complex and changeable operating conditions and working conditions, cavitation erosion and sediment wear damage of hydraulic machinery is a complex physical process, with the characteristics of multiphase, microscopic, transient, and random, involving complex gas-liquid, sediment wear, tribology, materials science, surface protection, and other multidisciplinary problems. The relevant theoretical modeling and experimental research are often very complex and difficult. Although some progress has been made, the relevant research is still inadequate. The development of the hydraulic machinery abrasion discipline has important theoretical research and engineering significance.

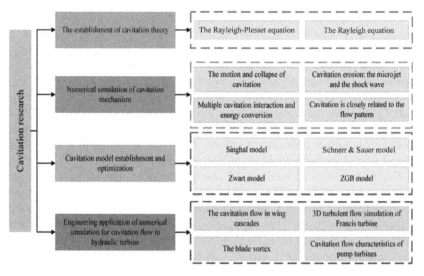

FIGURE 6.2 The flowchart of cavitation research.

Cavitation can be divided into five types according to its causes: hydraulic cavitation, oscillatory cavitation, acoustic cavitation, photoinduced cavitation, and non-phase variation cavitation [28]. The main type of cavitation in hydraulic machinery is hydraulic cavitation. As a common phenomenon in hydraulic machinery, it is of great significance to study cavitation characteristics of hydraulic machinery to ensure production. The experimental operation of the cavitation mechanism is difficult, and a high-speed camera is often used to photograph the cavitation process. However, the collapse time of cavitation bubbles is very short, and the process is difficult to capture. Therefore, the requirement of equipment is high, and the research cost is expensive. The numerical simulation method not only qualitatively and quantitatively analyzes a specific parameter and its variation rule with time but also carries out full-flow channel flow field simulation to predict the cavitation area within the fluid machinery. The working condition area that can be realized is larger than the experimental study. The study of cavitation flow in modern times is mainly carried out by numerical simulation and verified by experimental results.

Cavitation flow numerical simulation refers to the process of the evolution of pressure pulsation, force characteristics, and flow characteristics of the unit by discretizing the continuous fluid domain with the numerical calculation method and combining the governing equation based on cavitation theory. Then by changing the initial conditions, the numerical simulation of various working conditions is realized. With the development of computer simulation technology, the visual degree of numerical calculation for cavitation flow increases. The cavitation research is discussed from four perspectives, as shown in Fig. 6.2.

6.2.1 Establishment of cavitation theory

As early as in 1753, Euler had predicted that there would be a vacuum area in the interior of a rapidly moving liquid. In 1897, Barnaby and Parsons formally proposed the concept of "cavitation" of a liquid, based on the idea that the acceleration of a propeller to a certain extent would cause a drop in the speed of a ship. After 1906, Parsons took the lead in carrying out a large number of experimental studies on cavitation phenomenon, continuously enriching the theoretical connotation of liquid cavitation research, and laying the foundation of cavitation dynamics from the perspective of the force and motion on cavitation bubbles [29]. In 1917, Rayleigh [30] assumed that the research medium was an ideal fluid and proposed the Rayleigh equation for cavitation dynamics on the premise of ignoring viscosity and surface tension:

$$\frac{P_a(t) - P_\infty(t)}{\rho_f} = R\frac{d^2R}{dt^2} + \frac{3}{2}\left(\frac{dR}{dt}\right)^2, \tag{6.1}$$

where $P_a(t)$ and $P_\infty(t)$ represent the pressure of the cavitation bubble and the pressure at infinity from the cavitation bubble, respectively. ρ_f and R stand for the liquid density and the radius of cavitation bubble, respectively.

However, the assumptions made in the Rayleigh equation cannot fully reveal the mechanism of the whole process from the birth to the collapse of cavitation bubbles. To enhance the applicability of the equation, Plesset proposed the famous Rayleigh-Plesset equation based on the Rayleigh equation in 1949, which is expressed as follows:

$$\frac{P_a(t) - P_\infty(t)}{\rho_f} = R\frac{d^2R}{dt^2} + \frac{3}{2}(\frac{dR}{dt})^2 + \frac{4\upsilon_t}{R}\frac{dR}{dt} + \frac{2\sigma}{\rho_f R}, \tag{6.2}$$

where σ and υ_t represent surface tension and kinematic viscosity, respectively. $\frac{dR}{dt}$ and $\frac{d^2R}{dt^2}$ stand for the velocity and acceleration of the cavitation bubble wall, respectively.

Thus, the cavitation dynamics was born, which opened the way to explore cavitation mechanism from the perspective of studying the development process of cavitation bubbles.

6.2.2 Numerical simulation of the cavitation mechanism

Based on the theoretical support of cavitation dynamics, researchers use the numerical calculation method to simulate the movement of cavitation bubbles in the cavitation phenomenon so as to enrich the essential connotation of the cavitation mechanism. Through a large number of experiments, the periodic behaviors of initial birth, development, shedding, discharge, and collapse for cavitation bubbles have been observed. In the aspect of numerical simulation, Chen et al. [31] studied the phenomenon of cavitation flow around a rotating body; the calculation results correctly reflected the change of cavitation shape with cavitation number, and the return jet was captured at the end of the cavitation. Liu et al. [32] simulated local cavitation flow in viscous flow and found that the shape of cavitation was related to the Reynolds number and developed toward the hydrofoil end. Ji et al. [33] studied the cavitation shedding process of three-dimensional warping airfoil, and pointed out that the main body shedding and secondary shedding structures were caused by the influence of return jet and side jet. At that time, little was known about the underlying mechanism of the periodic behavior for cavitation bubbles. Min et al. [34] took the lead in conducting a detailed study on the periodic flow and mechanism of airfoil cavitation by adopting the mean-balanced cavitation flow model based on the positive pressure relation cavitation model. The results show that unsteady cavitation flow is a turbulent flow with large-scale eddy current structure, and the negative pressure gradient after the closed zone of the cavitation leads to the formation of vortex and the development of return jet.

Compared with other steps in the development of cavitation, various phenomena caused by the collapse of cavitation are directly related to cavitation on the wall. Researchers at home and abroad have conducted a lot of research on the collapse of cavitation near the wall. In 1970, Plesset and Chapman [35] carried out numerical simulation for the cavitation collapse process for the first time and made the following research hypothesis: the fluid is incompressible and nonviscous liquid, the ambient pressure and vapor pressure are constant, the pressure inside the entire bubble is uniform and does not contain permanent gas, the surface tension of the liquid is ignored, and the cavitation bubble is symmetrical spherical. The numerical calculation method adopted in the study is the finite difference method, which simulates the following two specific situations: on one hand, the initial state of spherical cavity collapse near the plane solid wall is simulated, on the other hand, the cavitation state when the distance from the wall to the nearest point of the cavitation cavity is exactly half of the radius of the cavitation bubble is simulated. The cavitation collapse in the two cases is shown in Fig. 6.3. The results show that the efflux from the cavitation to the wall is generated in the early stage of cavitation collapse, and the velocity of the efflux can explain the damage degree of wall cavitation erosion. Since the cavitation bubbles in the actual collapse process no longer maintain spherical symmetry, and there was less discussion on the nonspherical symmetry behavior of the cavitation bubbles at that time, only qualitative analysis can be used. However, its calculation process has low complexity and accuracy, which is not conducive to practical application.

In 1981, Hirt and Nichols [36] proposed the volume of fluid (VOF) method to solve the precision problem of the numerical calculation method. They believe that the Eulerian method is more reasonable than the Lagrangian method for the problem of free boundary and put forward three problems to be considered when dealing with free boundary: (1) discrete representation, (2) free boundaries evolve over time, and (3) boundary condition. The VOF method provides a simple and economical way to track free boundaries in two- or three-dimensional grids, and is more flexible and efficient than the finite-difference method in dealing with problems involving highly complex free surface configurations.

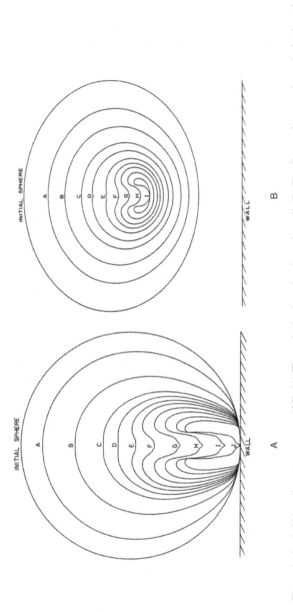

FIGURE 6.3 The cavitation bubble collapse in two cases [35]: (A) The cavitation boundary is on the wall. (B) The distance between the cavitation boundary and the wall surface is half of the cavitation diameter.

Since then, VOF method has become one of the commonly used numerical calculation methods in cavitation flow numerical simulation research and has been widely used in subsequent research.

Since Naudé and Ellis [37] experimentally verified the microjet phenomenon during cavitation collapse in 1961, and Plesset and Chapman used numerical calculation methods to make theoretical proof, microjets have been generally accepted as one of cavitation failure mechanisms. In the following decades, researchers at home and abroad have done a lot of experimental and theoretical research on the collapse of nonspherical cavitation. Lu and Huang [38] calculate the cavitation collapse near the rigid plane boundary wall in a solid-liquid two-phase fluid. Based on the assumption that the liquid phase is incompressible Newtonian fluid and the solid phase is spherical and its diameter is extremely small relative to the cavitation diameter, the Navier-Stokes equation of liquid motion in two-phase flow is solved without considering the mass diffusion effect and heat conduction effect of the cavitation. The process of cavitation collapse in viscous fluid and solid-liquid two-phase fluid is analyzed, respectively. The solid particles in the liquid will slow down the process of cavitation collapse and the microjet action, and with the increase of particle concentration, the mitigation effect become more obvious. When the particle concentration is maintained, the smaller the size is, the more obvious the mitigation of the cavitation collapse process will be. Because the collapse cavitation of the vacuole is not significantly different from the number level of sound velocity in the liquid, and because there is no energy exchange between the inside and outside of the vacuole, Li and Xiong [39] propose that it is necessary to consider the compressibility of the liquid. By analyzing the motion path of the cavitation bubble center near the rigid wall when a single spherical bubble collapses, the motion path of the bubble is decomposed into the relative motion along the radial direction and the motion of the bubble center along a certain path. The results show that the pressure difference between the upper and lower hemispheres of the bubble drives the movement of the bubble when it collapses, and the bubble center accelerates to the rigid side wall with the change of time. The study of the trajectory of the bubble center is helpful to find the collapse point, so as to analyze the failure effect of the maximum collapse pressure on the solid wall. Therefore, it is a new exploration angle in the process of cavitation mechanism research. He et al. [40] believed that surface tension should be included as one of the factors affecting cavitation and carried out numerical simulation for cavitation collapse near the solid wall. To avoid the numerical instability caused by overconcentration of markers, three different methods—median cubic interpolation, cubic spline interpolation, and cubic ENO interpolation—were used in the study. Based on the Bernoulli equation ignoring the cavitation gravity, the velocity and pressure of the cavitation outflow field were solved. The results showed that the surface tension played a significant role in the process of cavitation collapse.

As the commercial computational fluid dynamics (CFD) analysis software became mature, Li and Chen [41] carried out a numerical simulation for the collapse of the near-wall microbubble based on the Fluent software environment. In this study, the VOF model was used for unsteady calculations. When a single bubble collapsed, the process of accelerating the top bubble wall to the bottom until it penetrated the entire bubble and a high-speed microjet was formed to strike the wall when was simulated. The relationship between the distance between the cavitation and the wall and the strength of the microjet was also analyzed. The results showed that only when the cavitation was kept close enough to the wall can a high-speed jet be formed, and the high pressure generated by the jet was considered to be an important reason for the wall cavitation damage. This method provides an important experience for the subsequent research on the process of cavitation collapse. In addition, Xu et al. [42] used the full boundary element method to simulate the near-wall cavitation behavior of the elastic wall, and obtained the quantitative relationship between the distance of the wall and the wall deformation, as well as the influence of the distance of the wall on the jet flow. On this basis, the effect of the energy transfer relation of the cavitating jet on the particle velocity was further analyzed.

Cavitation erosion caused by cavitation collapse is not only caused by microjet theory [43] but also caused by shock wave. Laser cavitation shows good operability in the study of microjet pressure and shock wave pressure. Xing et al. [44] studied the mechanism of laser cavitation enhancement based on Fluent, and analyzed the release pressure when the cavitation collapsed by controlling the liquid properties, bubble wall distance, and laser energy parameter values. When 2A02 aluminum alloy is used, the jet and impact pressure are not enough to destroy the surface of the material. The results provide a new idea for an anticavitation effect in hydraulic machinery—that is, to choose different strengthening methods according to the properties of materials, and to improve the antifatigue strength of materials.

Compared with single cavitation collapse, the mechanism of multicavitation interaction and energy conversion is more complex. In 1994, Chahine [45] used a three-dimensional boundary element method model to study the multicavity behavior in a nonuniform flow field. It was found that the increase in the number of cavitation only extended the cavitation cycle without affecting the maximum size of the cavitation, and caused the central pressure of the cloud cavitation to increase sharply. In addition, the cavitation deformation and collapse were used to explain the noise of tip vortex cavitation. For the dynamic behavior of the vacuole group, the interface tracking method is the main research method. Chen et al. [46] analyzed the dynamic response of the cavitation bubble group and defined the low-frequency vibration in cavitation as the transformation process from the disorderly potential energy of a single cavitation to the ordered potential energy of a cavitation bubble group. Moreover, the frequency of a cavitation bubble group is much lower than that of a single cavitation. For a hydraulic system, the cavitation area is an excitation source of the overall dynamic behavior. Some studies believe that cloud cavitation is the main cause of cavitation [47–49]. To further explore the collapse mechanism of bubble groups on the mesoscale, Zhang et al. [50] analyzed it from the perspectives of single bubble collapse dynamics, bubble-bubble coupling and bubble-wall coupling, and different forms of energy conversion. Based on the OpenFOAM platform, under the conditions of considering the tension, viscosity, and compressibility at the model interface, the VOF model was selected to solve the governing equation of the flow field. The study on the single cavitation collapse process proved that the wall effect affected the cavitation in many aspects, such as the delay and deformation of single cavitation, strengthening the jet, restraining the release of pressure wave, and increasing the wall impact force, which was similar to the bubble interaction. In addition, the process of collapse near the wall of the bubble group was simulated numerically, and the influence of the expansion of bubble size and the change of bubble height on the collapse of bubble group and the change of energy distribution were analyzed, respectively. The energy conversion mechanism in the collapse of the bubble group was proposed for the first time.

The cavitation phenomenon is closely related to the flow pattern of water. The high-speed water flow in hydraulic machinery results in a sharp decrease in pressure, which leads to the primary cavitation. Therefore, the analysis of flow field structure in unsteady cavitation flow is essential. It is difficult to observe the complex eddy currents in the flow field in detail by experimental means, but the proper turbulence model can reflect cavitation more accurately in numerical simulation. Chumakov et al. [51] thought that the estimation of turbulence viscosity in the Reynolds average Navier-Stokes equations was too high, which affected the accuracy of cavity calculation. Chumakov et al. also proposed to use the large eddy simulation method to calculate compressible flows so as to more truly reveal the relevant structures of turbulence in cavitation calculation. Different from previous studies, Chumakov et al. pointed out that under the high-intensity turbulence effect, the shedding of cavitation was not only dominated by the back jet but also the development of turbulent pulsation caused the cavitation to break up. The relationship between the pressure pulsation in the far field caused by the periodic motion of the cavitation bubble and the second-order variation of the cavitation bubble volume was obtained by quantitative analysis.

6.2.3 Cavitation model establishment and optimization

Based on the Rayleigh-Plesset equation and the gradually improved cavitation mechanism, different scholars proposed various cavitation models for the actual engineering cavitation flow analysis. In commonly used commercial CFD software, three cavitation models are mainly provided: the Singhal model, Schnerr and Sauer model, and Zwart-Gerber-Belamri (ZGB) model [52]:

(1) *Singhal model.* The Singhal model takes into account such information as saturation vapor pressure, density, and surface tension. In addition, cavitation deformation and movement, pressure pulsation and velocity in

turbulent development, and composition of incompressible gas all have great influences on cavitation flow [53]. The phase transition rate of this model can be expressed as

$$R_e = C_e \frac{\sqrt{k}}{\sigma} \rho_l \rho_v \left(\frac{2}{3} \frac{P_v - P}{\rho_l} \right)^{\frac{1}{2}} (1 - f_v - f_g), \qquad (6.3)$$

$$R_c = C_c \frac{\sqrt{k}}{\sigma} \rho_l \rho_v \left(\frac{2}{3} \frac{P - P_v}{\rho_l} \right)^{\frac{1}{2}} f_v, \qquad (6.4)$$

where R_e and R_c are the liquid phase evaporation rate and vapor phase condensation rate, respectively. C_e and C_c represent the evaporation coefficient and condensation coefficient, respectively. As empirical constants, their values are generally 0.02 and 0.01. P and P_v represent the local absolute pressure and saturated vapor pressure, respectively; k is the turbulent kinetic energy and σ is the surface tension; and ρ_l and ρ_v are the densities of vapor and liquid, respectively. In addition, f_v, f_l, and f_g stand for the mass fraction of steam, liquid, and incompressible gas, respectively.

(2) *Schnerr and Sauer model.* Based on VOF technology, the Schnerr and Sauer model is often used to simulate NACA 0015 hydrofoil cavitation flow, which has a good effect on simulating cavitation characteristic effects such as the formation of cavitation clouds and refracted flows, and local hydrodynamic pressure peak caused by cavitation cloud collapse [54]. The transformation rate of this model can be expressed as

$$R_e = \frac{\rho_l \rho_v}{\rho} \alpha_v (1 - \alpha_v) \frac{3}{R_B} \sqrt{\frac{2}{3} \left(\frac{P_v - P}{\rho_l} \right)}, \qquad (6.5)$$

$$R_c = -\frac{\rho_l \rho_v}{\rho} \alpha_v (1 - \alpha_v) \frac{3}{R_B} \sqrt{\frac{2}{3} \left(\frac{P - P_v}{\rho_l} \right)}, \qquad (6.6)$$

where R_e and R_c are the liquid phase evaporation rate and vapor phase condensation rate, respectively; P and P_v represent the local absolute pressure and saturated vapor pressure, respectively; and ρ_l and ρ_v are the densities of vapor and liquid, respectively. R_B is the radius of cavitation bubble, and α_v is the volume fraction of the cavitation bubble.

(3) *Zwart model.* The Zwart model is proposed on the basis of the Kubota et al. [55] and Gerber [56] models, and now the ZGB model is commonly used. The ZGB model has good robustness in predicting three-dimensional cavitation flows [57], and its phase transition rate can be expressed as follows:

$$R_e = C_e \frac{3\alpha_{nuc}(1 - \alpha_v)\rho_v}{R_B} \sqrt{\frac{2 \max(P_v - P)}{3\rho_l}}, \qquad (6.7)$$

$$R_c = C_e \frac{3\alpha_v \rho_v}{R_B} \sqrt{\frac{2\max(P - P_v)}{3\rho_l}}, \qquad (6.8)$$

where R_e and R_c are the liquid phase evaporation rate and vapor phase condensation rate, respectively. C_e and C_c represent the evaporation coefficient and condensation coefficient, respectively. As empirical constants, their values are generally 50 and 0.01. P and P_v represent the local absolute pressure and saturated vapor pressure, respectively; ρ_l and ρ_v are the densities of vapor and liquid, respectively; and α_v is the gas core density.

In the numerical simulation application of hydraulic mechanical cavitation flow, the ZGB cavitation model has a better simulation effect on three-dimensional cavitation characteristics of the mixed-flow turbine than the Schnerr and Sauer model, with a wider application range and higher calculation accuracy [58]. However, the ZGB model is limited to some extent because it does not consider the noncondensable gas in the medium.

6.2.4 Engineering application of numerical simulation for cavitation flow in a hydraulic turbine

With the appearance of million-class units in hydropower stations, it is important to explore the hydraulic performance, cavitation performance, and energy performance of units. Researchers in the water conservancy and hydropower industry have conducted a large number of experimental and simulation studies on the cavitation flow of hydraulic machinery, and the results obtained are quite rich and have their own advantages and disadvantages. In this chapter, combined with the numerical simulation of cavitation flow applied in engineering practice, the defects and prospective achievements in each case are summarized. Meanwhile, the development direction of numerical simulation for cavitation flow in hydraulic machinery is prospected.

As early as the 1960s and 1970s, researchers have carried out studies on the cavitation flow in wing cascades. Nishiyama and Evanoff extended the single-wing linear solution to the confined cavitation flow problem of linear cascade. Brennen used the finite difference method to analyze axisymmetric flow problems. In 1983, Yamaguchi and Kato [59] proposed a nonlinear theoretical calculation method for the cavitation flow of thick wings, and the cavitiony shape was determined by using the adherent panel method on the premise of considering the viscous force on the boundary. Later, Johansen [60] improved the method of Yamaguchi and Kato and attempted to introduce it into the plane linear cascade. In 1998, based on the preceding theory, Ren and Chang [61] proposed a singularity solution applicable to the nonlinear theory for the cavitation flow field of arbitrary binary thick cascade, which realized the hydrodynamics solution of the ternary cavitation flow in the runner. Meanwhile, the ternary cavitation area on the runner blade of a Francis turbine was predicted. The experimental comparison proved that the theory was reliable for the prediction of the cavitation region of the hydraulic turbine and had important engineering practical significance for the research of cavitation flow for hydraulic machinery.

With the rapid development of CFD, three-dimensional turbulent flow simulation of the Francis turbine has been widely used. In 2009, Huang et al. [62] established a three-dimensional model of full flow channel with the Francis turbine of a power station as the prototype, and conducted steady and unsteady numerical simulation. In the steady calculation, the standard k-ε turbulence model was selected to solve the Navier-Stokes equation so as to obtain the internal flow field distribution of each component under stable operating conditions, and the vortex structure and cavitation performance in the turbine were predicted. Based on the steady calculation results, the RNG k-ε turbulence model is selected to calculate the unsteady turbulence under different conditions. The results showed that the turbulence at the vortex shell and the runner was complex, and especially the turbulent motion in the runner was intense. In this study, the dynamic and static interference of unit components was considered, and the relatively real flow field was simulated. Therefore, three-dimensional turbulence calculation has become an important direction for the calculation and analysis of hydraulic turbines.

In 2011, Huang et al. [63] further carried out a three-dimensional numerical simulation of cavitation turbulence field for a Francis turbine on the basis of the preceding research. The research scope was a full-flow channel model for a Francis turbine of a hydropower station, with a grid-scale unit number of about 2.02 million. By adding the Schnerr and Sauer cavitation model considering the influence of cavitation, the steady calculation was carried out in the multireference coordinate system to obtain the flow characteristics and distribution of cavitation in the flow passage for the hydraulic turbine under specific working conditions, and the region and degree of cavitation that had occurred. The results show that, compared with the condition with large opening, the cavitation area inside the runner is larger and the risk of blade fracture is greater under the condition with small opening, but the pressure pulsation of the draft tubes has less influence. The cavitation conditions of different parts under specific working conditions are analyzed, which provides a safety guarantee for the production and operation of the Francis turbine. Compared with the single-phase fluid model, the numerical simulation of a two-phase mixture can better reflect the real flow in the unit and has good engineering applicability. However, the limitation of this study is that three-dimensional cavitation flow steady calculation cannot simulate the whole process of cavitation initial generation, development, and collapse, and unsteady numerical calculation is not carried out.

A typical cavitation phenomenon—that is, the blade vortex—will occur in the runner chamber when the Francis turbine deviates from the rated operating conditions, as shown in Fig. 6.4. The resulting pressure vibration will affect the hydraulic stability of the unit and even damage the unit components. Kumar et al. [64] conducted a study on the blade vortex simulation of the rotor inside

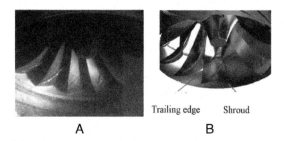

Trailing edge   Shroud

A   B

FIGURE 6.4   The shape of the leaf vortex [65]: (A) Experimental observation. (B) Numerical simulation.

the specific speed Francis turbine runner in the Fluent software environment. The study was based on the continuity equation and momentum equation, and the RNG k-ε turbulence model was applied. The results showed that under the condition of small flow, the pressure pulsation generated by the blade vortex and the vortex zone of the draft tube worked together, which greatly damaged the safe operation. There is almost no blade vortex under rated conditions. Under the condition of large flow, only the front of the runner blade appears in the passage vortex. In addition, based on the SST K-turbulence model and the Zwart cavitation model, Guo et al. [65] carried out numerical simulation of transient cavitation two-phase flow in a low-head Francis turbine model and verified the consistency with experimental results. The results showed that the cavitation in the runner of the blade vortex working area was periodically pulsating, and the water flow into the stern tube had a maximum velocity near the wall.

In recent years, as pumped-storage power plants have taken on increasingly important tasks in the power grid, the research on cavitation flow characteristics of pump turbines has become a hotspot in this field. Pump turbines often switch between two operating modes: pump operating mode and hydraulic turbine operating mode. The different operating modes cause differences in the internal flow field. For pump turbines, not only will cavitation occur in the runner chamber but also in the double-row cascade area under pump conditions, stall vortices are often formed, causing cavitation in the blocked flow path. However, there is some research on the cavitation characteristics of double-row cascade so far. In 2019, Zhai [66] conducted numerical simulation of cavitation flow in pump and hydraulic turbine conditions, respectively, to study the influence of environmental factors such as temperature and pressure on cavitation performance of the pump turbine. The control group was set by changing the flow rate, and the flow field in the double-row cascade, runner liquid flow angle, runner ring volume, and draft tube flow field were studied to analyze the energy characteristics. Then the variation of pressure field, blade load, and hydraulic loss under each cavitation coefficient was analyzed. The analysis on the characteristics of cavitation vortices in double-row cascades showed that the strength of stall vortices increased with the decrease of cavitation coefficient.

When considering the influence of temperature and pressure on the cavitation performance of the unit, the cavitation performance of the turbine was mainly affected by temperature and that of the pump was mainly affected by pressure. The distribution of hydraulic resources in China is uneven. The southwest region with rich hydraulic resources has a dangerous terrain, and the plateau low pressure for the pump turbine is a challenge. In addition, the severe cold climate in northeast China has also hindered the normal operation of pumped storage power stations. This study has considered the influence of environmental factors on the cavitation performance for the pump turbine, which is of great practical significance.

6.3 Intelligent evaluation and diagnosis technology

The fault diagnosis and condition assessment of hydropower units is a complex system engineering. Its purpose is to analyze and identify the current state and fault type of the unit systematically and accurately by obtaining the state information of the unit online on the premise of ensuring the safety of the unit so as to make a scientific and reasonable maintenance plan and to implement to improve the safety, availability, and economy of the unit. How to build a stable and efficient fault diagnosis and state assessment system for hydropower units, and further improve the management level of operation and maintenance of hydropower plants, is a hot topic of current research. Fault diagnosis and condition assessment is a subject integrating engineering mechanics, signal processing, artificial intelligence and model recognition, and system safety engineering. This chapter focuses on the theoretical research, commercial application system, and future development prospects related to the condition assessment and fault diagnosis of hydropower units. The structural relationship is shown in Fig. 6.5.

6.3.1 Current theoretical research hotspot

6.3.1.1 Research on intelligent fault diagnosis of hydropower units

The fault mechanism of hydropower units is complex, the fault types are various, and there are many fault coupling factors. These factors affect each other and are related to each other. A simple reasoning method is difficult to find out all the factors that lead to the fault accurately, and especially the contribution of different fault symptoms to the fault type discrimination is difficult to determine. In addition, due to the increasing complexity of the unit structure and the improvement of automation, it is impossible to rely solely on professional personnel to analyze and process the massive data obtained in the unit monitoring system. Therefore, it is necessary to improve the automation and intelligence of the fault diagnosis of hydropower units, and realize the efficient and reliable intelligent diagnosis of the equipment. In recent years, with the development of time-frequency

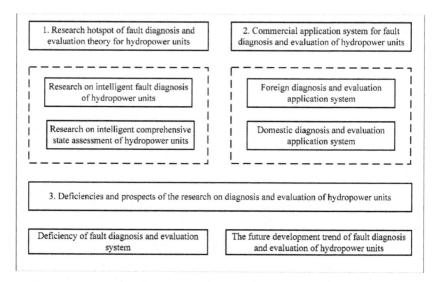

FIGURE 6.5 Logic block diagram of an intelligent fault diagnosis and evaluation of hydropower units.

signal processing technology, artificial intelligence, and other interdisciplinary subjects, scholars at home and abroad have put forward many effective fault diagnosis methods and achieved good application results. Generally speaking, there are two kinds of fault diagnosis methods: the data-driven fault diagnosis method and multi-information fusion fault diagnosis method. In the study of fault diagnosis of hydropower units, feature extraction is the key to fault diagnosis. By extracting useful equipment feature information from monitoring signals, feature parameters representing fault symptoms can be obtained, which provides the basis for fault diagnosis.

Research on fault feature parameter extraction.

Generally, the traditional feature extraction method of hydropower units is to extract the energy distribution features of the fault feature frequency band through statistical analysis in the time domain or calculation of the feature frequency band in the frequency domain. Due to the coupling of the hydropower unit structure and the variability of operating conditions, the actual measured vibration signals are complex and non-stationary [67]. In addition, the fault feature component is mixed with strong background noise and a large interference signal, showing a low signal-to-noise ratio. With the continuous development of signal processing technology, wavelet transform, empirical mode decomposition (EMD), and other nonstationary signal analysis methods and shaft orbit analysis methods have emerged one after another. These methods have laid a solid foundation for fault diagnosis and state evaluation of hydroelectric generating units:

(1) *Wavelet transform.* Wavelet transform uses different resolutions to approach the signal to be analyzed step by step, projects the signal into a set of orthogonal wavelet function subspace, and expands the signal on different scales so as to extract the characteristics of the signal in different frequency bands while retaining the original characteristics of the signal on each scale. Because of the good multiresolution characteristics of wavelet analysis, it is widely used in the fault feature extraction of hydropower units. Peng and Lui [68] proposed a method of using a wavelet packet to extract the fault features of hydropower units, and using support vector machine (SVM) as a classifier to diagnose the fault. Liu et al. [69] proposed a feature extraction method for vibration signals of hydropower units based on wavelet analysis and fuzzy relational symptom combination, and the extracted composite feature could reasonably describe the energy distribution of instantaneous signals. Gui and Han [70] defined the characteristic entropy of the wavelet packet by combining wavelet packet analysis and entropy, and used it as a characteristic parameter to characterize the severity of the fault of the tail water pipe to identify and estimate the degradation state of the tail water pipe. Peng et al. [71] used wavelet analysis and the correlation analysis method to analyze and study the cavitation erosion, crack, and collision wear state of the water turbine set, and obtained characteristic parameters suitable for turbine state monitoring and fault diagnosis. Shi et al. [72] combined the observation results of the cavitation state of the turbine model, carried out wavelet analysis of cavitation ultrasonic signals monitored under corresponding working conditions, and obtained the relationship between temporal and frequency domain changes of cavitation ultrasonic signals of the turbine.

(2) *Empirical mode decomposition.* Aiming at the defects of wavelet analysis, Huang et al. [73] proposed a new time-frequency analysis method—EMD, also known as Hilbert Huang transform—that can be used to analyze nonlinear nonstationary signals according to the local time-frequency decomposition of the signal, and has shown its advantages in the practical application of various mechanical fault diagnosis. Jia et al. [74] proposed the regression EMD method based on least squares SVM and applied the improved EMD to the fault diagnosis of the hydroelectric generating set. This method effectively suppressed the end effect and realized the accurate identification of the fault [74]. The method of integrating EMD and approximate entropy was applied to denoise the swing signal of hydrogenerator unit [75]. Xue et al. [76] used the method of EMD and index energy to effectively extract the dynamic characteristics of the pressure fluctuation signal of water turbine generator. Although a large number of research achievements have been made in the analysis of nonlinear

nonstationary signals, its basic mathematical theory has not been strictly proved, and there is no systematic and rigorous theoretical framework as the support [77].

(3) *Axis orbit*. The axis orbit is also an important feature of the failure of hydropower units. Many scholars have done a lot of research work on the automatic identification and classification of the axis trajectory. Qian et al. [78] proposed the purification of the axis trajectory by generalized particle swarm optimization. The analysis of the vibration signal retrieved from the purified trajectory map shows that the method has a good denoising effect. Wang et al. [79] introduced the improved image chain code to represent the axis track, making it have translation, scaling, and rotation without deformation, and classified it with probabilistic neural network. Zhang et al. [80] analyzed the fractal and geometric features of the axis track and proposed to use the box dimension, information dimension, correlation dimension in the fractal dimension, and the compactness and abundance in the geometric features to express the axis track eigenvector. Shi et al. [81] proposed a high-resolution spectrum for the purification of the axis track and accurately identified the axis track type through the moment feature and curve feature. Wang et al. [82] proposed a new method based on a novel pulse coupled neural network for the automatic identification of the axis trajectory, which has the characteristics of small calculation, fast speed, and high recognition rate.

The data-driven fault diagnosis method.

The purpose of fault diagnosis is to identify the current operation state of the unit, such as normal or fault, by monitoring and analyzing the operation data of the unit, then according to the fault status, further determine the location and extent of the fault, and develop appropriate unit maintenance strategy. Traditional fault diagnosis is mainly realized by artificial observation and analysis, which relies on necessary prior knowledge and expert experience, lacks the necessary theoretical and technical basis, and is difficult to ensure the accuracy of diagnosis, so it is greatly limited in practical application. With the continuous development and integration of artificial intelligence and big data analysis technology, more and more intelligent algorithms have been introduced into the field of fault diagnosis of hydropower units, and a series of research results has been achieved. The self-learning mechanism of the intelligent algorithm is used to realize unit fault diagnosis, reduce the artificial participation in the diagnosis process, effectively improve the diagnosis accuracy and calculation efficiency, and promote the intelligent construction process of a power station. Among them, the most widely used artificial intelligence diagnosis methods are the artificial neural network (ANN), SVM, and extreme learning machine (ELM):

(1) *Neural network*. As a typical representative in the field of artificial intelligence, the neural network has good nonlinear fitting ability and can fit any function in theory, which is suitable for processing massive sample data. ANN has received extensive attention in the field of mechanical equipment fault diagnosis because of its superior performance in fault tolerance, large-scale parallel processing, and self-learning [83]. Lu [84] proposed the method of vibration fault diagnosis of hydropower units based on the ant colony initialization wavelet neural network, combined with the advantages of ant colony algorithm to learn the parameters of wavelet neural network, overcome the defect that the algorithm is sensitive to the initial parameters, and obtain better diagnosis results. Li et al. [85] proposed a fault diagnosis method for hydropower units based on singular spectral entropy and a self-organizing feature map neural network for the nonstationary characteristics of vibration signals. Xie et al. [86], combined with the advantages of neighborhood rough set theory in reducing the redundancy of fault features, established a vibration fault diagnosis model of hydropower unit based on an improved neighborhood rough set and probabilistic neural network. According to the operation data of cavitation failure of hydropower units, Du et al. [87] adopted the integrated EMD to obtain the signal characteristics and input them into BPNN to accurately identify the types of cavitation failure under different guide vane opening.

The neural network has great potential in the field of equipment fault diagnosis. However, the neural network also has the following defects. First, the physical meaning of the neural network is not clear, and the data processing process is opaque, which belongs to the "black box" model. Second, the neural network training process needs a large number of training samples, which is difficult to meet the needs of small sample fault diagnosis of hydroelectric generating units.

(2) *Support vector machine.* SVM is a generalized linear classifier based on the principle of structural risk minimization proposed by Cortes and Vapnik [88]. SVM maps the sample data to a high-dimensional space, solves the maximum classification interval, and then obtains the optimal hyperplane. In addition, SVM is widely used in the study of fault diagnosis for all kinds of data samples, especially for small samples, due to its strong generalization ability of simple modeling theory. Zhang [89] combined the fuzzy sigmoid kernel function with SVM, proposed the improved multiclass fuzzy SVM method, and successfully solved the fault diagnosis problem of hydropower units. Peng et al. [90] applied the least square SVM and information fusion technology to diagnose the vibration fault of hydropower units and obtained a high diagnosis accuracy. Zhang et al. [91] used the method of variable mode decomposition to obtain the vibration fault characteristics of hydropower units, and used it as SVM input to realize the identification and diagnosis of unit fault modes. Cheng et al. [92] extracted the vibration signal features of hydropower units in different states, and used

SVM and a neural network classifier to diagnose the reduced dimension features. SVM is very suitable for small sample fault diagnosis. It has many advantages, such as high accuracy, strong robustness, and strong generalization ability. However, SVM also has some certain limitations. For example, it is not suitable for dealing with those uncertain systems. Therefore, it is difficult to realize the quantitative diagnosis that can better reflect the real fault state of the unit.

(3) *Extreme learning machine.* ELM is a single hidden layer feedforward neural network model proposed by Huang et al. ELM is based on the principle of minimum root mean square of error and seeks for the optimal weight coefficient of the output layer. It has superior approximation ability and classification performance, and is favored by many scholars. Xiao [93] introduced ELM into the fault diagnosis of hydraulic turbine generator set, proposed a two-stage evolutionary limit learning machine diagnosis algorithm, and achieved good diagnosis results. Luo [94] proposed a hybrid gravity search ELM model, which can realize the fault diagnosis of hydropower by integrating signal processing, multidimensional feature extraction, and parameter optimization.

Although ELM can complete the training quickly, as a neural network with single hidden layer structure, it is only suitable for small labeled datasets and cannot complete complex or unsupervised learning tasks. However, due to the weight and bias of the random initialization input, the parameters obtained by the network are not considered as the global optimal solution, which is prone to the problem of underfitting.

Research on diagnosis technology based on multi-information fusion

(1) *Expert system.* The expert system is an intelligent computer program system that can use the knowledge and experience of human experts in the field to infer and judge the fault process through a series of rules, simulate the decision-making process of human experts, and then realize fault location and give corresponding decision-making suggestions. The system contains a lot of expert knowledge related to the problem to be solved, which improves the efficiency and ability of effective analysis of the problem. Therefore, the intelligent analysis method based on the expert system has been widely used in the field of mechanical equipment fault diagnosis. According to the characteristics of hydropower units, Wang et al. [95] constructed an expert system for vibration fault diagnosis of hydropower units with a modularization idea, which realized the intelligent diagnosis of unit fault. By analyzing the operation characteristics of hydropower units, Mao et al. [96] collected the characteristics and causes of faults occurring during the operation of hydropower units, constructed a diagnosis knowledge base, and formed a fault diagnosis expert system for hydropower units so as to provide guidance for the operation and maintenance of hydropower units. On the basis of the HM9000 hydropower unit remote online monitoring and analysis system, Zhou et al. [97] developed an open expert system software platform for fault diagnosis of hydropower units, which fully integrated the diagnosis experience of experts in the field and realized accurate fault diagnosis with the construction of the Three Gorges Diagnosis Center.

The fault diagnosis method based on the expert system has some advantages in solving the problems that are difficult to describe with the mathematical model, but the establishment of an expert knowledge base and reasoning mechanism has always been the difficulty in the diagnosis process. In addition, to improve the analysis ability of the expert system, how to effectively maintain and update the knowledge base is also a problem worthy of in-depth study.

(2) *Fault tree analysis.* Fault tree analysis is a safe and reliable fault diagnosis and analysis technology. It takes the undesired fault condition of the system as the top event of fault analysis, decomposes it layer by layer according to the logical structure relationship of the system until all of the bottom events leading to the fault are found, and then links each event with its corresponding logical relationship to form a tree diagram of the relationship between the fault and the event. Li et al. [98] took the vibration fault of the hydro-generator unit as the research object, studied and analyzed the logical relationship between the components of the unit equipment in the structure, and established the fault tree of "vibration of hydro-generator unit." Hu and Xiao [99] transformed the existing fault tree structure of the generator fault model into the Bayesian network model, providing the basis for reliability design and fault diagnosis of a complex system. Liu [100] studied and constructed the fault tree model of a water pump turbine, calculated the fault tree model, obtained the fault of unit equipment and its probability, and effectively guided its safe and stable operation.

It is intuitionistic and simple to search the fault source by the fault tree method, but the key to fault tree diagnosis is to build on the correct fault tree structure, which is not always guaranteed to be correct in actual diagnosis. Once the fault tree is not built completely or correctly, the diagnosis result will become unreliable. In addition, fault tree analysis usually requires that the bottom event and top event of the system are determined—that is, there are only two states of normal and fault, so the fault tree cannot diagnose uncertain or fuzzy fault.

6.3.1.2   Research on intelligent comprehensive state assessment of hydropower units

In the operation process of hydropower units, the operation life is a dynamic gradual process from health performance degradation fault failure. Therefore, it is necessary to establish perfect health samples in the equipment health stage to realize the health assessment and performance degradation prediction in the later unit operation process. At present, the state assessment of hydropower units is generally based on national standards or industry regulations at home and abroad. The commonly used evaluation methods of unit operation health include the limit value evaluation standard, statistical evaluation standard, analogy evaluation standard, and fuzzy comprehensive state evaluation method.

The limit evaluation method.

The limit value evaluation standard of hydropower units is based on the national standards, industry specifications, and international standards issued by the International Standardization Association (ISO), among others, and by comparing the monitoring vibration value of the unit to see if it exceeds the limit value, it is regarded as the evaluation standard of unit performance. At present, GB/T7894-2001 fundamental technical specifications for hydro-generators and GB/T8564-2003 specification installation of hydraulic turbine generator units are mainly adopted in China, which stipulate the allowable vibration value of each part of the unit; DL/T817-2002 technological code for maintenance of vertical hydro-generators and GB/T8564-2003 specification installation of hydraulic turbine generator units have been in line with international standards in terms of vibration standards, the influence of rotating speed is also considered [101–104], and the detailed evaluation criteria are shown in Table 6.4.

GB/T11348.5-2002 mechanical vibration of nonreciprocating machine measurements on rotating shafts and evaluation criteria and GB/T15468-2006 fundamental technical requirements for hydraulic turbines further stipulate the evaluation limit of spindle swing. The operating speed and swing amplitude are divided into four areas: A, B, C, and D. Areas A and B can meet the long-term stable operation of the unit, area C is not suitable for long-term continuous operation, and area D is prohibited from operation. Its vibration can damage unit components [105], [106].

The statistical evaluation method.

The traditional limit value evaluation method considers that when the measured value of the unit operation monitoring data exceeds the threshold value or limit value, the unit status is abnormal. However, the condition monitoring data of hydropower units are affected by the fluctuation of operating conditions, the sensitivity of monitoring instruments, and random interference, so it is impossible to determine whether the performance of the equipment is actually degraded. As the monitoring data often have the statistical distribution characteristics that obey the normal distribution law, the alarm value can be determined according to 3σ (Leite criterion)—that is, the alarm limit. Therefore, Pan [107] proposed a statistical evaluation standard considering the distribution of monitoring data, using the reference value and alarm limit value established by a sufficient number of monitoring samples under the normal operation condition of the previous

TABLE 6.4 Allowable vibration values of various parts of the water turbine generator set.

Unit type	Project		Rated speed (n) r/min			
			$n < 100$	$100 < n < 250$	$250 < n < 375$	$375 < n < 750$
Vertical units	Water turbine	Head cover horizontal vibration	0.09	0.07	0.05	0.03
		Head cover vertical vibration	0.11	0.09	0.06	0.03
	Hydro-generator	Vertical vibration with thrust bearing bracket	0.08	0.07	0.05	0.04
		Horizontal vibration with guide bearing bracket	0.11	0.09	0.07	0.05
		Horizontal vibration of frame at stator core	0.04	0.03	0.02	0.02
		Stator core vibration (100 Hz, double amplitude)	0.03	0.03	0.03	0.03
Horizontal unit		Vertical vibration of each bearing	0.11	0.09	0.07	0.05
		Axial vibration of thrust support	0.10		0.08	
Bulb tubular unit		Radial vibration of each guide bearing	0.12		0.10	
		Radial vibration of bulb head	0.12		0.10	

Note: Vibration value refers to the double vibration value of the unit under various stable operation conditions except for over speed operation.

unit as the health sample of the characteristic quantity. If no abnormality occurs to the equipment—that is, the monitoring value exceeds the limit specified by the health sample—the probability of abnormal operation of the equipment is 99.73%. At the same time, to facilitate the application of engineering, when processing samples, the samples are divided according to the operation conditions in advance, and the monitoring samples after the division are statistically analyzed to get the ideal health samples. The solution of the reference value is to obtain the average value of a certain characteristic quantity in sample space (the characteristic quantity can be the average value, amplitude, frequency, etc., of the monitoring quantity) as the health value of its operation state through the monitoring data accumulated by the unit under normal operation for a long time, as shown in Eq. (6.9):

$$\bar{X} = \frac{(X_1 + X_2 + \cdots X_n)}{n}. \tag{6.9}$$

The analogy evaluation method.

In the statistical evaluation standard, although it overcomes factors such as the limitation of traditional evaluation standard of pumping units and sensitivity to the change of operating conditions, it only uses its own historical data instead of using the information between the same units for performance evaluation. The analogy evaluation standard is to comprehensively use the same monitoring measurement of different units when multiple units are operating under the same operating conditions. The comparison information of the monitoring data in different historical periods of a single unit and the comparison information of the monitoring data in different historical periods are used to evaluate the performance of the unit so as to accurately obtain the current operation status of the unit [108].

This method measures the distance between the health state and the current state. The common distance measures include Euclidean geometric distance and dynamic time warping distance. Euclidean geometric distance is the most basic and intuitive way. The expression of Euclidean geometric distance is very simple, which is to calculate the square root of the sum of the squares of two time series differences. For a group of unit health feature samples $X = \{x_1,\ldots, x_n\}$ and the current state feature samples $Y = \{y_1,\ldots, y_n\}$, the Euclidean geometric distance is

$$D(X, Y) = \left(\sum_{t=1}^{n} |x_t - y_t|^2\right)^{1/2}. \tag{6.10}$$

After defining the similarity distance measurement formula, the process of establishing the analogy evaluation strategy is shown in Fig. 6.6.

Because the analogy evaluation standard makes good use of the performance difference information between the same unit and the difference information of different historical stages of a single unit, it can well avoid the defects that

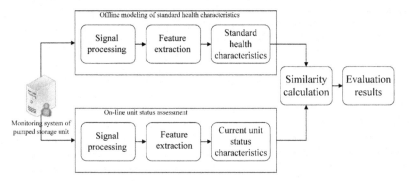

FIGURE 6.6   Flow of state assessment of a hydropower unit based on an analogy algorithm.

the current limit evaluation standard is sensitive to operating conditions and the degree of performance degradation is difficult to quantify.

6.3.2   Commercial intelligent evaluation and fault diagnosis systems

With the continuous development of power production processes, the complexity of hydropower units is getting higher and higher, and the coupling among water, machinery, electricity, and other factors is also getting stronger and stronger. There is an urgent need for an effective system to monitor, evaluate, and diagnose the real-time performance and status of hydropower unit equipment so as to take effective measures to ensure the normal, lasting, and stable operation of the equipment. Therefore, many universities and research institutes are carrying out research and development of fault diagnosis systems.

6.3.2.1   The foreign diagnosis and evaluation application system

Canada and the United States have significant advantages in large-scale power station online monitoring technology and system research. The ZOOM2000 online monitoring and diagnosis system developed by Canada VibroSystM for hydropower units can monitor the air gap and magnetic field strength of the generator stator and rotor, unit vibration, swing, pressure, temperature, and other unit states online, and has certain equipment performance evaluation and fault diagnosis functions [109]. The gems system jointly developed by Ontario Hydropower Bureau of Canada and Sri International under the American Academy of Electrical Sciences is a unit fault diagnosis system that can comprehensively diagnose the faults of the generator stator winding and core, excitation system, and seal oil system, among others. [110]. Westinghouse Electric used artificial intelligence technology for online monitoring of units, and gradually developed into a large-scale online monitoring and diagnosis system for power plants, and established a fault diagnosis center in Warrendo [111]. Many enterprises in Europe and Japan have also made some achievements in multiparameter integrated monitoring. Typical products, such as the CAMPASS system of B & K Company in Denmark, monitor vibration parameters and unit process parameters, and have relatively extensive fault diagnosis function [112].

6.3.2.2 The domestic diagnostic and evaluation application system

Due to the long-term use of planned maintenance system in China, the emphasis on online monitoring technology of equipment is not enough. Compared with foreign related research work, it started late. However, based on the introduction and absorption of foreign advanced technology, the development process is relatively rapid. Many domestic universities, scientific research institutions, and enterprises have developed quite effective online monitoring systems. After years of effort, the state monitoring and analysis system developed in China has played an important role in China's power industry and achieved good economic benefits:

(1) *HM9000: Comprehensive analysis system for condition monitoring of hydropower units.* China Hydropower Research Institute has developed the HM9000 hydropower unit remote online monitoring and analysis system (HM9000 system). Based on condition monitoring, the system combines local and remote monitoring to provide an efficient and intelligent operation condition evaluation platform for condition-based maintenance of hydropower plant equipment [113]. In addition, the HM9000 system also provides detailed data analysis tools that can analyze, process, and report the real-time or long-term status data in detail, and provide technical support for promoting the informatization and automation of power plant management. At present, the HM9000 monitoring comprehensive analysis system has been successfully applied in Taipingwan Hydropower Station in northeast China. Through long-time operation and use, the results of the HM9000 monitoring analysis system are compared with other independent monitoring results, and the accuracy and reliability of the HM9000 system are verified.

(2) *TN8000: Fault diagnosis system for condition monitoring and analysis of hydropower units.* The state monitoring analysis fault diagnosis system of the TN800 hydropower unit can collect and analyze state parameters such as vibration swing, pressure fluctuation, air gap, magnetic field strength, and partial discharge of the unit in real time, identify the state of the unit in time, and find the early symptoms of the fault. At the same time, it can make a judgment on the cause, severity, and development trend of the fault so as to avoid the occurrence of destructive accidents, and provide a solid technical basis for the realization of condition-based maintenance of the unit [114]. The TN8000 hydropower unit condition monitoring and analysis fault

diagnosis system has been applied to a large number of large and medium-sized hydropower units, such as the Three Gorges Left Bank Power Station, the Three Gorges Right Bank Power Station, and the Beijing Shisanling Pumped Storage Power Station, which play an important role in ensuring the safe and stable operation of the units.

(3) *PSTA2003: State detection and tracking analysis system.* The PSTA2003 state detection and tracking analysis system is a diagnosis and evaluation platform jointly developed by Tsinghua University and Beijing Orient General Electrical Technology Institute. Through online monitoring of the vibration, swing, pressure pulsation, air gap between the stator and rotor of the hydropower unit, and combining the monitoring information of the computer monitoring system, it uses various analyses, diagnosis strategies, and algorithms to establish a full-featured system. The tracking and analysis system with strong practicability has the functions of monitoring, alarming, state evaluation, and fault diagnosis, among others, which can grasp the state of the unit in real time, and provide strong technical support for safe operation, optimized scheduling, and maintenance guidance [115].

6.3.3 Application status and deficiency of current diagnosis systems

At present, most of the systems that have been put into operation in China are relatively mature in terms of unit condition monitoring technology, whereas in terms of unit fault diagnosis and intelligent evaluation, they only stay in the diagnosis and analysis of a certain link or part of abnormal state of the unit. In addition, the real time and accuracy of diagnosis are difficult to guarantee. Therefore, to achieve the application goal of hydropower unit fault diagnosis and transform science and technology into productivity, more in-depth research on fault diagnosis and evaluation technology is needed. In recent years, China has invested a lot in the field of intelligent assessment and fault diagnosis technology of hydropower equipment, and achieved a lot of research results. In theory, it has reached the international level, but there is still a certain gap in application technology with foreign countries.

6.3.3.1 Deficiency of the fault diagnosis and evaluation system

The fault diagnosis and evaluation system of hydropower units is a framework platform composed of a software and hardware system. Its diagnosis ability and application effect depend on the diagnosis knowledge converted into computer language, including the fault model, fault sample, and expert experience, among others. In the intelligent fault diagnosis system, every type of fault must be able to describe the fault according to the program rules of the diagnosis system, draw the fault reasoning or classification flowchart, give the corresponding symptom input and diagnosis result output, and input them into the knowledge base so that the computer system can realize the intelligent diagnosis. Therefore, the premise of the intelligent diagnosis system is to give the computer language description of the fault reasoning or classification clearly. Based on the limitations of the computer diagnosis system, there are three kinds of problems for the possible faults of the hydropower units:

(1) *Information island problem.* A large number of basic data used for fault diagnosis are scattered in different application systems in isolation, forming an information island. It will be difficult for a monitoring or diagnosis system to access a certain information subsystem data source. Therefore, it is an important measure to integrate the monitoring information, business management information, and other data of hydropower unit equipment, build a unified data platform, and provide a basic data and analysis platform for the fault system and auxiliary decision making.

(2) *Diagnosis knowledge management is closed and low utilization rate.* At present, the fault diagnosis system for hydropower generators lacks effective organization and management of expert knowledge. A large amount of experience knowledge and failure cases distributed in different business systems are difficult to integrate and share effectively. Valuable expert knowledge is not fully utilized, and it is difficult to play its true value. In addition, most existing diagnosis systems are closed. After the system is completed, it is difficult to supplement and enrich the diagnosis knowledge resources. Therefore, there is an urgent need for a service-oriented and developing network knowledge management platform to integrate the knowledge resources of different users in different places, and solve the equipment failure in a timely manner and collaboratively.

(3) *Failure prediction and health management is not available.* In view of the degradation of the performance and the gradual change of the failure of the hydroelectric unit equipment, fault prediction and health management technology can effectively change the current maintenance mode of the hydroelectric unit equipment by adopting prediction methods such as model- and data-based methods, fault prediction, fault management, planning maintenance, and auxiliary decision making. In other words, the maintenance mode can be changed from "breakdown maintenance" to "condition-based maintenance" and "predictive maintenance," which makes the fault diagnosis system of the hydropower unit change from passive diagnosis to an active prevention model.

6.3.3.2 The future development trend of fault diagnosis for hydropower units

The generation and development of hydroelectric unit fault is the coupling effect of many factors, such as hydraulic, mechanical, electromagnetic, and structural, which contain a lot of uncertain factors. It is difficult to get satisfactory diagnosis results by conventional diagnosis methods. At present, the research on the fault diagnosis system of hydropower units is still in the stage of continuous innovation and development. Many achievements have been made in the diagnosis theory and method, and the set of a perfect and expandable application system platform of diagnosis technology has not been formed. In the next step, we should focus on the application research of the fault diagnosis technology of hydropower units, and research and develop the practical diagnosis system of the project; its application prospect will be broad:

(1) At present, the online monitoring and fault diagnosis system of the unit can accurately perceive and identify certain faults, yet the fault of the hydropower unit has developed to the middle and late stage, the fault characteristics are obvious, and the performance of the components has been seriously degraded. Although the current diagnosis method can avoid the occurrence of major accidents in time, the unit has suffered serious wear and performance degradation at this time, and it often needs to be overhauled or updated in the end. The significance of predictive maintenance for fault diagnosis of hydropower units is lost. Therefore, it is necessary to carry out in-depth research on the feature extraction method of early fault of hydropower units under the background of strong noise, reveal the mapping relationship between different fault states and weak fault symptoms in the fault evolution process, carry out the early performance degradation evaluation system of hydropower units, grasp the fault state and its dynamic evolution process in time, and take timely remedial measures for the early fault. It is of great engineering significance to prevent degradation failure of group performance.

(2) Based on the massive data of the large-scale online monitoring system, combined with new signal processing technologies such as EMD and information entropy, an evolutionary search framework for nonstationary feature extraction and optimal feature selection is constructed, which can fully represent the nonstationary symptom set such as fault location and damage degree, accurately capture the location and impact degree of fault occurrence, and realize the transformation from qualitative diagnosis to definite diagnosis. The breakthrough of quantitative diagnosis reveals the occurrence, development, and evolution of the fault state of hydropower units, thus providing basic basis for the safety analysis, reliability evaluation, and life prediction of hydropower units.

(3) Hydraulic fault, mechanical fault, and electromagnetic fault occur in the form of parallel or cascading, and their fault characteristic signals are coupled with each other, which becomes a pattern recognition problem of multiple faults. However, the existing diagnosis system of intelligent hydropower unit is still weak in multifault concurrent diagnosis, and most of the intelligent methods need to meet the limitations of a large number of training samples and artificial parameters. Therefore, it is necessary to

build a multimodel hybrid intelligent diagnosis technology with complementary performance, through a small number of expert mark faults and a large number of online monitoring samples for self-learning semisupervised integrated diagnosis, to achieve the separation of multifault coupling features of hydropower units, and finally build an accurate fault diagnosis with rich knowledge, correct reasoning, and a reliable conclusion.

(4) At present, large and medium-sized hydropower stations have installed online monitoring and fault diagnosis systems for hydropower units. However, these systems all have the defects of the data island and single service. It will be a future research direction to build a multisource, heterogeneous, and massive information distributed diagnosis system and realize the cross-discipline, cross-basin, and cross-platform remote intelligent diagnosis. With the help of cloud computing technology with large-scale distributed computing and storage capacity, the knowledge integration and data sharing of water motor groups in the basin can be realized; with the increasing popularity of mobile terminals and wearable devices, the engineering and technical personnel of hydropower plants can obtain the performance information and corresponding maintenance strategies of unit equipment at any time and place, and complete the regular point inspection and condition monitoring seamless integration to achieve reliable, economic, and efficient operation of power plants.

6.4 Prognostics research

During the operation of the hydropower unit, condition monitoring obtains the operating parameters of the equipment under different working conditions through collection, measurement, and so on, and uses fault diagnosis technology to evaluate and diagnose it, then realizes the analysis, judgment, and prediction of the operation state of the hydropower unit. To ensure that hydroelectric mechanical equipment can reliably and effectively realize its function in a certain working environment and during a certain continuous working period, regular maintenance is changed to predictive maintenance, and fault prediction technology is the main research direction in the field in the future.

The state trend prediction technology is to make a judgment on the working state of the equipment by analyzing the historical trend of the unit state parameters. Its main contents include the following: (1) predict whether a failure may occur in the future, (2) estimate how long the failure will occur in the future, and (3) determine the probability of possible failure in the future. For decades, fault prediction has been applied in the field of rotating machinery. The main research focus is to predict the remaining useful life of the equipment—that is, to predict the remaining time until the failure occurs according to the current state of the equipment and historical maintenance records. Compared with fault diagnosis, research on fault prediction is relatively backward. The fault prediction of the hydroelectric generating set is also at the preliminary research stage.

At present, time series forecasting is a research hotspot in the field of state forecasting. Most of the forecasting of the state trend of unit equipment can be classified as time series modeling and forecasting. Through reasonable assumptions to analyze the historical data and real-time observation values of the unit equipment system, the corresponding time series dynamic relationship model can be established. On the basis of the model, the development trend of the time series can be predicted. Prediction theory can be divided into four categories: prediction based on the classic linear model, prediction based on intelligent technology, prediction based on fuzzy set theory, and combined prediction.

6.4.1 Prediction based on the classical linear time series model

In 1927, Yule was the first to study time series based on an autoregressive model. Based on this, Wold, a Swedish statistician, established a sliding smooth moving average model and a linear autoregressive moving average model of stationary time series. Box et al. [116] proposed the time series analysis method in "Time Series Analysis: Forecasting and Control" and gave an autoregressive integrated moving average model for handling nonstationary time series. This model is more representative Highlight and apply the more general linear model. Since then, based on this model, many experts and scholars have conducted in-depth explorations and concluded many models with outstanding influence, such as the conditional heteroscedastic ARCH model and the threshold autoregressive model, and apply it to the prediction of time series. Fu et al. [117] predicted the water demand of well-irrigated rice through the self-excitation threshold autoregressive model. The model has high prediction accuracy and small calculation amount. Jin et al. [118] optimized the threshold and autoregressive coefficients by the genetic algorithm to establish the genetic threshold. The autoregressive model has a very wide application value in the prediction process of various meteorological nonlinear time series.

The preceding prediction model has been extensively studied and widely used in the time series state prediction process, but many shortcomings remain, which are specifically manifested in the following three aspects. First, the prediction based on the classic linear model is mainly for the linear time series as the object of study, and in engineering practice, most of the data have strong nonlinearity, which makes the classic linear model based on the prediction result relatively rough. Second, for imperfect, missing, and ambiguous time series data, the classic linear model is more difficult to process. In addition, if there is less historical data, the linear model built based on statistical methods also has certain deficiencies in prediction. Third, the prediction results of the classic linear model have quantitative characteristics, but some of the actual phenomena often have ambiguity and randomness. For example, people often use fuzzy language such as "up" and "down" to describe the change of the stock index. When it comes to precise numerical values, nonprofessionals cannot deeply understand the meaning, and fuzzy semantics are relatively easy to understand.

6.4.2 Time Series Prediction Based on Intelligent Technology

In actual engineering, linear time series and nonlinear time series often coexist, and modeling linear or nonlinear time series alone is not applicable. Because of its outstanding adaptability, the time series forecasting model based on intelligent technology has become an ideal choice for solving time series forecasting problems. At present, time series prediction based on intelligent technology has become a research hotspot, specifically involving many intelligent technologies such as ANN, SVM, genetic algorithms, Bayesian networks, and information granularity:

(1) *Neural network-based prediction.* Since the late 1980s, neural networks have been widely used in the field of time series prediction, and many scholars have conducted relevant research and achieved many results [119–123]. Velasquez [119] proposed a new method for nonlinear time series prediction using an adaptive multidimensional neural-fuzzy inference system. Based on phase space reconstruction and the BP neural network, Pei [110] conducted real-time prediction of horizontal vibration and hydraulic conductivity of the upper frame of the turbine. Chen et al. [121] predicted the vibration trend of the lower frame of the turbine based on the LSTM network. He et al. [122] designed and trained the wavelet network based on the hierarchical evolutionary algorithm based on hierarchical structure chromosomes to predict the operation state of the hydraulic unit. Zhu and Bie [123] predicted the hydraulic turbine regulation control based on the model identified by the NNARX turbine regulation system. Sun et al. [124] took the Francis99 high-head Francis model turbine as the research object, and used the TBR model to predict its hydraulic performance and load. Based on the distributed entropy discrimination and ELM segmented prediction model, Zhang [125] effectively predicted the state trend of the turbine unit. Liu [126] proposed a multipart nonlinear trend prediction method based on variational modal decomposition and the convolutional neural network to predict the vibration trend of the top cover of the hydraulic turbine.
Neural networks are widely used in time series prediction mainly for the following reasons. First, neural networks are nonparametric prediction methods, and they do not need to make corresponding assumptions about research problems. This is extremely critical, because many time series data cannot be clearly generated from the internal data mechanism and cannot make assumptions in advance. Second, the neural network is a typical nonlinear model, and the time series data in the project are often nonlinear,

and the neural network can identify the nonlinear functional relationship well. Third, the theoretical approximation effect of the neural network is more prominent. As long as the number of hidden layers and neurons is sufficient, it can be approximated with arbitrary precision for any function. The neural network prediction model also has certain defects in the time series prediction process, which is mainly manifested in the following ways: (1) how to choose a reasonable neural network structure—that is, how to determine the number of input layer and hidden layer neurons—currently mainly depends on the specific situation; (2) the prediction results are not interpretable, and the prediction process is a "black box"; and (3) the prediction process takes a long time, and it is easy to fall into a local minimum.

(2) *SVM-based prediction.* In terms of data analysis, SVM is another important tool. In 1963, Vapnik [127] proposed such a machine learning method based on statistical learning theory. The SVM method can effectively deal with time series prediction and classification problems. In recent years, SVMs have been widely used in time series prediction. In the prediction process of SVMs, the nonlinear function is transformed into a linear problem through the kernel function, so the determination of the kernel function is the key that determines the accuracy and effectiveness of the prediction result. Research scholars at home and abroad have done a lot of work in the theoretical research and practical application of SVMs. In terms of theoretical research, the regression algorithms and principles of support vector machines are elaborated and explored, such as improving training algorithms and constructing kernel functions. In terms of practical application, it is widely used in many fields such as regression analysis and time series forecasting. Tang et al. [128] combined iterative error correction and least squares SVM to automatically obtain the optimal parameters of the prediction model from a simple and effective pattern search algorithm, and proposed a time series least squares support vector based on the iterative error correction prediction algorithm. Pei [120] constructed the SVR regression model created by the sample set training through the item space reconstruction method to predict the upper frame of the turbine and the hydraulic conductivity. Zhu [129] used the SVM model based on gray theory to predict the multinozzle oblique impact turbine. Peng [130] adopted data dependency based on Riemannian geometry to improve the method for RBF kernel function in SVM, and predicted the stability state of hydropower units. Hui [131] combined fuzzy clustering technology with SVM to predict the state of the turbine unit. Fu [132] proposed a prediction method based on aggregated EEMD and SVM regression to predict the vibration trend of hydroelectric generating units.

Compared with neural network technology, SVM adopts the principle of structural risk minimization, and its learning and generalization capabilities are more prominent. It can effectively solve problems such as local minimums and dimensional disasters, but it is more sensitive to missing data and larger samples.

(3) *Chaos theory.* The nonlinear prediction method based on chaos theory is another extremely important research field. Its theoretical basis is the phase space reconstruction theory proposed by Packard in 1980. The time series data is reconstructed into the corresponding nonlinear chaotic model through phase space reconstruction. Compared with traditional prediction, phase space reconstruction prediction can obtain more accurate prediction results by using the rich information in the time series. Such prediction methods include the Lyapunov exponential prediction method and the weighted zero-order local method, among others [133]. When making short-term predictions, the prediction results of the preceding methods are ideal, but because chaotic motion is particularly sensitive to initial conditions, it is difficult to achieve long-term accurate prediction of time series data. Therefore, many scholars consider improved chaotic time series prediction methods. Based on the orthogonal basis neural network, combined autocorrelation method, and Cao method, Li et al. [134] reconstructed the phase space of chaotic time series, transformed the model parameter optimization problem into the function optimization problem in multidimensional space, and realized the chaotic time sequence single-step and multistep prediction. In the study of some reservoir bank landslide displacement prediction process in the Three Gorges Reservoir area, based on the exponential smoothing method, the multivariable chaos model, and ELM, Huang et al. [135] proposed a better combined landslide displacement prediction model. Gao [136] comprehensively improved the SVM and chaos theory to predict the time series of water content and achieved good results. Wu [137] combined the RBF neural network prediction model with phase space reconstruction to perform single-step and multistep predictions on the extracted chaotic time series, thereby realizing the prediction of the operating state of the Francis turbine regulation system.

The relevant research on the prediction method of chaos theory provides important tools for the prediction of time series with chaotic characteristics, but it also has shortcomings: it is sensitive to the initial value and cannot achieve long-term prediction. Therefore, further research is needed on chaotic prediction theory and its combination with other prediction technologies to explore effective models with high prediction accuracy, fast convergence, and long-term prediction.

(4) *Bayesian network–based prediction method.* The Bayesian network is also widely used in time series data prediction. In 1988, Pearl proposed the Bayesian network, which is an important tool for exploring reasoning and uncertainty problems in the field of intelligence. The Bayesian network derives the probability information of the variable of interest according to the strength of dependence between the variables in the network, and gives

reasoning or judgment literature based on uncertain and incomplete information [138]. Regarding the application of Bayesian network for prediction, Wang [139] systematically described the time series model of the Bayesian statistical inference method, which verified the superiority of the prediction result after using the Bayesian inference method. Zhang et al. [140] applied the sparse Bayesian method to the prediction of chaotic time series. The sparse Bayesian method not only has the performance of SVM but also uses fewer kernel functions than SVM and achieves a better prediction effect. Das and Ghosh [141] incorporated spatial semantics as a form of domain knowledge into the standard Bayesian network and proposed a Bayesian network variant multivariate prediction method that is not easily affected by the uncertainty of parameter values. Based on the multi-information fusion optimization model, Xiao et al. [142] proposed a dynamic Bayesian network that can perform multistep prediction of time series and has high efficiency and robustness using the optimal theorem and dynamic Bayesian network. Bayesian network is a more effective method of uncertainty reasoning and forecasting in the fields of financial engineering, learning prediction, and statistical decision making, but its application is restricted due to its greater influence on the prior probability.

In addition to ANNs, SVMs, chaos theory, and Bayesian networks, there are other intelligent technologies in time series prediction, but these technologies are mostly used alone or in combination with other methods, scattered for local prediction of time series in the process. For example, based on time series data, knowledge rules in the form of IF-THEN are mined through rough sets to predict the change trend of time series [143]. Fuzzy decision trees are an effective classification technique in classification problems that can also be used in time series. Used in forecasting, this technology is similar to rough sets and uses the extracted prediction rules to predict time series [144]. In addition, in time series prediction, there are particle swarm optimization, the genetic algorithm, and dynamic programming algorithms [145–147].

The biggest advantage of intelligent technology is that it can capture the non-linear characteristics of time series data well so as to have a more comprehensive and profound understanding of time series data. In recent years, literature that uses intelligent methods and intelligent techniques to predict time series continues to appear, but there is currently no method applicable to any problem. Various methods have advantages and disadvantages, prompting many experts and scholars to continue to explore the prediction methods, improve, and optimize.

6.4.3   Fuzzy time series prediction based on fuzzy set theory

The prediction results of the classic linear model have quantitative characteristics, but some of the actual phenomena often have ambiguity and randomness. For example, people often use fuzzy language such as "up" and "down" to describe the change of the stock index. For this reason, Zadeh put forward the theory of fuzzy sets in 1965. The membership of fuzzy sets for each element is not only a binary relationship but also a subordinate relationship. The emergence of fuzzy set theory allows people to mathematically deal with the fuzzy phenomena in real life, which is of great significance to production and scientific research. According to the proposed fuzzy set theory, Zadeh established a prediction model to deal with ambiguity problems, which also laid the foundation for the establishment of subsequent fuzzy time series prediction models [148], [149]. Based on the theory of fuzzy sets, Song and Chissom integrated the concept of fuzzy sets into the analysis of time series in 1994, defined the fuzzy time series, and established a prediction model based on the fuzzy time series. The model mainly has four steps: (1) define and divide the discourse: collate data, determine, and divide the discourse; (2) definition of fuzzy sets and fuzzification of data: define fuzzy sets in the discourse domain, and fuzzify sample data based on fuzzy sets; (3) construct fuzzy prediction rules: establish fuzzy relationships of sample data, and construct fuzzy prediction rules according to all fuzzy relationships; (4) prediction and defuzzification: prediction is made by fuzzy rules, and the prediction result is obtained by defuzzifying the prediction result.

Since then, more and more scholars have devoted themselves to the theory and application of fuzzy time series, mainly including the improvement of the traditional model from the division of the discourse domain and the extraction of fuzzy rules, and the comparison with the classical time series analysis, and achieved fruitful results.

For example, Zhu [150] combined fuzzy system theory and the RBF neural network to propose a method for fault prediction of hydroelectric generating units based on constrained generalized dynamic fuzzy neural network, which achieves accurate fault prediction for generating units. Liang and Ni [151] discussed the fuzzy random variables and their digital characteristics of the cavitation system, and proposed a fuzzy random theory prediction method for the cavitation state. Zou [152] took the prediction of the performance deterioration trend of pumped storage units as a breakthrough and, based on the improved interval type 2 FCR clustering algorithm, proposed the interval type 2 fuzzy interval prediction model of the performance deterioration trend of pumped storage units, using a multi-index evaluation system, to achieve multiscale assessment of accuracy and reliability of prediction intervals. Jia et al. [153] proposed the characteristic prediction method of electro-hydraulic servo valve based on a rough set and adaptive neuro-fuzzy inference system, which has reference value for the state prediction of hydroelectric generating the unit transmission mechanism.

In view of the shortcomings of the classical linear time series forecasting model, it can be solved by constructing the time series fuzzy forecasting model through fuzzy set theory. The fuzzy time series prediction model can effectively solve the situation that the time series contains fuzzy language variables or uncertain language variables.

6.4.4 Combination forecast

With the development of prediction methods and prediction theories, time series prediction models are also emerging. However, at present, there is no one method that can effectively solve all prediction problems. Any traditional single statistical model has its own shortcomings, and each has its own different scope and conditions of application. Therefore, while studying and improving the single prediction method, many scholars have studied the combined prediction— that is, using two or more prediction methods to make predictions, and effectively combining the individual prediction results to obtain the final prediction results. The idea of the combined prediction model was first proposed by Bates and Granger [154] in the 1960s, and was widely used in the field of time series prediction. Compared with the single model, the combined forecasting model can effectively reduce the system risk while ensuring a higher model effect.

After the combination forecast model was put forward, it immediately received the attention of many experts and scholars, and actively invested in its research. From the current point of view, the combination forecast model has achieved stable development with a high rate and high quality. The publication of a series of combined forecasting model albums in the authoritative academic *Journal of Forecasting* in the field of international forecasting embodies the recognition of the combined forecasting technology by various experts and scholars and thus greatly enhances its position in forecasting.

At present, domestic and foreign scholars have mainly proposed the following combined forecasting methods, such as the least variance method, unconstrained least squares method, constrained least squares method, Bayes method, combined prediction method based on different criteria and norms, and recursive combined prediction method. Among the preceding various combined forecasting methods, the least variance method is the most practical and theoretical research, and most of them use absolute error as the criterion to calculate the weight vector of the combined forecasting method. However, in addition to the absolute error, the index for evaluating prediction accuracy can also be used to reflect the prediction accuracy index using correlation indicators. Therefore, the current research on combination forecasting methods is not perfect, and further research is needed.

Based on the time series prediction model, there are different fields of demand in social production and life, and the perspectives of domestic and foreign scholars on time series research are different. The existing time series models can be divided into many categories, such as parametric and nonparametric models, linear and nonlinear models, and single and combined prediction models. Here, according to the relationship between the combined prediction value and each research method, the different combined prediction models are divided into two categories: linear and nonlinear combined prediction:

(1) *Linear combination prediction.* Linear combination prediction uses the weighted average method to predict the results obtained by all single prediction methods. In this process, the design of weights when combining single methods is the core. There are many specific methods for determining weights. For example, Hao [155] used the genetic optimization algorithm to determine combination weights. Li and Qin [156] used the iterative optimization algorithm to determine combination weights. Xu and Hui [157] used the quadratic planning algorithm to determine the combination weight and so on. Compared with the single prediction method, although the linear combination prediction effect is more prominent, because the model is too simple, it is often difficult to deal with complex problems.

(2) *Nonlinear combination prediction.* Due to the limitations of linear combination forecasting methods, people have begun to pay attention to nonlinear combination forecasting methods in recent years. At present, some scholars have carried out research on nonlinear combination prediction through comprehensive wavelet theory, SVM, and the genetic algorithm, among others.

Aiming at the problem of hydropower generating unit trend prediction, Xue et al. [158] proposed a hybrid prediction model based on energy entropy reconstruction and support vector regression (SVR). First, the fast integrated empirical mode decomposition algorithm is used to decompose the detection signal into multiple components; each component is reconstructed based on the energy entropy theory; and the generated reconstructed intrinsic modal functions are used as input. The optimal SVR is obtained by training model parameters, which is used to predict the development trend of the unit status. The hybrid forecasting model has high forecasting accuracy, which provides certain guidance for the formulation of unit operation and maintenance strategy. An et al. [159] used the EEMD method to decompose the state degradation trend vector of the hydropower unit, and the resulting components were used to select the chaos model or the gray model for trend prediction. Finally, the component reconstruction was used to realize the state prediction of the unit. In such methods, how to build an appropriate prediction model based on the characteristics of the timing signal corresponding to the device status characterization is the key to accurately predict the status trend. Fu et al. [160] proposed a hydroelectric generating unit vibration trend prediction model based on optimal variational mode decomposition and SVR to better predict the hydroelectric generating unit vibration trend. The study conducted a predictive test on the vibration monitoring data of a large Francis hydropower unit and conducted a comparative analysis. The results show that the model can effectively predict the vibration trend of the hydropower unit. Cheng et al. [161] proposed the combination of wavelet band analysis and gray prediction theory to predict the fault of hydropower units based on the neural network "energy-fault" mapping relationship. Taking the hydraulic turbine shaft swing signal as an example, this method is used to extract and predict feature information, which shows that the combination of wavelet energy extraction and gray prediction theory is more effective in predicting vibration feature information, which provides a new idea for fault prediction. Taking into account the advantages of the gray prediction method and Markov random theory, Zhang et al. [162] designed a method for predicting the state trend of energy conversion equipment combining Markov theory and gray correlation analysis. The experimental results show that the method has a good prediction effect. In general, nonlinear combination forecasting, combining the advantages of wavelet theory, support vector machine, genetic algorithm and other methods, breaks through the limitations of linear combination forecasting, and improves the accuracy and speed of forecasting. It is foreseeable that nonlinear combination forecasting will be a hot research topic in the future forecasting field.

References

[1] L. Lu, Z.N. Peng, X. Wang, L. Zhu, X.L. An, J. Liu, Development in the field of hydraulic machinery research, Journal of China Institute of Water Resources & Hydropower Research 16 (5) (2018) 442–450.

[2] L. Lu, Z.X. Gao, L.P. Pan, S.P. Ma, Development and review of hydraulic electromechanical research field in the past 50 years, Journal of China Institute of Water Resources & Hydropower Research 6 (4) (2008) 299–307.

[3] M.K. Padhy, R.P. Saini, A review on silt erosion in hydro turbines, Renewable & Sustainable Energy Reviews 12 (7) (2008) 1974–1987.

[4] F. Wang, X.-Q Li, J.-M Ma, M. Yang, Y.-L Zhu, Experimental investigation of characteristic frequency in unsteady hydraulic behaviour of a large hydraulic turbine, Journal of Hydrodynamics 21 (1) (2009) 12–19.

[5] Y. Yang, D.L. Jones, C. Liu, Recovery of rectified signals from hot-wire/film anemometers due to flow reversal in oscillating flows, Review of Scientific Instruments 81 (2010) 0151041.

[6] M.A. Ardekani, F. Farhani, Experimental study on response of hot wire and cylindrical hot film anemometers operating under varying fluid temperatures, Flow Measurement & Instrumentation 20 (4-5) (2009) 174–179.

[7] H.E. Albrecht, M. Borys, N. Damaschke, C. Tropea, Laser Doppler and Phase Doppler Measurement Techniques. Springer, 2003.

[8] N.A. Worth, T.B. Nickels, Time-resolved volumetric measurement of fine-scale coherent structures in turbulence, Physical Review E 84 (2011) 02530122.

[9] M.J. Murphy, R.J. Adrian, PIV through moving shocks with refracting curvature, Experiments in Fluids 50 (4SI) (2011) 847–862.

[10] L. Adrian, R.J. Adrian, J. Westerweel, Particle Image Velocimetry, Cambridge University Press, Cambridge, UK, 2010.

[11] W.T. Su, Research on Internal Flow Stability of Large Francis Turbine Mode, Harbin Institute of Technology, Harbin, China, 2014.

[12] D. Stefan, S. Houde, C. Deschenes, Numerical investigation of flow in a runner of low-head bulb turbine and correlation with particle image velocimetry and laser doppler velocimetry measurements, Journal of Fluids Engineering 141: (2019) 0914039.

[13] M. Benisek, I. Bozic, B. Ignjatovic, The comparative analysis of model and prototype test

results of Bulb turbine, IOP Conference Series: Earth & Environmental Science 12 (1) (2010) 012091.

[14] Y.-K. Cheng, Development of Tubular Model Turbine Test Device, Harbin Institute of Technology, Harbin, China, 2017.

[15] X.Y. Chen, Q. Sun, J.P. Sun, Overview of the research status of dynamic stress characteristics of hydraulic turbine runners, Hydropower Automation & Dam Monitoring 37 (05) (2013) 57–61.

[16] Y.-L. Xu, Experimental Research on Stress Test Model of Francis Pump Turbine Runner, Harbin Institute of Technology, 2017.

[17] L.H. Liu, L.Y. Zheng, Development of hydrodynamic stress measuring instrument, Journal of Huazhong University of Science & Technology (Natural Science Edition) (10) (2005) 29–31.

[18] X. Huang, J. Chamberland-Lauzon, C. Oram, A. Klopfer, N. Ruchonnet, Fatigue analyses of the prototype Francis runners based on site measurements and simulations, in IOP Conference Series-Earth and Environmental Science, N. Desy, et al., N. Desy, et al. Editors. 2014.

[19] L.-P. Pan, Research on Dynamic Stress Testing Technology of Large Hydraulic Turbine Runner, Tsinghua University, Beijing, China, 2005.

[20] X. Liu, Y. Luo, Z. Wang, A review on fatigue damage mechanism in hydro turbines, Renewable & Sustainable Energy Reviews 54 (2016) 1–14.

[21] F.-E.H. Mullker F, Efficient monitoring and diagnosis for hydro power plants, International Journal on Hydropower & Dams 23 (5) (1996) 31–34.

[22] A. Molki, L. Khezzar, A. Goharzadeh, Measurement of fluid velocity development in laminar pipe flow using laser Doppler velocimetry, European Journal of Physics 34 (5) (2013) 1127–1134.

[23] X. Zhao, Research on Turbine Efficiency Test System, Huazhong University of Science and Technology, Wuhan, China, 2011.

[24] K. Xiao, Three Gorges Hydropower Plant No. 21 Turbine Stability Test, Huazhong University of Science and Technology, Wuhan, China, 2011.

[25] Z. Huang, Design and Implementation of Multi-Channel Data Acquisition System Based on DSP and FPGA, Huazhong University of Science and Technology, Wuhan, China, 2018.

[26] Y.H. Chang, Z.H. Gui, F.W. Lu, L.N. Fan, Analysis of current status of no-load stability study of pumped storage units, Hydropower & Pumped Storage 2 (5) (2016) 70–75.

[27] H. Bian, G. Liu, J. Feng, B. Gao, H. Yang, Design of multichannel data acquisition system based on Ethernet. In *Proceedings of the 2017 IEEE 17th International Conference on Communication Technology (ICCT)*. 692--695.

[28] S.C. Li (Ed.), Cavitation of Hydraulic Machinery, World Scientific, 2000.

[29] X.-W. Luo, B. Ji, Y. Tsujimoto, A review of cavitation in hydraulic machinery, Journal of Hydrodynamics, Series B 28 (3) (2016) 335–358.

[30] L. Rayleigh, On the pressure developed in a liquid during the collapse of a spherical cavity, London, Edinburgh, & Dublin Philosophical Magazine & Journal of Science 34 (1917) 94–98.

[31] X. Chen, C.-J. Lu, J. Li, Z.-C. Pan, The wall effect on ventilated cavitating flows in closed cavitation tunnels, Journal of Hydrodynamics, Series B 20 (2008) 561–566.

[32] C.-J. Liu, S. Wang Y, J.-B. Liu, Numerical simulation of cavitation flow around two-dimensional hydrofoil, Journal of Naval University of Engineering (2008) 5.

[33] B. Ji, X.W. Luo, X.-X. Peng, Y.-L. Wu, Three-dimensional large eddy simulation and vorticity analysis of unsteady cavitating flow around a twisted hydrofoil, Journal of Hydrodynamics, Series B (04) (2013) 24–33.

[34] Z. Min, Z.H. Zhang, H.B. Luo, D.K. Zhu, Numerical simulation and analysis of turbine guide vane airfoil based on CFD. Journal of Lanzhou University of Technology 2012;3:47–50.

[35] M.S. Plesset, R.B. Chapman, Collapse of an Initially Spherical Vapor Cavity in the Neighborhood of a Solid Boundary, California Institute of Technology, Pasadena, CA, 1970.

[36] C.W. Hirt, B.D. Nichols, Volume of fluid (VOF) method for the dynamics of free boundaries, Journal of Computational Physics 39 (1981) 201–225.

[37] C.F. Naudé, A.T. Ellis, On the mechanism of cavitation damage by non-hemispherical cavities collapsing in contact with a solid boundary, Journal of Basic Engineering 83 (4) (1961) 648–656.

[38] L. Lu, J.-T. Huang, Experimental investigation of bubble collapse near a rigid boundary in solid-liquid two-phase fluid, Hydro-Science & Engineering 4 (1991) 339–347.

[39] W. Li, C.-K. Xiong, Analysis of the center moving path of a spherical bubble broken near a rigid wall, Journal of Xihua University (Natural Science Edition) 1 (2006) 16.

[40] J. He, X.M. Liu, J. Lu, X. Ni, Surface tension effects on the growth and collapse of cavitation bubbles near a rigid boundary, Chinese Journal of Lasers 36 (2) (2009) 342–346.

[41] J. Li, H.S. Chen, Numerical simulation of micro bubble collapse near solid wall in fluent environment, Tribology 28 (4) (2008) 311–315.

[42] M. Xu, C. Ji, J. Zou, X.D. Ruan, X. Fu, Particle removal by a single cavitation bubble, Science China Physics, Mechanics & Astronomy 04 (2014) 78–83.

[43] M. Kornfeld, L. Suvorov, On the destructive action of cavitation, Journal of Applied Physics 15 (6) (1994) 495–506.

[44] Q. Xing, L. Zhang, S. Li, J. Lu, Effect of laser shock processing on electrochemical corrosion behavior of 2A02 aluminum alloy, Corrosion Science & Protection Technology 25 (5) (2013) 402–405.

[45] G.L. Chahine, Strong interactions bubble/bubble and bubble/flow, Bubble Dynamics and Interface Phenomena. Fluid Mechanics and Its Applications, Vol. 23, Springer (1994).

[46] H. Chen, S.-C. Li, Z.-G. Zuo, S. Li, Direct numerical simuiation of bubble-cluster's dynamic characteristics, Journal of Hydrodynamics 20 (2008) 689–695.

[47] K.H. Kato, H. Yamaguchi, M. Tanimura, Y. Tagaya, Mechanism and control of cloud cavitation, Journal of Fluids Engineering 236 (4) (1997) 788–794.

[48] Y. Tomita, A. Shima, Mechanisms of impulsive pressure generation and damage pit formation by bubble collapse, Journal of Fluids Engineering 169 (1986) 535–564.

[49] S.N. Buravova, Y.A. Gordopolov, Bubble-induced cavitation effect upon solid surface, Technical Physics Letters 36 (8) (2010) 717–719.

[50] J. Zhang, L. Zhang, J. Deng, Numerical study of the collapse of multiple bubbles and the energy conversion during bubble collapse, Water 11 (2) (2019) 247.

[51] S. Chumakov, D. Cook, F. Ham, U. Iben, Large Eddy Simulation of cavitation in turbulence, APS Division of Fluid Dynamics Meeting Abstracts (2012) A3:006.

[52] F. Hong, J. Yuan, B. Zhou, Y. Shi, Assessment of improved Schnerr-Sauer model in cavitation simulation around a hydrofoil, Journal of Harbin Engineering University 037 (007) (2016) 885–890.

[53] A.K. Singhal, M.M. Athavale, H. Li, Y. Jiang, Mathematical basis and validation of the full cavitation model, Journal of Fluids Engineering 124 (3) (2002) 617–624.

[54] G.H. Schnerr, J. Sauer, in: Physical and numerical modeling of unsteady cavitation dynamics, *Proceedings of the 4th International Conference on Multiphase Flow*, 2000.

[55] A. Kubota, H. Kato, H. Yamaguchi, A new modelling of cavitation flows: A numerical study of unsteady cavitation on a hydrofoil section, Journal of Fluid Mechanics 240 (1992) 59–96.

[56] A.G. Gerber. A CFD model for devices operating under extensive cavitation conditions. In *Proceedings of the 2002International Mechanical Engineering Congress and Exhibit*.

[57] P.J. Zwart, A.G. Gerber, T. Belamri, A two-phase flow model for predicting cavitation dynamics, Proceedings of the 5th International Conference on Multiphase Flow (2004).

[58] J. Yang, L.J. Zhou, Z.W. Wang, Numerical investigation of the cavitation dynamic parameter in a Francis turbine draft tube with columnar vortex rope, Journal of Hydrodynamics, Series B 31 (2019) 939.

[59] H. Yamaguchi, H. Kato, in: On application of nonlinear cavity flow theory to thickfoil sections, In *Proceedings of the 2nd International Conference on Cavitation*, 1983.

[60] T.K. Johansen, Description of a computer program, Capro, for the calculation of cavitation on two-dimensional hydrofoils, Computer Programs (1979).

[61] J. Ren, J.S. Chang. A nonlinear singularity solution for the cavitaiton flow in a 2D thick profile annular cascade. In *Proceedings of the 1996 International Conference on Fluid Machinery and Fluid Engineering*.

[62] J.F. Huang, L.X. Zhang, S.H. He, Numerical simulation of 3-D steady and unsteady flows in whole flow passage of a Francis hydro-turbine, Proceedings of the CSEE 29 (2) (2009) 87–94.

[63] J.F. Huang, L.X. Zhang, J. Yao, L. Cao, Numerical simulation of 3D cavitation turbulent flow in Francis hydro-turbine passage on mixture modeling technology, Proceedings of the CSEE 31 (32) (2011) 115–121.

[64] G.K. Kumar, T. Tanaka, N. Yamaguchi, T. Taniwaki, K. Miyagawa, W. Takahashi, Influence of guide vane clearance on internal flow of medium-specific speed Francis turbine, IOP Conference Series: Earth & Environmental Science 240 (2) (2019) 022056.

[65] P.C. Guo, Z.N. Wang, X.Q. Luo, Y.L. Wang, J.L. Zuo, Flow characteristics on the blade channel vortex in the Francis turbine, IOP Conference Series: Materials Science & Engineering 129: (2016) 012038.

[66] G.-H. Zhai, Research on the Cavitation Characteristics of Pump Turbine Under the Influence of Environmental Factors, Xi'an University of Technology, Xi'an, Beilin, China, 2019.

[67] J. Chen, Z. Li, J. Pan, G. Chen, Y. Zi, J. Yuan, B. Chen, Z. He, Wavelet transform based on inner product in fault diagnosis of rotating machinery: A review, Mechanical Systems & Signal Processing (2016) 70–71 1-35.

[68] W. Peng, X. Luo, Research on vibrant fault diagnosis of hydro-turbine generating unit based on wavelet packet analysis and support vector machine, Proceedings of the CSEE 26 (24) (2007) 164–168.

[69] Z. Liu, J. Zhou, Y. Zhang, M. Zou, RBFNN fault diagnosis based on complex feature extraction of hydropower unit, Automation of Electric Power Systems 31 (11) (2007) 87–91.

[70] Z. Gui, F. Han, Neural network based on wavelet packet-characteristic entropy for fault diagnosis of draft tube, Proceedings of the CSEE 25 (4) (2005) 99–102.

[71] Z. Peng, Y. He, Q. Lu, W. Lu, Using wavelet method to analyse fault features of rub rotor in generator, Proceedings of the CSEE 23 (5) (2003) 75–79.

[72] H. Shi, C. Li, Y. Bi, Study and application of method for analysis on cavitation characteristics of hydro-turbine based on acoustic detection, Water Resources & Hydropower Engineering 39 (9) (2008) 75–77.

[73] N.E. Huang, Z. Shen, S.R. Long, M.C. Wu, H.H. Shih, Q. Zheng, N.-C. Yen, C.C. Tung, H.H. Liu, The empirical mode decomposition and the Hilbert spectrum for nonlinear and

non-stationary time series analysis. *Proceedings of the Royal Society A: Mathematical, Physical & Engineering Sciences* 454 (1971) (1998) 903–995.

[74] R. Jia, X. Wang, Z.-H. Cai, L. Zhang, Hilbet-Huang transform based on least squares support regression machine and its application in the fault diagnosis of hydroelectric generation unit, Proceedings of the CSEE 26 (22) (2007) 128–133.

[75] X. An, L. Pan, F. Zhang, De-noising of hydropower unit throw based on EEMD and approximate entropy, Journal of Hydroelectric Engineering 34 (4) (2015) 163–169.

[76] Y. Xue, X. Luo, H. Wang, Dynamic characteristic extraction of the draft tube vortex based on EMD multi-scale feature entropy. Transactions of the CSAE, 2011, 27(3):210–214.

[77] N.E. Huang, Introduction to the Hilbert-Huang transform and its related mathematical problems, Interdisciplinary Mathematics 5: (2005) 1–26.

[78] Y. Qian, H. Zhang, D. Peng, F. Xia, Orbit purification of generator unit based on a new generalized particle swarm optimization method, Proceedings of the CSEE 32 (2) (2012) 130–137.

[79] C. Wang, J. Zhou, P. Kou, Z. Luo, Y. Zhang, Identification of shaft orbit for hydraulic generator unit using chain code and probability neural network, Applied Soft Computing 12 (1) (2012) 423–429.

[80] Z. Zhang, S. Xue, Y. Zhang, Intelligent fault diagnosis method of rotating machinery based on characteristic parameters, Journal of Vibration, Measurement & Diagnosis 29 (3) (2009) 256–260.

[81] D. Shi, W. Wang, P. Unsworth, L. Qu, Purification and feature extraction of shaft orbits for diagnosing large rotating machinery, Journal of Sound & Vibration 279 (3) (2005) 581–600.

[82] C. Wang, J. Zhou, H. Qin, C. Li, Y. Zhang, Fault diagnosis based on pulse coupled neural network and probability neural network, Expert Systems with Applications 38 (11) (2011) 14307–14313.

[83] A. Jami, P. Heyns, Impeller fault detection under variable flow conditions based on three feature extraction methods and artificial neural networks, Journal of Mechanical Science & Technology 32 (9) (2018) 4079–4087.

[84] N. Lu, Multiwavelets Based Vibration Feature Extraction and Fault Diagnosis Methods for Hydro-Turbine Generating Unit, Wuhan University, Wuhan, China, 2014.

[85] H. Li, M. Jiao, X. Yang, L. Bai, Fault diagnosis of hydroelectric sets based on EEMD and SOM neural networks, Journal of Hydroelectric Engineering 36 (2017) 83–91.

[86] L. Xie, J. Lei, F. Xu, Vibrant fault diagnosis for hydro-turbine generating unit based on improved neighborhood rough sets and PNN, Journal of Shanghai University of Electric Power 32 (2016) 181–187.

[87] Y. Du, J. Zhou, Y. Shan, S. Li, Y. Xu, W. Jiang, Cavitation fault diagnosis of hydropower unit based on EMD-BPNN, Water Resources & Power 36 (3) (2018) 157–160.

[88] C. Cortes, V. Vapnik, Support vector machine, Machine Learning 20 (3) (1995) 273–297.

[89] X. Zhang, Hybrid Intelligent Fault Diagnosis Method Integrating Support Vector Machines for Hydroelectric Generator Units, Huazhong University of Science and Technology, Wuhan, China, 2012.

[90] W. Peng, P. Guo, X. Luo, Vibration fault diagnosis of hydroelectric unit based on LS-SVM and information fusion technology, Water Resources & Power (2007) 137–142.

[91] X. Zhang, W. Chen, Y. Yang, T. Li, Vibration fault diagnosis of hydroelectric generators based on VMD decomposition and support vector machine, Power System & Clean Energy 33 (2017) 134–138.

[92] X. Cheng, Q. Chen, W. Wang, Y. Zheng, D. Guo, Q. Lou, Fault diagnosis of hydroelectric generating sets based on multidimensional features and multiple classifiers, Journal of Hydroelectric Engineering 38 (04) (2019) 181–188.
[93] J. Xiao, Performance Degradation Evaluation and Intelligent Fault Diagnosis for Hydropower Generator Unit, Huazhong University of Science and Technology, Wuhan, China, 2014.
[94] M. Luo, Study on Vibration Fault Diagnosis and Trend Prediction of Hydroelectric Generator Units, Huazhong University of Science and Technology, Wuhan, China, 2017.
[95] Q. Wang, T. Yang, R. Shen, B. Yang, Fault caused by vibration diagnosis expert system for a pump storage group, Journal of Vibration & Shock (07) (2012). 167--170, 179
[96] C. Mao, H. Liu, X. Li, J. Wu, H. Pan, J. Wang, H. Qian, Fault diagnosis expert system of hydropower unit based on knowledge base, Huadian Technology 37 (2015). 25–28, 32, 78
[97] Y. Zhou, S. Tang, L. Pan, Design and development of HM9000ES fault diagnosis expert system, Journal of China Institute of Water Resources & Hydropower Research 12 (2014) 104–108.
[98] S. Li, Research of Vibration Fault Diagnosis for Hydropower Unit Based on Fault-Tree Analysis, Nanchang Institute of Technology, Nanchang, China, 2015.
[99] Y. Hu, Z. Xiao, Fault diagnosis analysis of fault tree based on Bayesian network in hydropower units, China Rural Water & Hydropower (8) (2017).
[100] Y. Liu, Research and Application of Pump Turbine Fault Diagnosis Technology Based on Fault Tree Analysis, Huazhong University of Science and Technology, Wuhan, China, 2019.
[101] GB/T7894-2001Fundamental technical specifications for hydro generators. General Administration of Quality Supervision, Inspection and Quarantine of the People's Republic of China, 2001.
[102] GB/T8564-2003Specification installation of hydraulic turbine generator units. General Administration of Quality Supervision, Inspection and Quarantine of the People's Republic of China, 2003.
[103] DL/T817-2002Technological code for maintenance of vertical hydro-generators. General Administration of Quality Supervision, Inspection and Quarantine of the People's Republic of China, 2002.
[104] GB/T8564-2003Specification installation of hydraulic turbine generator units. General Administration of Quality Supervision, Inspection and Quarantine of the People's Republic of China, 2003.
[105] GB/T11348.5-2002Mechanical vibration of non-reciprocating machines measurements on rotating shafts and evaluation criteria. General Administration of Quality Supervision, Inspection and Quarantine of the People's Republic of China, 2002.
[106] GB/T15468-2006Fundamental technical requirements for hydraulic turbines. General Administration of Quality Supervision, Inspection and Quarantine of the People's Republic of China, 2006.
[107] L. Pan, Research on Fault Diagnosis System of Hydropower Unit Based on Health Assessment and Deterioration Trend Forecast, China Institute of Water Resources and Hydropower, Beijing, 2013.
[108] W. Zhu, Research on Methods of Fault Diagnosis, Prognostics and Condition Assessment for Hydroelectric Generator Units, Huazhong University of Science and Technology, Wuhan, 2016.
[109] W. Yu, Research on Condition Monitoring and Intelligent Fault Diagnosis System of Large Hydroelectric Generator Units, Central South University, Changsha, Hunan, China, 2007.

[110] S. Wang, Research on the Theory and Application Technology of Condition Monitoring, Signal Analysis and Fault Diagnosis of Turbo Generator Set, Southeast University, Nanjing, Jiangsu, China, 2000.

[111] J. Yu, DCS-distributed control system and its application status, Industrial Instrumentation & Automation (4) (1993) 43–47.

[112] J. Wang, Compass system for prediction and analysis of machine equipment and its application, Northeast Electric Power Technology (11) (1995) 16–19.

[113] Z. Gui, L. Pan, S. Tang, Y. Zhou, F. Sun, Development and application of HM9000 hydropower unit condition monitoring and comprehensive analysis system, Hydropower Automation & Dam Monitoring 31 (6) (2008) 32–35.

[114] W. Sun, Y. Zheng, Application of TN8000 condition monitoring and fault diagnosis system in Pengshui hydropower plant, Central China Electric Power 2 (2013).

[115] Z. Tao, B. Liu, J. Li, Application of PSTA2003 condition monitoring and analysis diagnosis system in mixed flow unit, Hydropower Plant Automation 31 (3) (2010) 34–38.

[116] G.E.P. Box, G.M. Jenkins, G.C. Reinsel, et al. Time series analysis: forecasting and control. John Wiley & Sons, 2015.

[117] Q. Fu, F.L. Wang, Y. Sun, Self-exciting threshold auto-regressive model (SETAR) to forecast well-irrigated rice water requirements, Journal of Experimental Botany 54 (2003) 41–42.

[118] J.L. Jin, X.H. Yang, B.M. Jin, J. Ding, Application of genetic threshold auto-regressive model to forcasting meteorological time series, Journal of Tropical Meteorology 17 (4) (2001) 415–422.

[119] D. Velasquez, Adaptive multidimensional neuro-fuzzy inference system for time series prediction, IEEE Latin America Transactions 13 (8) (2015) 2694–2699.

[120] Z. Pei, Prediction of Hydropower Generating Unit Vibration Based on Artificial Intelligence, Huazhong University of Science and Technology, Wuhan, China, 2016.

[121] C. Chen, X. Li, W.Y. Cui, Hydraulic turbine operation status detection based on LSTM network prediction, Journal of Shandong University (Engineering Science) 49 (3) (2019) 39–46.

[122] Y.Y. He, F.L. Chu, B.L. Zhong, Wavelet network based on hierarchical evolving algorithm for machine condition forecasting, Journal of Tsinghua University (Science & Technology) 42 (6) (2002) 754–757.

[123] D.-L. Zhu, X.-B. Bie, Research on the NNARX identification and prediction control of the hydraulic turbine system, Automation & Instrumentation (12) (2017) 228–230 233.

[124] L. Sun, P. Guo, Q. Ma, X. Zheng, X. Luo, Hydraulic performance prediction for high-head Francis turbine based on TBR mode, Transactions of the CSAE 35 (7) (2019) 62–69.

[125] B. Zhang, Study on Intelligent Fault Diagnosis and State Tendency Prediction of Hydroelectric Generator Units Based on Time-Frequency Analysis and Nonlinear Entropy, Huazhong University of Science and Technology, Wuhan, China, 2019.

[126] H. Liu, Study on Multiple Information Fault Diagnosis and State Trend Forcasting for Hydroelectric Generator Unit, Huazhong University of Science and Technology, Wuhan, China, 2019.

[127] V.N. Vapnik, Statistical Learning Theory, Wiley, New York, NY, 1998.

[128] Z.-J. Tang, F. Ren, T. Peng, W.-B. Wang, A least square support vector machine prediction algorithm for chaotic time series based on the iterative error correction, Acta Physica Sinica 63 (5) (2014) 050505.

[129] Y.X. Zhu, Study on the performance prediction of multi-nozzle Turgo impulse turbines based on support vector machine, Large Electric Machine & Hydraulic Turbine (6) (2013) 55–58.

[130] B. Peng, Hydropower Generating Unit's Fault Diagnosis Based on Improved SVM and Information Fusion, Huazhong University of Science and Technology, Wuhan, China, 2008.

[131] L. Hui, Research on Wireless Viberation Monitoring and Fault Diagnosis System Base on SVM for Hydroelectric Generating Set, Xi'an University of Technology, Xi'an, Beilin, 2009.

[132] L. Fu, Vibration Signal Processing and Intelligent Fault Diagnosis of Hydroelectric Generator Units, Huazhong University of Science and Technology, Wuhan, China, 2016.

[133] Y. Li, S. Chen, Improved maximal Lyapunov exponent chaotic forecasting method based on Markov chain theory, Computer Science 43 (04) (2016) 270–273.

[134] R.G. Li, H.L. Zhang, W.H. Fan, Y. Wang, Hermite orthogonal basis neural network based on improved teaching-learning-based optimization algorithm for chaotic time series prediction, Acta Physica Sinica 64 (20) (2015) 200506.

[135] F.M. Huang, K.L. Yin, B.B. Yang, X. Li, L. Liu, X.L. Fu, X.W. Liu, Step-like displacement prediction of landslide based on time series decomposition and multivariate chaotic model, Earth Science 43 (3) (2018) 887–898.

[136] F. Gao, Time series prediction based on chaotic least squares support vector machines, Mathematics in Practice & Theory 48 (8) (2018) 239–244.

[137] J. Wu, Nonlinear State Forecasting and Stability Control of Hydro Turbine Regulation System, Northwest A&F University, 2019.

[138] S.B. Zheng, Q.W. Zhong, L.L. Peng, A simple method of residential electricity load forecasting by improved Bayesian neural networks. Mathematical Problems in Engineering 2018, 4276176.

[139] A. Wang, Optimization of time series prediction model based on Bayesian statistical inference, Journal of Taiyuan University (Natural Science Edition) 37 (04) (2019) 19–24.

[140] X.D. Zhang, F. Chen, J. Gao, T.J. Fang, Sparse Bayesian and its application to time series forecasting, Control & Decision 21 (5) (2006) 585–588.

[141] M. Das, S.K.Ghosh. semBnet, A semantic Bayesian network for multivariate prediction of meteorological time series data, Pattern Recognition Letters 93 (2017) 192–201.

[142] Q.K. Xiao, L. Xing, G. Song, Time series prediction using optimal theorem and dynamic Bayesian network, Optik 127 (23) (2016) 11063–11069.

[143] J.Y. Yan, J.X. Hua, Time series forecast of GMDH neural network based on rough set, Journal of Computer Application 29 (2009) 179–181.

[144] L. Liu, X.P. He, Model of time series forcasting based on GA and fuzzy decision tree, Computer Engineering & Design 29 (19) (2008) 5044–5046.

[145] X.L. Gong, L.S. Kong, C.L. Yuan, H.Q. Xiao, J.H. Liu, Nonlinear time series prediction model based on particle swarm optimization B-spline network, Journal of Electronic Measurement & Instrumentation 31 (12) (2017) 1890–1895.

[146] J. Zhong, G. Dong, Y.M. Sun, Z.Y. Zhang, Y.Q. Wu, Application of the nonlinear time series prediction method of genetic algorithm for forecasting surface wind of point station in the South China Sea with scatterometer observations, Chinese Physics B 25 (11) (2016) 167–173.

[147] D. Zheng, S. Wang, Q. Meng, Dynamic programming track-before-detect algorithm for radar target detection based on polynomial time series prediction, IET Radar Sonar & Navigation 10 (8) (2017) 1327–1336.

[148] K. Bisht, S. Kumar, Fuzzy time series forecasting method based on hesitant fuzzy sets, Expert Systems with Application 64 (2016) 557–568.

[149] B.P. Joshi, S. Kumar, Intuitionistic fuzzy sets based method for fuzzy time series forecasting, Cybernetics & Systems 43 (1) (2012) 34–47.

[150] W.-L. Zhu, Research on Method of Fault Diagnosis, Prognostics and Condition Assessment for Hydroelectric Generator Units, Huazhong University of Science and Technology, Wuhan, China, 2016.

[151] C. Liang, H.G. Ni, Prediction of cavitation state using fuzzy random theory, Journal of Chengdu University of Science & Technology (Engineering Science Edition) (1995) 15–19.

[152] W. Zou, Prediction of Pumped Storage Unit Performance Deterioration Trend Based on Type 2 Fuzzy Model Identification, Huazhong University of Science and Technology, 2019.

[153] Z.Y. Jia, J.W. Ma, F.J. Wang, W. Liu, Characteristics prediction method of electro-hydraulic servo valve based on rough set and adaptive neuro-fuzzy inference system, Chinese Journal of Mechanical Engineering 23 (2) (2010) 200–208.

[154] J.M. Bates, C.W.J. Granger, The combination of forecasts, Journal of the Operational Research Society 20 (4) (1969) 451–468.

[155] Y. Hao, The best linear prediction for complex time series, Journal of Taiyuan Normal University (Natural Science Edition) 16 (3) (2017) 23–25.

[156] A.G. Li, Z. Qin, Adaptive local linear predicting chaotic time series, System Engineering Theory & Practice 24 (6) (2004) 67–71.

[157] Y. Xu, H. Chen, Local linear prediction improvement of chaotic time series, Journal of Hangzhou Teachers College (Natural Science Edition) 6 (5) (2007) 337–341.

[158] X.M. Xue, S.Q. Cao, C.S. Li, State trend prediction of hydropower units based on energy entropy reconstruction and support vector regression, Water Resources & Power 37 (9) (2019) 139–142 135.

[159] X.L. An, L.P. Pan, F. Zhang, Y.J. Tang, State degradation assessment and nonlinear prediction of hydropower units, Power System Technology (5) (2013) 1378–1383.

[160] W.L. Fu, J. Zhou, Y. Zhang, Y. Zheng, Prediction of vibration trend of hydropower units based on OVMD and SVR, Journal of Vibration & Shock 35 (8) (2016) 36–40.

[161] B.Q. Cheng, F.Q. Han, Z.H. Gui, Application of grey prediction theory based on wavelet in fault prediction of hydropower units, Power System Technology 29 (13) (2005) 40–44.

[162] D. Zhou, Z. Yu, H. Zhang, S. Weng, A novel grey prognostic model based on Markov process and grey incidence analysis for energy conversion equipment degradation, Energy 109 (2016) 420–429.

Chapter 7

Fault detection and fault identification in marine current turbines

Tianzhen Wang, Zhichao Li and Yilai Zheng
School of Logistic Engineering, Shanghai Maritime University, China

Since the 21st century, in the face of energy shortage, environmental degradation, and climate change, more and more countries have turned their attention to clean and renewable energy [1–3]. The theoretical power generation of global ocean energy is 2000 to 4000 Twh/year, and 0.02% of ocean energy can meet the global energy demand [4–6]. The theoretical estimation value of current energy is about 10^8 kW in the world, and the energy of current energy is more concentrated, predictable, relatively stable, and not affected by day and night, and can be used for continuous power generation [7, 8]; the marine current turbine (MCT) is installed underwater, does not occupy valuable land resources, and has no noise or visual pollution [9, 10]. Many countries are increasing the development of marine current energy all over the world [11–13]. However, because the MCT is installed in the sea, the cost of transportation and installation is high, and it is not easy to detect and maintain [14–16]. At the same time, compared with the maintain land power generation equipment, the MCT is affected by many factors, such as sharp change of instantaneous velocity, seawater corrosion, attached organisms, surge, and turbulence [17–19], which accelerates the wear and corrosion of devices, resulting in higher mechanical loss and more likely to cause turbine, control system, sensor, and other failures [20]. There are many types of failures, and the fault features and forms are not the same in the MCTs [21]. The current fault diagnosis methods are not fully suitable for the diagnosis of the MCT due to the mismatch of fault model, the lack of fault data, and the lack of obvious fault characteristics in the time or frequency domain [22, 23]. Effective detection and diagnosis technology have important practical significance in promoting the actual project implementation of the condition-based maintenance of the MCT, ensuring the safe and efficient operation of the MCT and reducing maintenance costs.

Fault Diagnosis and Prognosis Techniques for Complex Engineering Systems. DOI: 10.1016/B978-0-12-822473-1.00005-7

There are some faults caused by the attachment of blades in MCTs. The attachment degree diagnosis is an important domain for MCT research [25–27]. Thus, it is important to detect the degree of blade attachment in time. There are two fault detection methods introduced by an electrical signal in this chapter:

(1) *Hilbert transform–based imbalance fault detection method using the stator voltage.* In this method, the voltage signal of the MCT is transformed into instantaneous voltage frequency and average voltage frequency by Hilbert transform (HT). Then the imbalance fault frequency is extracted by cubic spline interpolation. Last, the imbalance fault is detected though the wavelet time-frequency spectrum after the imbalance fault frequency is processed by wavelet transform (WT).

(2) *The wavelet threshold denoising–based imbalance fault detection method using the stator current.* First, the parameters are set by wavelet threshold denoising, then the features of the imbalance faults are extracted by HT and principal component analysis (PCA). Since the electrical signal cannot be used to diagnose the uniform faults and symmetrical faults, there are two identification methods introduced by the image signal in this chapter.

(3) *The identification method of blade attachment based on the sparse autoencoder and softmax regression.* Here, the image signal is used for fault diagnosis. The sparse autoencoder is used to extract the image features, and softmax regression is used to classify the different degrees of attachment.

(4) *The identification method of blade attachment based on depthwise separable convolutional neural network.* There are three parts: the first is to set reasonable labels for uniform and symmetrical faults diagnosis, then conduct nearest-neighbor interpolation to compress data, and then extract image features through a convolutional neural network (CNN) (Fig. 7.1).

7.1 The HT-based detection method

7.1.1 Problem description

As shown in Fig. 7.2, if an imbalance fault happens in the MCT system, there would be an equivalent mass generated that would rotate with the blades [22]. The torque generated by this equivalent mass acting on the shaft could be expressed as follows:

$$T_{\text{im}}(t) = F_d r_u \sin(\omega_m t), \tag{7.1}$$

where F_d means the downward synthetic force generated by the equivalent mass and r_u means the distance from the equivalent mass to the bearing center. There

FIGURE 7.1 Attachment of marine current turbine after 6 months of launching. From Keenan et al. [24].

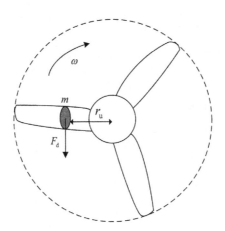

FIGURE 7.2 Influence of imbalance fault on the MCT.

is $\omega_m = 2\pi f_m$, where f_m means the turbine rotation frequency. The varieties of the rotation frequency affected by the imbalance fault could be expressed as

$$\Delta\omega_m = \frac{\rho g V - mg}{J\omega_m} r_u \cos(\omega_m t). \tag{7.2}$$

The current signal gotten from the MCT, which is in the fault state, could be expressed as follows:

$$i_s(t) = I_m \cos(p(\omega_m t + \Delta\omega_m t) + \gamma), \tag{7.3}$$

where I_m means the amplitude, γ means the initial angle, and p means the number of pole pairs. When there is $\Delta\omega_m = 0$, it means that the MCT is under the healthy condition.

If the turbine turning period is regarded as a working period of the MCT system, the average value of the current frequency could be p times the turbine rotation frequency [28]. The average value of the current frequency signal could be calculated by

$$\bar{f}_e = p \times f_m. \tag{7.4}$$

The current frequency variety could be calculated as

$$\Delta f_e = f_e - \bar{f}_e, \tag{7.5}$$

where f_e means the instantaneous frequency signal. With Eq. (7.5), the varieties of the current frequency could be calculated. The imbalance fault characteristic signal f_{fault} could be estimated by reducing the interference from Δf_e.

7.1.2 The HT-based detection method

To extract the fault characteristic signal and detect whether an imbalance fault happens, a HT-based detection method is given. In this method, HT, cubic spline interpolation, and WT are used. HT is used to calculate the instantaneous current frequency signal f_e and its average signal $\bar{f}_e$. Cubic spline interpolation is used to interpolate f_e and $\bar{f}_e$, and calculate the fault characteristic signal f_{fault}. WT is used to analyze the fault characteristic signal f_{fault} and draw the time-frequency spectrum to detect whether an imbalance fault happens in the MCT system.

When the influences of waves and fault are weak, the current signal could be considered as a signal of the narrow band. Therefore, HT could have good performance in calculating the instantaneous frequency signal [29].

The HT of the current signal $I(t)$, which is defined as $y(t)$, could be calculated by

$$y(t) = H[I(t)] = \frac{1}{\pi} \int_{\infty}^{-\infty} \frac{I(\tau)}{t - \tau} d\tau. \tag{7.6}$$

A complex conjugate pair could be formed by $I(t)$ and $y(t)$ as follows:

$$z(t) = I(t) + jH[I(t)] = A(t)e^{j\phi(t)}, \tag{7.7}$$

where $A(t)$ means the amplitude and $\varphi(t)$ means the phase. $A(t)$ and $\varphi(t)$ could be expressed as follows:

$$A(t) = |z(t)| = \sqrt{I^2(t) + H^2[I(t)]}, \phi(t) = \arctan\left[\frac{y(t)}{I(t)}\right]. \tag{7.8}$$

The instantaneous frequency signal of the current signal could be further calculated as

$$f_e(t) = \frac{1}{2\pi} \times \frac{d\phi(t)}{dt}. \tag{7.9}$$

With Eq. (7.4,), the average frequency signal $\bar{f}_e$ could be estimated.

As analyzed earlier, the fault characteristic signal f_{fault} could be estimated by decreasing the interference signal from Δf_e. Interpolation for f_e and $\bar{f}_e$ could reduce the interference. Replacing f_e and $\bar{f}_e$ in Eq. (7.5) with the interpolation results, the current frequency varieties could be estimated. There would be little interference in the estimated stator current frequency varieties. Therefore, the estimated stator current frequency varieties could be regarded as the fault characteristic signals. Two main parts to extract the imbalance fault characteristic signals are given:

(1) *Cubic spline interpolation.* Cubic spline interpolation could construct polynomials based on the known value of sampling frequency. The details of this method can be found in the work of Voltz and Webster [30].

   With the length N of the original stator current signal $I(n)$, the sampling interval T_s, the instantaneous current frequency f_e, and the average current frequency $\bar{f}_e$, the new N-point sequences $f_e^*(n)$ and $\bar{f}_e^*(n)$ could be reconstructed by using the cubic spline interpolation.

(2) *Imbalance fault characteristic signal extraction.* Using Eq. (7.5), replacing f_e and $\bar{f}_e$ with the interpolation results $f_e^*(n)$ and $\bar{f}_e^*(n)$, the current frequency varieties could be expressed as

$$\Delta f_e^* = f_e^*(n) - \bar{f}_e^*(n), \tag{7.10}$$

Δf_e^* could be considered as the fault characteristic signal, then there is $f_{\text{fault}} = \Delta f_e^*$.

Define $\psi(t)$ as a square-integrable function, then there is $\psi(t) \in L^2(R)$, and $\psi(t)$ meets the wavelet permissibility condition:

$$\int_{-\infty}^{\infty} |\omega|^{-1} |\hat{\psi}(\omega)|^2 d\omega < \infty, \tag{7.11}$$

where $\psi(t)$ means the mother wavelet function and $\hat{\psi}(\omega)$ means the Fourier transform of $\psi(t)$.

For the extracted fault characteristic signal $f_{\text{fault}}(t)$, its continuous WT could be calculated as

$$W(a, b) = |a|^{-1/2} \int_{-\infty}^{\infty} f_{\text{fault}}(t) \psi^* \left(\frac{t - b}{a} \right) dt, \tag{7.12}$$

where $W(a, b)$ means the wavelet coefficient, $\psi^*(t)$ means the conjugate function of the mother wavelet function $\psi(t)$, a means the scale factor, and $a \in R$, $a \neq 0$, whereas b means the translation factor, $b \in R$.

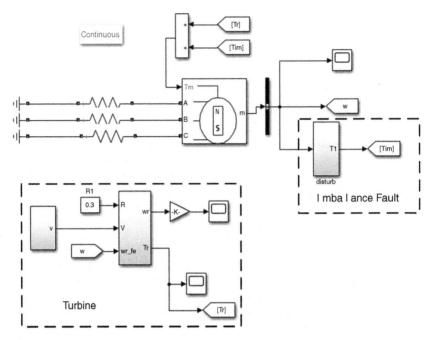

FIGURE 7.3 Simulation model of the MCT.

Complex Morlet wavelet could have good localization and symmetry performances in both the time domain and the frequency domain. The complex Morlet wavelet is chosen as the mother wavelet in this method.

The complex Morlet wavelet could be calculated as follows:

$$\psi_0 = \frac{1}{\sqrt{\pi f_b}} e^{2i\pi f_c t} e^{2t^2/f_b}, \tag{7.13}$$

where f_b means the wavelet bandwidth, f_c means the wavelet center frequency, and ψ_0 could meet the permissibility condition when $2\pi f_c \geq 5$.

Process the fault characteristic signal f_{fault} by using the WT and the time-frequency spectrum of WT could be drawn. The WT time-frequency spectrum could show the fault characteristic frequency clearly. The steps of this method can be found in the work of Yan et al. [31].

7.1.3 Simulation results and analysis

A model of the MCT system is established by MATLAB/Simulink (Fig. 7.3). The sampling frequency of this simulation model is 1 kHz. This simulation model contains the turbine part and generator part: the turbine part mainly uses the input water flow velocity information to calculate the mechanical torque; the generator part mainly uses torque, speed, and other information to calculate the stator current. The imbalance fault could be simulated by adding an imbalance torque at the input of torque.

After using the HT-based detection method, the fault characteristic signal could be extracted. The raw stator current signals under different blade states and their time-frequency spectrums of WT are given in Fig. 7.4 and Fig. 7.5, respectively. These figures show the detection results of the HT-based detection method (wavelet basis: complex Morlet, 4-4; center frequency, 4 Hz). The water flow velocity is about 1.1 to 1.2 m/s, and the equivalent mass is set to 200 g. Fig. 7.4 gives the stator current signal under the healthy condition and its detection result using this method. Fig. 7.5 displays the stator current signal under the imbalance fault condition and its detection result. The fault characteristic frequency under the imbalance fault condition would appear at the turbine rotation frequency, which is about 1.9 Hz. The simulation results validate that the HT-based detection method could easily detect whether an imbalance fault happens.

To make the imbalance fault detection easier, a HT-based detection method is given to detect whether an imbalance fault happens in the MCT system. This method could be implemented in the following three steps: (1) convert the stator current signal into the instantaneous current frequency signal and the average current frequency signal; (2) apply the cubic spline interpolation to decrease the interference information and estimate the fault characteristic signal, and (3) detect whether an imbalance fault happens in the MCT system by the WT. The simulation results have proved that the HT-based detection method has good performance in detecting the imbalance fault of the MCT system.

7.2 The wavelet threshold denoising–based dectection method

7.2.1 Problem description

In the underwater environment, there would be an interference signal included in the current signal of the MCT system, which is caused by turbulence and waves [32]. In the past few years, many researchers have given many strategies to decrease the interference signal and detect whether the imbalance fault happens. However, some strategies do not take the influences that are caused by different water flow velocities into account. Most of these strategies detect the fault by finding the harmonic that stands for the imbalance fault. Therefore, the detection performance could be affected by the turbine rotation frequency. In the underwater environment, the velocity of water flow varies continuously [33], which indicates that the turbine rotating frequency is continuously changing. The spectrum of the fault signature signals in the same blade condition under different water flow velocities is given in Fig. 7.6. This figure shows that the frequencies of the imbalance fault characteristic signals are changing with the water flow velocity, and their amplitudes are different in different water flow velocities. It

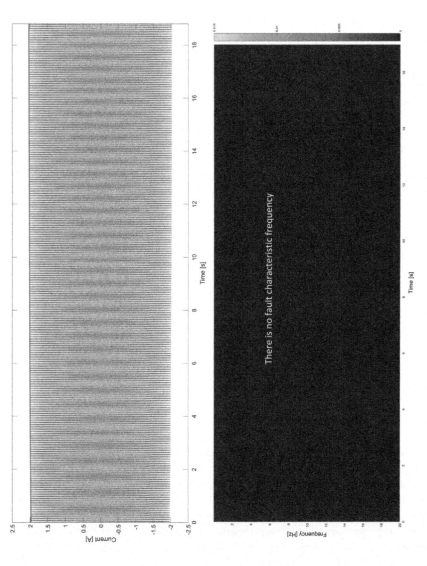

FIGURE 7.4 Stator current signal under the healthy condition and its time-frequency spectrums of WT.

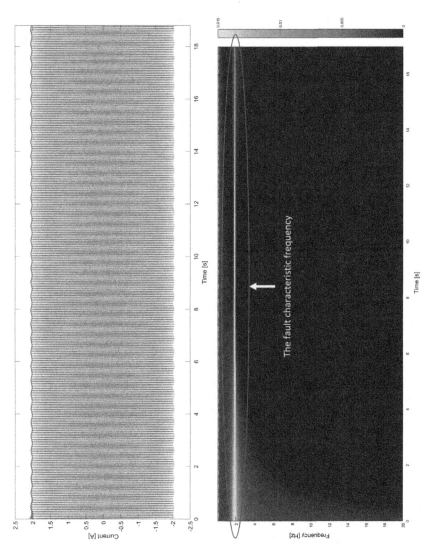

FIGURE 7.5 Stator current signal under an imbalance fault condition and its time-frequency spectrums of WT.

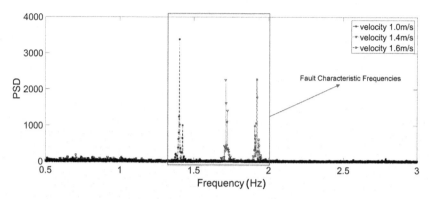

FIGURE 7.6 Spectrum of the fault characteristic signals under different water flow velocities.

is not easy to detect whether an imbalance fault happens by setting the control limit [34]. Moreover, the interference signal, which is generated by turbulence and wave, would vary with the water flow velocity. Therefore, the stability of the detection needs to be improved.

7.2.2 The wavelet threshold denoising–based detection method

At the aim of solving the preceding problems, a method called *wavelet threshold denoising–based detection* is given. This method consists of three steps: the parameters offline setting, the HT, and PCA-based detection, and by using this strategy, the fault could be found easily and the stability of this method in different water flow velocities is good:

(1) *Parameters offline setting.* For the purpose of decreasing the interference signal from the stator current signal, wavelet threshold denoising is chosen because of its good effect in decreasing noise. During the fault detection process of the MCT system, there may be some sudden changes due to interference, which could be considered as noise. With the calculated threshold, the wavelet coefficients, which denote the interference part, could be processed. In addition, the interference signal could be decreased. The details of this wavelet threshold denoising method can be found in the work of Liu et al. [35].

In the process of this denoising method, there are three crucial parameters: the number of decomposition layers, the wavelet basis function, and the threshold selection function. These parameters could affect the denoising performance [36]. An offline method is given to choose suitable parameters, and it consists of two parts, which are shown as follows:

(a) *The number of decomposition layers.* There is a certain correlation between every decomposition layer and the sampling frequency F_s. Define the number of the layers to be j, then the bandwidth of the minimum band could be expressed as $B_j = F_s/2^{j+1}$. In the actual MCT system, the first harmonic is mainly used to detect whether an imbalance fault occurs [37]. For the purpose of detecting the imbalance fault of the MCT system, the bandwidth of the minimum band needs to meet this following requirement:

$$B_j \geq 2f, \tag{7.14}$$

where f means the mean value of the frequency signal. Replacing the B_j in Eq. (7.14), the number of layers could be expressed by $j \leq \log_2[F_s/(4f)]$. To decrease the useless information from the original stator current signal in the decomposition process, the number of layers could be set as follows:

$$j = \log_2\left(\frac{F_s}{4f}\right), \tag{7.15}$$

where * means to round down the *.

(b) *The wavelet basis function and the threshold selection function.* Different groups of parameters are chosen to process the historical stator current signal $x(n)$ and decrease the interference signal in it. Haar, Db4, Coif4, and Sym4 are chosen as the different wavelet basis functions. Sqtwolog, rigrsure, heursure, and minimaxi are chosen as the different threshold selection functions. In addition, these parameters could be combined into different parameter groups in turn. To analyze the performance of these groups and choose the suitable group, the signal-to-noise ratio (SNR), the mean square error (MSE), and the correlation coefficient (CORR) are selected as the evaluation standard of denoising performance [38].

The SNR could show how much the noise is removed. After using the i-th group ($i = 1, 2, 3,..., I$, where I is the number of these groups), the SNR_i could be expressed as follows:

$$SNR_i = 10\log_{10}\left(\frac{\frac{1}{N}\sum_{n=1}^{N} Y_i^2(n)}{\frac{1}{N}\sum_{n=1}^{N}[x(n) - Y_i(n)]^2}\right), \tag{7.16}$$

where $Y_i(n)$ denotes the result after using the i-th group. With the MSE, the deviation of the raw signal and the signal after denoised could be reflected. This criterion could be expressed as follows:

$$MSE_i = \frac{1}{N}\sum_{n=1}^{N}(x(n) - Y_i(n))^2, \tag{7.17}$$

where MSE_i denotes the MSE after using the i-th group. The CORR could show the correlation between two signals. The $CORR_i$ after using

TABLE 7.1 Parameter group setting rules for wavelet threshold denoising.

Different situations	Parameter group setting results
$i_{snr} \neq i_{mse} \neq i_{corr}$	The i_{corr}-th parameter group should be set as the final group.
$a = i_{snr} = i_{mse} \neq i_{corr}$ or $a = i_{snr} = i_{corr} \neq i_{mse}$ or $a = i_{mse} = i_{corr} \neq i_{snr}$	The a-th parameter group is selected.
$b = i_{snr} = i_{mse} = i_{corr}$	The b-th parameter group is selected.

the i-th group could be expressed as

$$CORR_i = \frac{\sum_{n=1}^{N} (x(n) - \bar{x})\left(Y_i(n) - \bar{Y}_i\right)}{\sqrt{\sum_{n=1}^{N} (x(n) - \bar{x})^2}\sqrt{\sum_{n=1}^{N} \left(Y_i(n) - \bar{Y}_i\right)^2}}, \tag{7.18}$$

where $\bar{x}$ indicates the mean value of $x(n)$, whereas $\bar{Y}_i$ indicates the mean value of $Y_i(n)$. With these criteria, the denoising results of different parameter groups could be analyzed. The best result would meet: the smallest SNR, the smallest MSE, and the largest CORR. By comparing the results using different groups, a suitable group could be set. The suitable parameter group corresponding to each criterion could be selected by

$$[M_{snr}, i_{snr}] = \min(SNR_i), \tag{7.19}$$

$$[M_{mse}, i_{mse}] = \min(MSE_i), \tag{7.20}$$

$$[M_{corr}, i_{corr}] = \max(CORR_i), \tag{7.21}$$

where M_{snr}, M_{mse}, and M_{corr} are optimal results of these criteria. In application, the results may not meet all of the preceding requirements (which means that i_{snr}, i_{mse}, and i_{corr} could be different). The parameter group of the wavelet threshold denoising that meets as many requirements mentioned earlier as possible is chosen.

There are three different conditions in the relationship between i_{snr}, i_{mse}, and i_{corr}: when i_{snr}, i_{mse}, and i_{corr} are different from each other, the i_{corr}-th group is chosen to make sure there is enough useful information retained for the detection method to detect whether an imbalance fault occurs; when two of i_{snr}, i_{mse}, and i_{corr} are the same (the value is a), the a-th group is chosen to make sure that enough interference signals are decreased; when i_{snr}, i_{mse}, and i_{corr} are the same (the value is b), the b-th group is chosen to make sure the denoising performance is good. The parameter group setting rules are given in Table 7.1.

(2) *The instantaneous frequency estimation based on HT.* The HT is selected to calculate the instantaneous frequency signal of the current signal. Details of the HT method are shown in Section 7.1.2.

The instantaneous frequency signal f_e, which is calculated by HT, could be regarded as the fault characteristic signal. With the fault characteristic signal, the imbalance fault could be detected. After analyzing the instantaneous frequency signal in the frequency domain, the signal $s(k)$ could be obtained (k indicates the number of the frequency bands of the instantaneous frequency signal). By using the signal $s(k)$, the imbalance fault could be detected.

(3) *PCA-based detection.* For the purpose of decreasing the dimensions of signals and detecting whether there is an imbalance fault, a PCA-based detection method [39] is given. With PCA, the dimensions of the $s(k)$ could be decreased and a reference model could be built. The statistics and their control limits could be estimated. This PCA-based detection method could be divided into the following three steps:

(a) *Data normalization.* The $s(k)$ is processed first. Use all of the $s(k)$ acquired in different water flow velocities to form a new matrix $S \in \mathbf{R}^{q \times k}$ (where q denotes the number of samples). The S could be normalized as follows:

$$S^* = \frac{S - \bar{S}}{\sqrt{\mathrm{Var}(S)}}, \tag{7.22}$$

where $\bar{S}$ denotes the mean value of S, whereas $\mathrm{Var}(S)$ denotes the variance value of S.

(b) *The data matrix model.* The matrix S^* could be defined as follows:

$$S^* = TP^T, \tag{7.23}$$

where the matrix $T = [t_1, t_2, \ldots, t_k] \in \mathbf{R}^{q \times k}$ contains the transformed variables, the vectors $t_i \in \mathbf{R}^q$ denote the principal components (PCs), and the matrix $P = [p_1, p_2, \ldots, p_k] \in \mathbf{R}^{k \times k}$ contains the orthogonal vectors $p_i \in \mathbf{R}^k$. The corresponding covariance matrix could be defined as C and could be calculated as follows:

$$C = S^{*T}S^*/(q-1) = P\Lambda P^T, \tag{7.24}$$

where $PP^T = P^TP = I_k$. $\Lambda = \mathrm{diag}\{\lambda_1, \lambda_2, \ldots, \lambda_k\}$ ($\lambda_1 \geq \lambda_2 \geq \ldots \geq \lambda_k$) is the diagonal matrix, and it includes the eigenvalues that are related to the k PCs. The matrix I_k denotes the identity matrix. At the aim of decreasing the amount of computation, the amount of PCs (defined as l) has to be reduced. To determine the value of l, the cumulative percent variance (CPV) is chosen and could be calculated as follows:

$$CPV(l) = \frac{\sum_{i=1}^{l} \lambda_i}{\sum_{i=1}^{k} \lambda_i} \times 100\%. \tag{7.25}$$

The S^* could be calculated by

$$S^* = TP = [\hat{T}\tilde{T}][\hat{P}\tilde{P}]^T, \qquad (7.26)$$

where the matrix $\hat{T} \in \boldsymbol{R}^{q \times l}$ contains l retained PCs, the matrix $\tilde{T} \in \boldsymbol{R}^{q \times (k-l)}$ contains $k - l$ PCs, the matrix $\hat{P} \in \boldsymbol{R}^{q \times l}$ contains l retained eigenvectors, and the matrix $\tilde{P} \in \boldsymbol{R}^{q \times (k-l)}$ contains $k - l$ eigenvectors. Eq. (7.26) could be defined as follows:

$$
\begin{aligned}
S^* &= \hat{T}\hat{P}^{\mathrm{T}} + \tilde{T}\tilde{P}^{\mathrm{T}} \\
&= S^*\hat{P}\hat{P}^{\mathrm{T}} + S^*\left(I_k - \hat{P}\hat{P}^{\mathrm{T}}\right) \\
&= \hat{S^*} + \tilde{S^*},
\end{aligned} \qquad (7.27)
$$

where $\hat{S^*}$ denotes the S^* in the principal subspace, whereas $\tilde{S^*}$ denotes the projection of S^* in the residual subspace.

(c) *Statistics and control limit.* Two kind of statistics are selected: the T^2 and the Q statistics [40]. These two statistics could be calculated as follows:

$$T^2 = ||S^*||_2^2 = S^{*\mathrm{T}}P\Lambda^{-1}P^T S^* \le T_\alpha^2, \qquad (7.28)$$

$$Q = ||\tilde{S^*}||_2^2 \le \delta_\alpha^2, \qquad (7.29)$$

where T_α^2 means the limit for the T^2 statistic, and δ_α^2 means the limit for the Q statistic. When the newly calculated statistic exceeds the limit, it means that an imbalance fault has happened.

This detection method could be divided into two parts: offline training and online fault detection. In the first part, the historical current signal is used. The parameters, which are needed in the denoising process, could be selected. The reference model of the healthy MCT system is built. In the second part, the parameters, which are selected in the offline training part, could be used to decrease the interference signals. Whether an imbalance fault happens could be detected by comparing the calculated statistics with their corresponding control limits.

7.2.3 Simulation results and analysis

The fault detection performance of this fault detection method is verified by the MCT simulation model built in Section 7.1.3. The equivalent mass of 200 g is set, the center of mass is 0.1 m away from the turbine axis, and the load is set to 50 Ω:

(1) *Detection results under the same water flow velocity.* At the aim of showing the advantages of this method, simulations under different blade conditions in the same velocity of water flow are used.

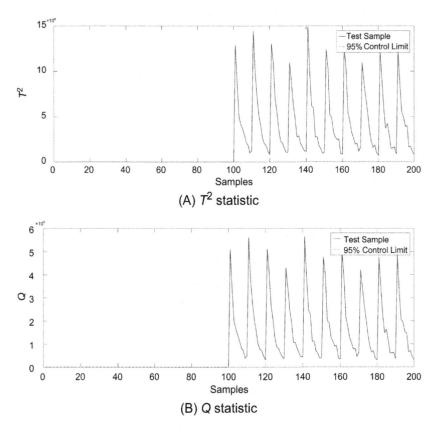

(A) T^2 statistic

(B) Q statistic

FIGURE 7.7 Simulation detection results under the same water flow velocity.

In the simulation, select 200 signal groups under the healthy condition and 100 signal groups under the fault condition in the same velocity of water flow (around 1.2 m/s). The length of every signal group is 3000. A total of 100 signal groups were chosen from the healthy current signals to compose the offline training data (the size of the data is 100*3000). The rest of the signals compose the online test data (the size of the data is 200*3000).

The frequency value of the historical healthy current signal is about 9.2 Hz (the turbine rotation frequency is about 1.15 Hz). The number of wavelet decomposition layers j is chosen as 5. Coif4 and rigrsure are set as the wavelet basis function and the threshold selection function. The contribution rate of PCs is set to 95%. By calculating the T^2 and Q statistics, the imbalance fault detection results are given in Fig. 7.7. It can be seen from this figure that both the T^2 statistic and the Q statistic could get good performance in detecting whether an imbalance fault happens. By using the T^2 statistic, the rate of missed detection and the false alarm rate are 0% and

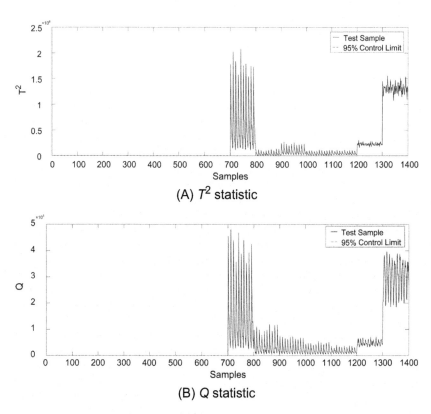

(A) T^2 statistic

(B) Q statistic

FIGURE 7.8   Simulation detection results under different water flow velocities.

0%, respectively; by using the Q statistic, the rate of missed detection and the false alarm rate are 0% and 3%, respectively.

(2) *Detection results under different water flow velocities.* At the aim of proving the stability of this method under different water flow velocities, 200 signal groups under the healthy condition and 100 signal groups under the fault condition in different velocities of water flow were selected (those water flow velocities are about 1.2, 1.4, 1.6, 1.7, 2.0, 2.2, and 2.4 m/s). The length of every signal group is 3000. A total of 100 signal groups were chosen from the healthy current signals to compose the offline training data (the size of the data is 700*3000). The rest of the signals compose the online test data (the size of the data is 1400*3000).

The frequency value of the historical healthy current signal is about 10 to 16 Hz (the turbine rotation frequency is about 1.25 to 2 Hz). The number of wavelet decomposition layers j is chosen as 4. Coif4 and rigrsure are set as the wavelet basis function and the threshold selection function. The contribution rate of PCs is set to 95%. Fig. 7.8 gives the simulation fault detection results (the water flow velocity conditions and health states are given in Table 7.2). It could be found from this figure that both the T^2 statistic and the Q statistic could be used to detect whether an imbalance fault happens. By using the T^2 statistic, the rate of missed detection and the false alarm rate are 0% and 6.29%, respectively; by using the Q statistic, the rate of missed detection and the false alarm rate are 0% and 8.43%, respectively.

7.2.4 Experimental results and analysis

At the aim of evaluating the detection performance and stability of the wavelet threshold denoising–based detection method, an experimental platform [41] is used. The number of pole pairs of the MCT is 8. The sampling frequency of this platform is 1 kHz. Emulate the fault by adding ropes to the blade of the MCT:

(1) *Detection results under the same water flow velocity.* In the experiment, we selected 150 signal groups under the healthy condition and 100 signal groups under the fault condition in the same velocity of water flow (around 1.0 m/s). The length of every signal group is 3000. A total of 50 signal groups were chosen from the healthy current signals to compose the offline training data (the size of the data is 50*3000). The rest of the data compose the online test data (the size of the data is 200*3000).
The frequency value of the historical healthy current signal received from this experimental platform is about 11.7 Hz (the turbine rotation frequency is about 1.5 Hz). The number of wavelet decomposition layers j is chosen as 4. Coif4 and rigrsure are set as the wavelet basis function and the threshold selection function. The contribution rate of PCs is set to 95%. By calculating the T^2 and Q statistics, the detection results are given in Fig. 7.9. It can be seen from this figure that the Q statistic could get good performance in detecting the imbalance fault. The rate of missed detection and the false alarm rate are 0% and 5.5%, respectively.

(2) *Detection results under different water flow velocities.* In the experiment, we selected 150 signal groups under the healthy condition and 100 signal groups under the fault condition in different velocities of water flow (the water flow velocities are about 1.0, 1.1, 1.2, 1.3, and 1.4 m/s). The length of every signal group is 3000. A total of 50 signal groups were chosen from the healthy current signals to compose the offline training data (the size of the data is 250*3000). The rest of the signals compose the online test data (the size of the data is 1000*3000).
The frequency value of the historical healthy current signal received from this experimental platform is about 10 to 16 Hz (the turbine rotation frequency is about 1.25 to 2 Hz). The number of wavelet decomposition layers j is chosen as 4. Coif4 and rigrsure are set as the wavelet basis function and the threshold selection function. The contribution rate of PCs is set

TABLE 7.2 Water flow velocity conditions and health states of simulation samples.

Samples	1-100	101-200	201-300	301-400	401-500	501-600	601-700	701-800	801-900	901-1000	1001-1100	1101-1200	1201-1300	1301-1400
Velocity condition	1.2 m/s	1.4 m/s	1.6 m/s	1.8 m/s	2.0 m/s	2.2 m/s	2.4 m/s	1.2 m/s	1.4 m/s	1.6 m/s	1.8 m/s	2.0 m/s	2.2 m/s	2.4 m/s
Health state	Healthy state							Fault state						

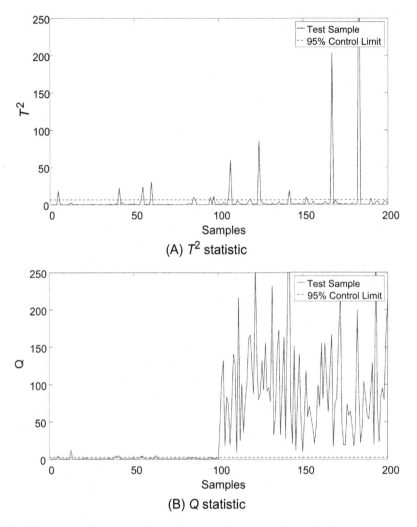

FIGURE 7.9 Experimental detection results under the same water flow velocity.

to 95%. Fig. 7.10 gives the experimental fault detection results (the water flow velocity conditions and health states are shown in Table 7.3). It can be seen from this figure that the imbalance fault could not be detected by using the T^2 statistic. By using the Q statistic, the performance of the imbalance fault detection could satisfy the requirements of fault detection. The rate of missed detection and the false alarm rate are 4.6% and 0.8%, respectively.

The wavelet threshold denoising–based detection method is a nonintrusive method that does not need to intrude on the equipment. This method could be

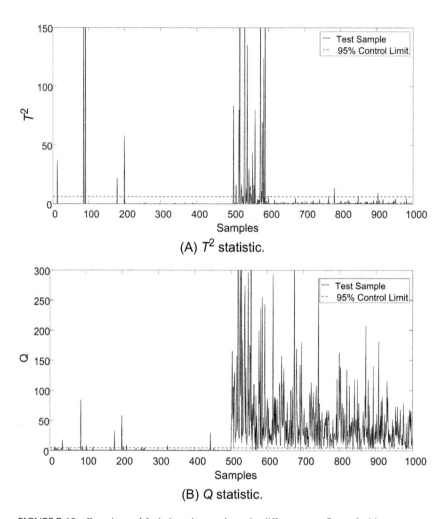

(A) T^2 statistic.

(B) Q statistic.

FIGURE 7.10 Experimental fault detection results under different water flow velocities.

TABLE 7.3 Water flow velocity conditions and health states of experimental samples.

Samples	1-100	101-200	201-300	301-400	401-500	501-600	601-700	701-800	801-900	901-1000
Velocity condition	1.0 m/s	1.1 m/s	1.2 m/s	1.3 m/s	1.4 m/s	1.0 m/s	1.1 m/s	1.2 m/s	1.3 m/s	1.4 m/s
Health state	Healthy state					Fault state				

implemented in the following three steps: (1) use the wavelet threshold denoising to decrease the interference signal under different water flow velocities; (2) use the HT to calculate the instantaneous frequency signal; and (3) apply the PCA and detect whether an imbalance fault happens in the MCT system by comparing the statistics with their corresponding control limits. Both simulation and experimental results verify that this method could detect whether an imbalance fault happens in the MCT system. In addition, this method has good performance of stability under different water flow velocities.

7.3   The identification method of blade attachment based on the sparse autoencoder and softmax regression

7.3.1   Problem description

According to Zhang et al. [12,41], methods are proposed based on electrical signal to diagnose imbalance faults of the MCT. However, the deformation of the blade also affects MCT power production. This kind of fault is resulted in by uniform and symmetrical attachment. In this part, the output voltages of the MCT are sampled under health condition and uniform attachment; fast Fourier transform is used to analyze the output voltage, and the analysis results are presented in Fig. 7.11. The frequency behavior of uniform attachment is better in terms of harmonic components distribution and amplitude. This results in the difficulties to accurately diagnose the deformation of the blade by an electrical signal.

We choose the MCT image signal as a fault diagnosis signal in this part. The onshore lighting condition is significantly better than the underwater environment, as there is natural light. However, underwater cameras have to depend on the artificial light, which results in difficulties of feature extraction due to poor visibility [42]. According to the preceding reason, the feature extraction method is significant for attachment degrees diagnosis based on image recognition.

7.3.2   The recognition method based on the sparse autoencoder and softmax regression

This method consists of four steps as shown in Fig. 7.12. In the training step, part of the data is divided into unlabeled data and labeled data. In step 1, the convolution kernels are pretrained by using the sparse autoencoder, where preprocessed unlabeled images are fed into the sparse autoencoder [43]; in step 2, the convolved features are calculated by making the convolution between trained convolution kernels and the labeled images; in step 3, obtain the pooled features by using average pooling operation, where convolutional features are as the input of average pooling operation; finally, in step 4, use the softmax classifier to diagnose the faults category, where the input of the softmax [44] classifier is pooled features:

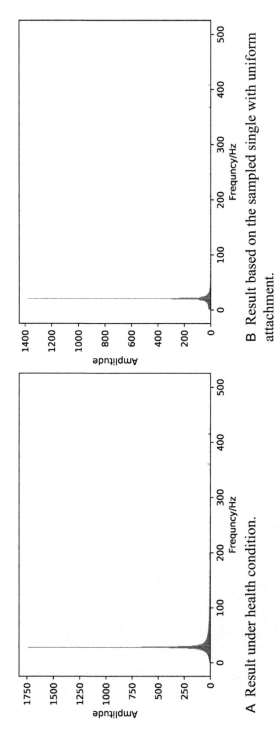

A Result under health condition.

B Result based on the sampled single with uniform attachment.

FIGURE 7.11 Fast Fourier transform results based on the output voltage of MCT under different conditions.

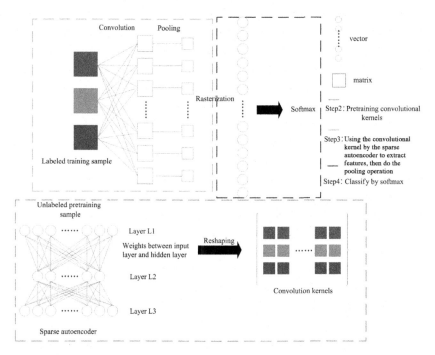

FIGURE 7.12   Frame of the method based on the sparse autoencoder and softmax regression.

(1) *Data preprocessing.* The unlabeled images are used to extract patches for training the sparse autoencoder. We extracted 500 patches of 20 × 20 from each unlabeled image, which are arranged in matrix $X_{unlabel} = \left[x^1_{unlabel}, \ldots, \ldots\right]$, where $x^k_{unlabel}$ is the k-th column of $X_{unlabel}$. $X^*_{unlabel}$ is obtained by removing the mean of $X_{unlabel}$. We calculate the covariance matrix of $X^*_{unlabel}$. Then we calculate the eigenvectors of the covariance matrix of $X^*_{unlabel}$ and the diagonal matrix of eigenvalues. Finally, the zero-phase component (ZCA) whitening [45] is used to calculate matrix $X_{whitening}$:

$$x^{*k}_{unlabel} = x^k_{unlabel} - \frac{1}{m}\sum_{i=1}^{m} x^i_{unlabel}, \tag{7.30}$$

$$C_X = \frac{1}{m}X^*_{unlabel}\left(X^*_{unlabel}\right)^T, \tag{7.31}$$

$$X_{whitening} = U(S + I)^{-\frac{1}{2}}X^*_{unlabel}, \tag{7.32}$$

where $x^{*k}_{unlabel}$ is the k-th column of $X^*_{unlabel}$, C_X is the covariance matrix of $X^*_{unlabel}$, m is the number of samples, S is the diagonal matrix of eigenvalues and U is the eigenvectors of C_X, and ε is the regularization parameter.

(2) *Pretraining convolutional kernels based on the sparse autoencoder.* When training the CNN, convolutional kernels and parameters of softmax are

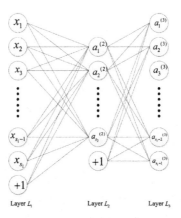

Layer L_1 Layer L_2 Layer L_3

FIGURE 7.13   SA neural network structure.

simultaneously trained. In this part, convolutional kernels and parameters of softmax classifier are asynchronously trained. In addition, the sparse autoencoder is used to train convolutional kernels.

The sparse autoencoder neural network is shown in Fig. 7.13. The sparse autoencoder has three layers: the input layer (L_1), hidden layer (L_2), and output layer (L_3), where "+1" is the threshold. Input of the sparse autoencoder equals ideal output, which means that the sparse autoencoder can learn features by encoding and decoding training data. Assuming the preprocessed input matrix $X_{whitening} = [x^1, ..., x^{80000}]$, where x^k is the k-th column of $X_{whitening}$, $x^k \in \mathbb{R}^n$, and $n = 1200$ is the number of pixels of each patch. Input $Z^{(2)}$ of the action function of the hidden layer is obtained by making matrix multiplication between input matrix $X_{whitening}$ and $W^{(1)}$. Output $A^{(2)}$ is obtained by using the action function of the hidden layer. The detail is shown in Eq. (7.33) and Eq. (7.34). $W_{ji}^{(1)}$, for $i = 1, ..., s_1, j = 1, ...,$ s_2, denotes the weight connecting the i-th element of L_1 and the j-th element of L_2. $b^{(1)}$ denotes the input threshold in L_2. The activation function of L_2 is the sigmoid function, where $s_1 = 1200$ is the number of elements of L_1 and $s_2 = 800$ is the number of elements of L_2. Input $Z^{(3)}$ of the action function of L_3 is obtained by making matrix multiplication between input matrix $Z^{(2)}$ and $W^{(2)}$. Output $A^{(3)}$ is obtained by using the action function of L_3. The detail is shown in Eq. (7.35) and Eq. (7.36). $W_{ij}^{(2)}$, for $i = 1, ..., s_3, j = 1, ...,$ s_2, denotes the weight connecting the j-th element of L_2 and the i-th element of L_3, and the activation function of L_3 is the proportional function, where $s_3 = 1200$ is the number of elements of L_3. After having trained the sparse autoencoder, the convolutional kernels are obtained by reshaping $W^{(1)}$:

$$z_j^{(2)} = \sum_{i=1}^{S_1} W_{ji}^{(1)} x_i + b_j^{(1)}, \tag{7.33}$$

$$a_j^{(2)} = f_1\left(z_j^{(2)}\right) = \frac{1}{1 + \exp\left(-z_j^{(2)}\right)}, \tag{7.34}$$

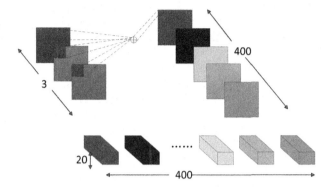

FIGURE 7.14 Convolution operation.

$$z_i^{(3)} = \sum_{j=1}^{S_2} W_{ij}^{(2)} a_j^{(2)} + b_i^{(2)}, \tag{7.35}$$

$$a_i^{(3)} = f_2\left(z_i^{(3)}\right) = t z_i^{(3)}, \tag{7.36}$$

where x_i is the i-th element of vector x; $z_j^{(2)}$ and $a_j^{(2)}$ are the output of the linear transformation and the activation function in the j-th elements of L_2, respectively; $z_i^{(3)}$ and $a_i^{(3)}$ are the output of the linear transformation and the activation function in the i-th element of L_3, respectively; and t is the proportionality coefficient.

(3) *Feature extraction method based on convolution and pooling.* In this step, the convolutional feature maps are obtained by trained convolution kernels, and the pooled features are extracted by pooling convolutional feature maps. Different feature values are calculated at each location in the labeled image by making convolution between each labeled image and the pre-trained kernels. The convolution operation is shown in Fig. 7.14. A pixel and its neighboring pixels are multiplied by the convolutional kernel. Then the convolved feature is obtained by adding the product. This operation will be repeated many times until every pixel of the image is covered.

The pooling operation is the process of extracting the features from the output of a convolution, which is applied to reduce the size of convolutional feature maps. The pooling operation is shown in Fig. 7.15. This will also follow the same process of sliding over the convolved features with a specified pool size. Two types of pooling are available: max pooling and average pooling. Because the features are not complex, the average pooling operation is used in this part [46].

(4) *A classifier based on softmax regression.* According to attachment degrees, the different labels are set. After obtaining the pooling features, the pooling features are fed into the softmax classifier. The detailed formula is shown in

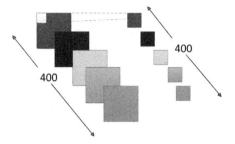

FIGURE 7.15 Pooling operation.

TABLE 7.4 Diagnostic category label.

Attachment degree	Classifier labels
(0%, 0%, 0%–5%)	1
(0%, 0%, 5%–20%)	2
(0%, 0%, 20%–40%)	3
(0%, 0%, 40%–60%)	4
(0%, 0%, 60%–90%)	5
(0%, 60%–90%, 60%–90%)	6
(60%–90%, 60%–90%, 60%–90%)	7

Eq. (7.37), where θ is a parameter of softmax and x is the input of softmax. The output is obtained by Eq. (7.37), where $y^{(i)}$ is the label of $x^{(i)}$ and $p(y^{(i)} = 1|x^{(i)}; \theta)$ is the probability that x is classified to category 1.

$$
h_\theta\left(x^{(i)}\right) = \begin{bmatrix} p(y^{(i)} = 1|x^{(i)}; \theta) \\ p(y^{(i)} = 2|x^{(i)}; \theta) \\ \vdots \\ p(y^{(i)} = k|x^{(i)}; \theta) \end{bmatrix} = \frac{1}{\sum_{j=1}^{k} e^{\theta_j x^{(i)}}} \begin{bmatrix} e^{\theta_1 x^{(i)}} \\ e^{\theta_2 x^{(i)}} \\ \vdots \\ e^{\theta_k x^{(i)}} \end{bmatrix}. \tag{7.37}
$$

7.3.3 Experimental results and analysis

In this experiment, each category of images will be sampled from four different configurations as shown in Fig. 7.16. Sampled images consist of unlabeled pretraining images, labeled training images, and the testing images. The label and datasets information are shown in Table 7.4 and Table 7.5. The model parameters and model are shown in Table 7.6, and Table 7.7.

The parameters of the sparse autoencoder and softmax have been trained repeatedly 20 times. In addition, Gaussian noises are added in the training sample and test sample. The convolution kernels shape of the compared CNN

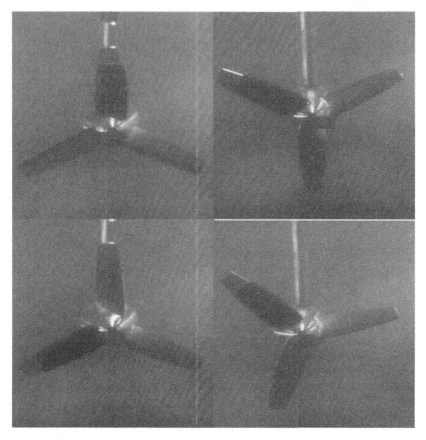

FIGURE 7.16   Single blade with different degrees of attachment.

TABLE 7.5 Detail of datasets.

Dataset	Number
Unlabeled pretraining image	160
Labeled training image	420
Testing image	280

is consistent with kernels trained by the sparse autoencoder. However, kernels values of the compared CNN are different for the sparse autoencoder because the kernels and parameters of softmax are trained simultaneously.

Due to harsh working conditions, the MCT image is chosen as the diagnostic signal to diagnose the different biological attachment degrees. The parameters sparse autoencoder is reshaped and used as a convolution kernel to extract

TABLE 7.6 Parameters for the method.

Parameters	Value
ε	0.1
M	80,000
λ_1	0.003
β	3
ρ	0.1
λ_2	0.0001
Hidden size	400
T	1

TABLE 7.7 Diagnostic results based on different methods.

Diagnosis method	Average accuracy
SA + softmax	98.214%
CNN	97.500%

features and softmax regression is used to classify attachment degree. In addition, this part simultaneously compares the training method (CNN). The experimental results show that the identification method based on the sparse autoencoder and softmax regression is effective to diagnose the different attachment degrees and shows higher accuracy.

7.4 The identification method of blade attachment based on depthwise separable CNN

7.4.1 Problem description

The limited light source is a disadvantage of the underwater environment. Meanwhile, the tide can create a rapid current up to 7 m/s [47], which causes MCT to rotate at high speed. Because of the preceding two reasons, images taken from the underwater environment are blurry and dim. An instance of an image taken from the underwater environment is shown in Fig. 7.17. In the worst situation, these images can be unrecognizable by a human. In addition, static and rotating MCT images are presented in Fig. 7.18. Motion blur has a great effect on MCT images.

Zheng et al. [48] use the sparse autoencoder to extract features, which causes insufficient feature extraction in harsh working conditions. In their work [48], 500 patches are extracted from each unlabeled sample (the number of unlabeled samples is 160). Therefore, when the number of unlabeled samples is large, the

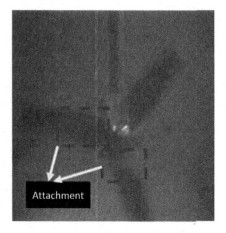

FIGURE 7.17 Image taken from abominable environment.

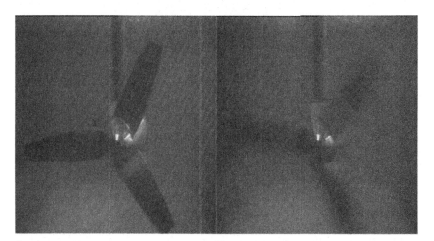

A Static MCT.   B Rotating MCT.

FIGURE 7.18 Static and rotating images taken from an abominable environment.

sparse autoencoder with one hidden layer cannot fit the distribution. Meanwhile, after multiple experiments, we find that some networks based on depthwise separable convolution [49] cannot extract feature information under harsh working conditions. After multiple feature extraction, the feature maps lose information. For example, a blurry image taken from the abominable environment is used as input to a trained MobileNetV2 [50], and the feature maps of the first and seventh convolutional block are shown in Fig. 7.19. The MCT features information of the first convolutional block can be presented in the heat map, which means that the first convolutional block can extract features. However, the feature maps of the seventh convolutional layer only show some simple colors, which means that MobileNetV2 cannot effectively extract and keep the MCT features.

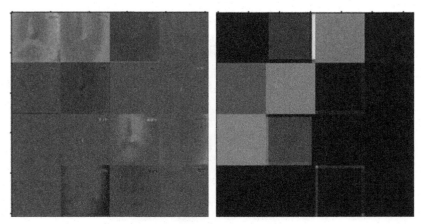

A Feature maps of 1st convolutional block. B Feature maps of 7th convolutional block.

FIGURE 7.19 Feature maps with information loss.

7.4.2 The recognition method based on depthwise separable CNN

This method consists of three steps. In step 1, the images are processed by nearest-neighbor interpolation and normalization; in step 2, the features are calculated by using CNN; in step 3, use the softmax classifier to diagnose the faults category, where the input of the softmax classifier is convolutional features.

(1) *Preprocessed data based on nearest-neighbor interpolation and normalization processing.* To have high diagnostic accuracy and efficiency, images are preprocessed before feature extraction. This step includes nearest-neighbor interpolation and normalization processing. First, the nearest-neighbor interpolation is used to resize images, where m and n are raw image size, and m_d and n_d are input sizes of the network. The detailed information of X and $\hat{X}$ is presented in Eq. (7.38) and Eq. (7.39). The formula of nearest-neighbor interpolation is shown in Eq. (7.40):

$$X = \begin{bmatrix} x_{11} & \cdots & x_{1n} \\ \vdots & \ddots & \vdots \\ x_{m1} & \cdots & x_{mn} \end{bmatrix},$$

$$\hat{X} = \begin{bmatrix} \hat{x}_{11} & \cdots & \hat{x}_{1n_d} \\ \vdots & \ddots & \vdots \\ \hat{x}_{m_d 1} & \cdots & \hat{x}_{m_d n_d} \end{bmatrix},$$

$$\hat{X}_{i_d, j_d} = X_{i_s \left\lfloor \frac{m_d}{m} \right\rfloor || , j_s \left\lfloor \frac{n_d}{n} \right\rfloor}, \tag{7.40}$$

where i_s, j_s, i_d, and j_d are the row and column index of X and $\hat{X}$, respectively. Meanwhile, $\lfloor\rfloor$ is a floor function that can map a raw image index to a resized image index. Eq. (7.40) is a value of the matrix indexing operation. According to the indices of two matrices, each value of $\hat{X}$ is obtained by mapping the raw image value to a corresponding resized image value. Each resized image will be made as the normalization. Per channel (three channels) is removed mean μ and divided by the standard deviation σ. The detailed formulas are shown next.

$$\mu = \frac{1}{m_d n_d} \sum_{i_d=1}^{m_d} \sum_{j_d=1}^{n_d} \hat{X}_{i_d, j_d}, \tag{7.41}$$

$$\sigma = \sqrt{[2]} \frac{1}{m_d n_d} \sum_{i_d=1}^{m_d} \sum_{j_d=1}^{n_d} (\hat{X}_{i_d, j_d} - \mu)^2, \tag{7.42}$$

$$\widehat{\hat{X}} = \frac{\hat{X} - \mu}{\sigma}. \tag{7.43}$$

(2) *Feature extraction for attachment image.*

(a) *Feature extraction based on depthwise separable convolution.* The 3×3 kernels can capture the multiple vision notions by using the least amount of parameters. For reducing the number of parameters, kernel size greater than 3×3 is not chosen in this method. Meanwhile, a 1×1 kernel can be used to efficiently change the number of channels of feature maps [51, 52]. Therefore, the architecture built by stacking 1×1 and 3×3 convolutions can use a smaller number of parameters to extract features effectively.

Meanwhile, assuming the kernel size is $D_K \times D_K$, the number of channels of the input and output of convolution are M and N, respectively, and K is the size of one feature map of output. The number of multiplications for a traditional convolutional layer is $D_K \times D_K \times N \times M \times K \times K$. By using depthwise separable convolution, the number of multiplications is $M \times (D_K \times D_K \times K \times K + N)$ when the output size is the same as that of traditional convolution. After this operation, the 1×1 convolution is used to change the number of channels. According to Fig. 7.20, the number of parameters of depthwise separable convolution and 1×1 convolution is $D_K \times D_K \times M + N \times M$; however, the number of parameters of traditional convolution is $D_K \times D_K \times M \times N$. Compared with traditional convolution, depthwise separable convolution makes faster feature extraction. Meanwhile, depthwise separable convolution uses fewer parameters to extract features, which means that depthwise separable convolution can be applied to extract features of the MCT by mobile devices.

f is used to denote a convolution operation. X_c and X_{last} (X_{last} can be $\widehat{\hat{X}}$ or outputs of activation function) are used to denote the convolution

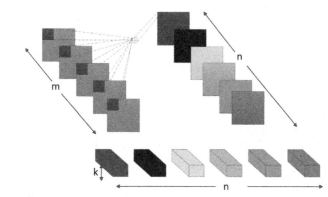

A Traditional convolution kernels.

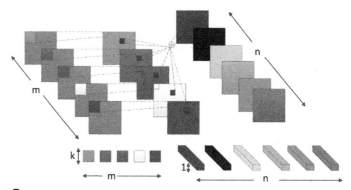

B Depthwise separable convolution kernels.

FIGURE 7.20 Different convolution operations.

output and input, respectively.

$$X_c = f(X_{last}).\qquad(7.44)$$

The feature maps X_c are calculated by depthwise separable convolution, and then the standardized features are calculated by batch normalization processing.

(b) *Fitting distribution based on batch normalization.* Batch normalization [53] is used to extract the fault features effectively, which makes the network better fit the distribution. The inputs are values of X_{ci} over a mini-batch $\mathcal{B} = \{X_{c1} \ldots X_{ck}\}$ and γ, β are trained parameters to be learned where each X_{ci} is the output of the last convolutional layer. First, the mini-batch mean and variance are computed, followed by mini-batch normalization. Finally, mini-batch is scaled and shifted by γ, β.

$$\mu_{\mathcal{B}} = \frac{1}{k}\sum_{i=1}^{k} X_{ci},\qquad(7.45)$$

$$\sigma_B^2 = \frac{1}{k} \sum_{i=1}^{k} (X_{ci} - \mu_B)^2, \tag{7.46}$$

$$\hat{X}_{ci} = \frac{X_{ci} - \mu_B}{\sqrt{\sigma_B^2 + \epsilon}}, \tag{7.47}$$

$$BN_{\gamma,\beta}(X_{ci}) = Y_{ci} = \gamma \hat{X}_{ci} + \beta. \tag{7.48}$$

Batch normalization $BN_{\gamma,\beta}$ effectively fixes the distribution of MCT images data.

(c) *Selection of activation function.* After batch normalization, the output of batch normalization is fed into the activation function. In addition, depthwise separable convolution is used to extract convolutional features in MobileNet [54], which sacrifices the feature quality (information is lost) for fast feature extraction. Meanwhile, if 1×1 convolution is used to change the number of channels from majority to minority, the information will be lost when ReLU is used as the activation function [55]. Therefore, linear activation function and ReLU6 are used to keep information and improve the robustness [50], where ReLU6 is used to make nonlinear activation for the outputs of batch normalization Y_{ci}. The detail formula is presented in Eq. (7.49), where y_{ci} is one of the elements in Y_{ci}.

$$ReLU6(Y_{ci}) = \begin{cases} 6, & y_{ci} > 6 \\ y_{ci}, & 6 \geq y_{ci} > 0 \\ 0, & y_{ci} \leq 0 \end{cases} \tag{7.49}$$

(3) *A classifier based on softmax regression.* After multiple feature extractions, diagnosis features are obtained by a linear transformation, and the detailed formula of softmax is shown in the following. Suppose θ is a parameter matrix and x is the input of softmax. The input of the normalized exponential function is obtained by making matrix multiplication between θ and x. The normalized exponential function is applied to each element $z^{(i)}$. Then the probability of the corresponding category is calculated by dividing the sum of all of these exponentials. The detail is shown in Eq. (7.50) and Eq. (7.51), where $y^{(i)}$ is the label of $z^{(i)}$ and $p(y^{(i)} = k|z^{(i)})$ is a probability that x is classified to category k.

$$z = \theta x \tag{7.50}$$

$$h\left(z^{(i)}\right) = \begin{bmatrix} p(y^{(i)} = 1|z^{(i)}) \\ p(y^{(i)} = 2|z^{(i)}) \\ \vdots \\ p(y^{(i)} = k|z^{(i)}) \end{bmatrix} = \frac{1}{\sum_{j=1}^{k} e^{z_j^{(i)}}} \begin{bmatrix} e^{z_1^{(i)}} \\ e^{z_2^{(i)}} \\ \vdots \\ e^{z_k^{(i)}} \end{bmatrix}. \tag{7.51}$$

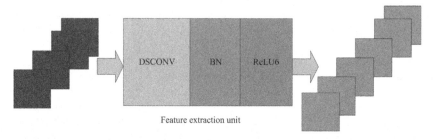

Feature extraction unit

FIGURE 7.21   Feature extraction unit.

(4) *Architecture with high diagnostic speed and high accuracy.* For high diagnostic speed and accuracy, we use a network based on MobileNet (a depthwise separable convolutional network) and experimentally choose the most suitable image size. The base feature extraction unit of the network based on MobileNetV1 is shown in Fig. 7.21, which consists of depthwise separable convolution, batch normalization, and ReLU6. The whole network consists of 14 feature extraction units and a softmax classifier. We use five-time subsampling to reduce the size of the feature map, and the detailed information is shown in Table 7.8. Convolution and depthwise separable convolution with two strides are used to reduce the size of the feature map the first four times, and the global average pooling operation is used to reduce size in the final time.

7.4.3   Experimental analysis

In this experiment, data is sampled from the environment illuminated by an artificial light source. In addition, each category will be sampled with 2400 images for a total of 19,200 images. The instances of the environment by the artificial light source are shown in Fig. 7.22. The data consists of the training set and test set. The training set and test set include 6400 and 12,800 images, respectively.

Because electrical signals cannot diagnose symmetrical faults, symmetrical faults (20%–40%, 20%–40%, 40%–60%) are set in this part. Table 7.9 shows the detailed percentage of attachment degree and corresponding classification labels.

Since the input size affects the diagnostic speed and accuracy, we find the most suitable input size for each CNN by multiple experiments. The results are shown in Table 7.10, where size: 256 denotes that the image sizes m_d and n_d are 256. The best input size is 128 for ResNet50, 96 for MobileNet and InceptionResNetV2. At the same time, the best performance numbers of each model are marked in bold.

Because the useful network is based on MobileNet, the residual unit is not used in the network. The network based on MobileNetV1 is still affected by the

TABLE 7.8 Detail of the architecture based on MobileNetV1.

Type/Stride	Kernel shape
Conv/s2	$3 \times 3 \times 3 \times 32$
DSCONV/s1	$3 \times 3 \times 32 + 32 \times 64$
DSCONV/s2	$3 \times 3 \times 64 + 64 \times 128$
DSCONV/s1	$3 \times 3 \times 128 + 128 \times 128$
DSCONV/s2	$3 \times 3 \times 128 + 128 \times 256$
DSCONV/s1	$3 \times 3 \times 256 + 256 \times 256$
DSCONV/s2	$3 \times 3 \times 256 + 256 \times 512$
DSCONV/s1	$3 \times 3 \times 512 + 512 \times 512$
DSCONV/s1	$3 \times 3 \times 512 + 512 \times 512$
DSCONV/s1	$3 \times 3 \times 512 + 512 \times 512$
DSCONV/s1	$3 \times 3 \times 512 + 512 \times 1024$
DSCONV/s1	$3 \times 3 \times 512 + 512 \times 512$
DSCONV/s1	$3 \times 3 \times 512 + 512 \times 512$
DSCONV/s2	$3 \times 3 \times 512 + 512 \times 1024$
DSCONV/s2	$3 \times 3 \times 1024 + 1024 \times 512$
Global Avg Pool/s1	Pooling operation
FC/s1	$1024 \times$ number of categories
Softmax/s1	Classifier

FIGURE 7.22   Data sampled from the environment illuminated by the artificial light source.

TABLE 7.9 Diagnostic category label in the work environment.

Percentage of the area occupied by attachment	Softmax classifier labels
0%, 0%, 0%	0
0%–20%, 20%–40%, 0%	1
0%–20%, 20%–40%, 40%–60%	2
0%–20%, 40%–60%, 0%	3
0%–20%, 0%, 0%	4
20%–40%, 20%–40%, 40%–60%	5
20%–40%, 0%, 0%	6
40%–60%, 0%, 0%	7

TABLE 7.10 Experimental results with different CNNs.

| CNN | Accuracy | | | | | | |
	Size: 256	Size: 224	Size: 192	Size: 160	Size: 128	Size: 96	Size: 64
ResNet50	83.84%	76.55%	80.38%	81.57%	**89.04%**	85.82%	57.74%
MobileNet	86.94%	81.43%	85.14%	88.93%	91.19%	**93.97%**	92.21%
Inception-ResNetV2	83.86%	86.54%	84.51%	85.88%	84.37%	**93.14%**	

TABLE 7.11 Experimental results with the number of layers in MobileNetV1.

No. of layers	Accuracy
13	93.02%
14	93.13%
15	93.97%
16	94.09%
17	94.25%
18	93.83%

vanishing gradient, which results in limited network depth. Therefore, we study the effect of the number of convolutional layers on the performance by multiple experiments. The accuracy of 15-layer MobileNet [54] is used as the benchmark, and 17-layer MobileNet presents the best accuracy; the average accuracy results are shown in Table 7.11.

TABLE 7.12 Performance indicator results with different CNNs.

CNN	No. of floating-point operations	No. of parameters	Accuracy
ResNet50	47,053,088	23,606,153	89.04%
CNN based on MobileNetV1 (17-layer MobileNet)	7,505,631	3,778,760	94.25%
InceptionResNetV2	108,549,709	54,350,569	93.14%

Performance indicators include accuracy and the number of floating-point operations and parameters; the detail is shown in Table 7.12. Compared with other networks, the network based on MobileNetV1 has higher accuracy.

To extract distinct features without sacrificing efficiency, we use data compression and normalization to process the raw image taken from the abominable working conditions. To overcome the difficulty of blurry and dim image feature extraction, an effective feature extraction method is chosen that consists of three parts: (1) set label including imbalance and symmetrical attachment faults; (2) use nearest-neighbor interpolation and normalization to preprocess images; and (3) extract image features through CNN-based depthwise separable convolution. The identification method of blade attachment based on depthwise separable CNN has four advantages: (1) it has high diagnostic accuracy and speed, (2) it is suitable for underwater working conditions without a natural light source, (3) it has effective imbalance and symmetrical attachment fault diagnoses, and (4) it is robust into the recognition of blurred images.

7.5 Conclusion and future works

An MCT's rotor and blade are often affected by attachment, which leads to an imbalance fault. Thus, the attachment degree diagnosis is an important domain for MCT research. To detect the imbalance faults more easily and reduce the influence, a HT-based detection method is introduced in this chapter. The results of simulation show that this method is useful to detect the imbalance faults based on the voltage signal for the direct-drive MCT.

To decrease the interference signals, which are generated by turbulence and waves in different velocities of water flow, a wavelet threshold denoising–based detection method is introduced to detect the imbalance fault for MCTs. This method can detect the imbalance fault automatically and has good stability in different velocities of water flow. The experimental results in different velocities of water flow with Q statistics have shown satisfactory imbalance fault detection with false alarm and false-negative rates less than 1% and 5%, respectively.

Only an imbalance fault can be detected based on the electrical signal, and if the attachment is evenly distributed on the blades, it is difficult to find the faults.

The identification method of blade attachment based on the sparse autoencoder and softmax regression are introduced to monitor whether the blade is attached by benthos and then to determine its corresponding degree of attachment. The experimental results show that this method is useful to classify the different degrees of biological attachment. To classify the percentage of area occupied by attachment, the identification method based on depthwise separable CNN is introduced in this chapter. The diagnostic accuracy and efficiency are high speed in this method, which is suitable for an underwater environment with strong currents and complex spatiotemporal variability, and it is effective uniform and symmetrical attachment fault diagnosis; meanwhile, it is robust into the recognition of blurred pictures under high-speed rotation.

It will be better to combine image features and electrical features to diagnose uniform faults and symmetrical attachment faults in the future.

References

[1]   Z. Ren, Y. Wang, H. Li, X. Liu, Y. Wen, W. Li., A coordinated planning method for micrositing of tidal current turbines and collector system optimization in tidal current farms, IEEE Transactions on Power Systems 34 (1) (2018) 292–302.

[2]   Y. Dai, Z. Ren, K. Wang, W. Li, Z. Li, W. Yan, Optimal sizing and arrangement of tidal current farm, IEEE Transactions on Sustainable Energy 9 (1) (2017) 168–177.

[3]   O.A.L. Brutto, M.R. Barakat, S.S. Guillou, J. Thiébot, H. Gualous, Influence of the wake Effect on electrical dynamics of commercial tidal farms: Application to the Alderney Race (France), IEEE Transactions on Sustainable Energy 9 (1) (2017) 321–332.

[4]   M.R. Barakat, B. Tala-Ighil, H. Chaoui, H. Gualous, Y. Slamani, D. Hissel, Energetic macroscopic representation of a marine current turbine system with loss minimization control, IEEE Transactions on Sustainable Energy 9 (1) (2017) 106–117.

[5]   S.B. Chabane, M. Alamir, M. Fiacchini, R. Riah, T. Kovaltchouk, S. Bacha, Electricity grid connection of a tidal farm: An active power control framework constrained to grid code requirements, IEEE Transactions on Sustainable Energy 9 (4) (2018) 1948–1956.

[6]   Z. Li, N. Maki, T. Ida, M. Miki, M. Izumi, Comparative study of 1-MW PM and HTS synchronous generators for marine current turbine, IEEE Transactions on Applied Superconductivity 28 (4) (2018) 1–5.

[7]   G.L. Wick, W.R. Schmitt, R. Clarke, Harvesting Ocean Energy (1981).

[8]   P.A. Lynn, Electricity from Wave and Tide: An Introduction to Marine Energy, John Wiley & Sons, West Sussex, UK, 2013.

[9]   H.T. Pham, J.M. Bourgeot, M. Benbouzid, Fault-tolerant finite control set-model predictive control for marine current turbine applications, IET Renewable Power Generation 12 (4) (2017) 415–421.

[10]   Z. Ren, H. Li, W. Li, X. Zhao, Y. Sun, T. Li, F. Jiang, Reliability evaluation of tidal current farm integrated generation systems considering wake effects, IEEE Access 6: (2018) 52616–52624.

[11]   Z. Zhou, M. Benbouzid, J.F. Charpentier, F. Sciuller, T. Tang, Developments in large marine current turbine technologies—A review, Renewable and Sustainable Energy Reviews 71 (2017) 852–858.

[12] M. Zhang, T. Tang, T. Wang. Multi-domain reference method for fault detection of marine current turbine. In *Proceedings of IECON 2017—The 43rd Annual Conference of the IEEE Industrial Electronics Society*, 2017. IEEE, Los Alamitos, CA, 2017. 8087–8092.

[13] R. Rosli, E. Dimla. A review of tidal current energy resource assessment: Current status and trend. In *Proceedings of the 5th International Conference on Renewable Energy: Generation and Applications (ICREGA)*, 2018. IEEE, Los Alamitos, CA, 34–40.

[14] X. Yang, N. Liu, P. Zhang, Z. Guo, C. Ma, P. Hu, X. Zhang, The current state of marine renewable energy policy in China, Marine Policy 100: (2019) 334–341.

[15] A. Uihlein, D. Magagna, Wave and tidal current energy–A review of the current state of research beyond technology, Renewable & Sustainable Energy Reviews 58: (2016) 1070–1081.

[16] A. Mérigaud, J.V. Ringwood, Condition-based maintenance methods for marine renewable energy, Renewable & Sustainable Energy Reviews 66 (2016) 53–78.

[17] T. Flanagan, J. Maguire, C.M. Ó'Brádaigh, P. Mayorga, A. Doyle, Smart affordable composite blades for tidal energy, Proceedings of the 11th European Wave and Tidal Energy Conference (EWTEC) (2015) 6–11.

[18] M. Mueller, R. Wallace, Enabling science and technology for marine renewable energy, Energy Policy 36 (12) (2008) 4376–4382.

[19] J.M. Walker, K.A. Flack, E.E. Lust, M.P. Schultz, L. Luznik, Experimental and numerical studies of blade roughness and fouling on marine current turbine performance, Renewable Energy 66: (2014) 257–267.

[20] B. Polagye, B. Van Cleve, A. Copping, K. Kirkendall, Environmental effects of tidal energy development, Proceedings of the Tidal Energy Workshop (2011).

[21] T. Wang, J. Qi, H. Xu, Y. Wang, L. Liu, D. Gao. Fault diagnosis method based on FFT-RPCA-SVM for cascaded-multilevel inverter. ISA Transactions 2016;60: 156–163.

[22] M. Zhang, T. Wang, T. Tang, M. Benbouzid, D. Diallo, Imbalance fault detection of marine current turbine under condition of wave and turbulence, in: Proceedings of ECON 2016—The 42nd Annual Conference of the IEEE Industrial Electronics Society, IEEE, Los Alamitos, CA, 2016, pp. 6353–6358.

[23] A.N. Einrí, G.M. Jónsdóttir, F. Milano. Modeling and control of marine current turbines and energy storage systems. *IFAC-PapersOnLine* 2091;52(4):425–430.

[24] G. Keenan, C. Sparling, H. Williams, F. Fortune. *SeaGen Environmental Monitoring Programme: Final Report*. Marine Current Turbines, Northern Ireland, UK, 2011.

[25] W. Li, H. Zhou, H. Liu, Y. Lin, Q. Xu, Review on the blade design technologies of tidal current turbine, Renewable & Sustainable Energy Reviews 63: (2016) 414–422.

[26] H. Titah-Benbouzid, M.E.H. Benbouzid, Biofouling issue on marine renewable energy converters: A state of the art review on impacts and prevention, International Journal on Energy Conversion 5 (3) (2017) 67–78.

[27] X. Sheng, S. Wan, L. Cheng, Y. Li, Blade aerodynamic asymmetry fault analysis and diagnosis of wind turbines with doubly fed induction generator, Journal of Mechanical Science & Technology 31 (10) (2017) 5011–5020.

[28] X. Gong, W. Qiao, Bearing fault diagnosis for direct-drive wind turbines via current-demodulated signals, IEEE Transactions on Industrial Electronics 60 (8) (2013) 3419–3428.

[29] H. Talhaoui, A. Menacer, A. Kessal, A. Tarek, Experimental diagnosis of broken rotor bars fault in induction machine based on Hilbert and discrete wavelet transforms, International Journal of Advanced Manufacturing Technology 95 (1–4) (2018) 1399 1408.

[30] M. Voltz, R. Webster, A comparison of kriging, cubic splines and classification for predicting soil properties from sample information, Journal of Soil Science 41 (3) (1990) 473–490.

[31] R. Yan, R.X. Gao, X. Chen, Wavelets for fault diagnosis of rotary machines: A review with applications, Signal Processing 96 (2014) 1–15.

[32] H. Chen, N. At-Ahmed, M. Machmoum, M.E.H. Zam, Modeling and vector control of marine current energy conversion system based on doubly salient permanent magnet generator, IEEE Transactions on Sustainable Energy 7 (1) (2015) 409–418.

[33] H.T. Pham, J.M. Bourgeot, M.E.H. Benbouzid, Comparative investigations of sensor fault-tolerant control strategies performance for marine current turbine applications, IEEE Journal of Oceanic Engineering 43 (4) (2017) 1024–1036.

[34] Z. Li, T. Wang, Y. Wang, Y. Amirat, M. Benbouzid, D. Diallo, A wavelet threshold denoising-based imbalance fault detection method for marine current turbines, IEEE Access 8 (2020) 29815–29825.

[35] Z. Liu, Z. He, W. Guo, Z. Tang, A hybrid fault diagnosis method based on second generation wavelet de-noising and local mean decomposition for rotating machinery, ISA Transactions 61 (2016) 211–220.

[36] A.K. Bhandari, D. Kumar, A. Kumar, G.K. Singh, Optimal sub-band adaptive thresholding based edge preserved satellite image denoising using adaptive differential evolution algorithm, Neurocomputing 174 (2016) 698–721.

[37] X. Gong, W. Qiao, Imbalance fault detection of direct-drive wind turbines using generator current signals, IEEE Transactions on Energy Conversion 27 (2) (2012) 468–476.

[38] H.T. Chiang, Y.Y. Hsieh, S.W. Fu, K.H. Hung, Y. Tsao, S.Y. Chien, Noise reduction in ECG signals using fully convolutional denoising autoencoders, IEEE Access 7 (2019) 60806–60813.

[39] M.Z. Sheriff, M. Mansouri, M.N. Karim, H. Nounou, M. Nounou, Fault detection using multiscale PCA-based moving window GLRT, Journal of Process Control 54 (2017) 47–64.

[40] M. Mansouri, M.Z. Sheriff, R. Baklouti, M. Nounou, H. Nounou, A.B. Hamida, N. Karim, Statistical fault detection of chemical process-comparative studies, Journal of Chemical Engineering & Process Technology 7 (1) (2016) 282–291.

[41] M. Zhang, T. Wang, T. Tang, M. Benbouzid, D. Diallo, An imbalance fault detection method based on data normalization and EMD for marine current turbines, ISA Transactions 68 (2017) 302–312.

[42] G. Hou, Z. Pan, B. Huang, G. Wang, X. Luan, Hue preserving-based approach for underwater colour image enhancement, IET Image Processing 12 (2) (2017) 292–298.

[43] A. Ng, Sparse autoencoder, CS294A Lecture Notes 72 (2011) 1–19.

[44] B. Xin, T. Wang, T. Tang, A deep learning and softmax regression fault diagnosis method for multi-level converter, in: *Proceedings of the IEEE 11th International Symposium on Diagnostics for Electrical Machines, Power Electronics, and Drives (SDEMPED)*, 2017, IEEE, Los Alamitos, CA, 2017, pp. 292–297.

[45] V.D. Krsman, A.T. Sarić, Bad area detection and whitening transformation-based identification in three-phase distribution state estimation, IET Generation, Transmission & Distribution 11 (9) (2017) 2351–2361.

[46] Y. LeCun, L. Bottou, Y. Bengio, P. Haffner, Gradient-based learning applied to document recognition, Proceedings of the IEEE 86 (11) (1998) 2278–2324.

[47] H. Chen, T. Tang, N. Aït-Ahmed, M.E.H. Benbouzid, M. Machmoum, M.E.H Zaïm, Attraction, challenge and current status of marine current energy, IEEE Access 6 (2018) 12665–12685.

[48] Y. Zheng, T. Wang, B. Xin, T. Xie, Y. Wang, A sparse autoencoder and softmax regression based diagnosis method for the attachment on the blades of marine current turbine, Sensors 19 (4) (2019) 826.

[49] F. Chollet. Xception: Deep learning with depthwise separable convolutions. In Proceedings of the 2017 IEEE Conference on Computer Vision and Pattern Recognition. 1251–1258.

[50] M. Sandler, A. Howard, M. Zhu, A. Zhmoginov, L.C. Chen. MobileNetB2: Inverted residuals and linear bottlenecks. In Proceedings of the 2018 IEEE Conference on Computer Vision and Pattern Recognition. 4510–4520.

[51] K. He, X. Zhang, S. Ren, J. Sun. Deep residual learning for image recognition. In Proceedings of the 2016 IEEE Conference on Computer Vision and Pattern Recognition. 770–778.

[52] C. Szegedy, W. Liu, Y. Jia, P. Sermanet, S. Reed, D. Anguelov, D. Erhan, V. Vanhoucke, A. Rabinovich. Going deeper with convolutions. In Proceedings of the 2015 IEEE Conference on Computer Vision and Pattern Recognition. 1–9.

[53] S. Ioffe, C. Szegedy. Batch normalization: Accelerating deep network training by reducing internal covariate shift. arXiv:1502.03167, 2015.

[54] A.G. Howard, M. Zhu, B. Chen, D. Kalenichenko, W. Wang, T. Weyand, M. Andreetto, H. Adam. MobileNets: Efficient convolutional neural networks for mobile vision applications. arXiv:1704.04861, 2017.

[55] A. Howard, M. Sandler, G. Chu, L.C. Chen, B. Chen, M. Tan, W. Wang, et al. Searching for MobileNetV3. In Proceedings of the 2019 IEEE International Conference on Computer Vision. 1314–1324.

Chapter 8

Quadrotor actuator fault diagnosis and accommodation based on nonlinear adaptive state observer

Sicheng Zhou[a], Kexin Guo[a], Xiang Yu[a,b], Lei Guo[a,b] and Youmin Zhang[c]
[a] School of Automation Science and Electrical Engineering, Beihang University, China. [b] Beijing Advanced Innovation Center for Big Data-Based Precision Medicine, Beihang University, Beijing, China. [c] Department of Mechanical, Industrial, and Aerospace Engineering, Concordia University, Montreal, Quebec, Canada

Financial Support: This research was supported by the National Natural Science Foundation of China (No. 61833013, 61973012 and 61903019), the Program for Changjiang Scholars and Innovative Research Team (no. IRT 16R03), Zhejiang Lab Fund (2019NB0AB08), China Postdoctoral Science Foundation (no. 2019M660404), Zhejiang Provincial Natural Science Foundation (no. LQ20F030006), and NSERC.

8.1 Introduction

In recent years, unmanned aerial vehicles (UAVs) have been widely used because of the huge potential in military and civilian applications [1–3], such as traffic monitoring, recognition and surveillance, and search and rescue operations in hostile environments [4, 5], especially quadrotor UAV. With the rapid development of UAVs, the importance of fault-tolerant control (FTC) is increasing. Generally, faults of quadrotor UAVs can be classified into actuator faults, sensor faults, and component faults [6]. The occurrence of actuator faults can deteriorate the tracking performance and stability of the closed-loop system [7].

FTC design methods can be essentially classified into passive and active approaches [8, 9]. With respect to passive FTC, an adaptive FTC strategy with consideration of input saturation is presented against actuator faults and external disturbances [10]. In the work of Yu and Jiang [11], a robust nonlinear controller is developed to handle the disturbances and faults by combining the sliding

Fault Diagnosis and Prognosis Techniques for Complex Engineering Systems. DOI: 10.1016/B978-0-12-822473-1.00002-1

mode control and backstepping control techniques. In the work of Avram et al. [12], an adaptive FTC is designed to guarantee asymptotic convergence of the altitude and attitude tracking errors in the presence of multiple actuator faults and modeling uncertainties. Focusing on active FTC, to actively compensate actuator faults, a fault detection and diagnosis (FDD) module is necessary. In the past few years, researchers have paid tremendous attention to FDD design, resulting in different kinds of methods, such as sliding mode observer methods [13], the high-gain observer approach [14, 15], Kalman filter–based estimations [16], and the adaptive observer-based method [17]. Additionally, a moving horizon estimator and an unscented Kalman filter are compared to examine the FDD performance of quadrotor UAV [18]. In the work of Aguilar-Sierra et al. [19], a polynomial observer is proposed to diagnose actuator faults in a small-scale quadrotor UAV. In the work of Yu and Jiang [20], a hybrid FTC system that combines the merits of passive and active FTC systems is proposed to accommodate the partial actuator failures. It is also noteworthy that the neural network adaptive techniques have been exploited at the active FTC design stage in recent years [21, 22].

In this chapter, we focus on the design of active FTC subject to time-varying actuator faults of a quadrotor UAV. The developed FTC system includes the fault detection module, fault diagnosis module, and accommodation unit. Before the fault occurs, the fault detection module is exploited to monitor the state residual of the quadrotor UAV. Once the fault is detected, the fault diagnosis is thereby activated to identify fault amplitude and estimate unknown fault parameters. Based on the fault estimation, the accommodation unit adjusts the control signal to guarantee the tracking performance of the quadrotor UAV. Our major contributions are briefly stated as follows:

(1) An adaptive fault detection threshold is proposed to determine the fault occurrence in the presence of model uncertainties and external disturbances. The fault can be detected if the state residual generated by a nonlinear state observer exceeds the threshold.

(2) Four nonlinear adaptive state observers (NASO) with respect to four rotors are developed in the fault diagnosis module. In comparison with sliding mode observer and high-gain observer, NASO has better tracking performance for time-varying faults and can locate the fault rotor accurately as well.

(3) An accommodation unit is proposed to adjust the control signal without changing the original control architecture. With the proposed approach, some complicated FTC design can be avoided and there is no effect on the parameters of the baseline controller.

This chapter is organized as follows. Section 8.2 presents the dynamic model of a quadrotor UAV and time-varying actuator faults model. The FTC design algorithm is described in Section 8.3, including the fault detection module, fault

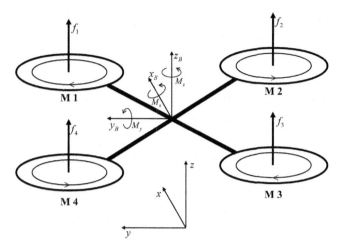

FIGURE 8.1   Structure of the quadrotor and frames.

diagnosis module, and accommodation unit. Moreover, the results of numerical simulation and flight test are illustrated to validate the effectiveness and applicability of the proposed FTC scheme in Section 8.4. Finally, Section 8.5 concludes the chapter.

8.2   Mathematical model of a quadrotor

This section presents the nonlinear quadrotor model and the actuator fault model, which provides a basis for the proposed active FTC design.

8.2.1   The nonlinear quadrotor model

Generally, the dynamic model includes translation and rotation equations. As shown in Fig. 8.1, the Body Frame (BF) is assumed to be at the center of gravity of the quadrotor, where the x-axis (x_B) is pointing head, the y-axis (y_B) is pointing left, and the z-axis (z_B) is pointing upward, and the Inertial Frame (IF) is assumed to be at the take-off point of the quadrotor.

The transformation of vectors from the BF to the IF can be expressed as

$$
R_{EB}
= \begin{bmatrix} \cos\psi\cos\theta & -\sin\psi\cos\phi + \cos\psi\sin\theta\sin\phi & \sin\psi\sin\phi + \cos\psi\sin\theta\cos\phi \\ \sin\psi\cos\theta & \cos\psi\cos\phi + \sin\psi\sin\theta\sin\phi & -\cos\psi\sin\phi + \sin\psi\sin\theta\cos\phi \\ -\sin\theta & \cos\theta\sin\phi & \cos\phi\cos\theta \end{bmatrix}
$$

$$(8.1)$$

where $\Omega_b = [\phi \quad \theta \quad \psi]^T$ represents body-axis pitch, roll, and yaw angle.

TABLE 8.1 Parameters for quadrotor dynamics

Parameter	Meaning
$p_E = [x \ y \ z]^T$	Position of the quadrotor in the IF
$v_E = [u \ v \ w]^T$	Linear velocity of the quadrotor in the IF
$\omega = [p \ q \ r]^T$	Angular rates of the quadrotor in the BF
$[\xi_v \ \xi_\omega]^T$	Model uncertainties in the translational and rotational dynamics
$[J_x \ J_y \ J_z]^T$	Moment of inertia in the BF
$[M_x \ M_y \ M_z]^T$	The rolling torque, the pitching torque, and the yawing torque
F_m	The thrust
m	Mass of the quadrotor
g	The gravitational acceleration

The nonlinear quadrotor dynamics considered in this chapter can be described as

$$\dot{p}_E = v_E, \tag{8.2}$$

$$\dot{v}_E = \frac{1}{m} R_{EB}(\Omega_b) \begin{bmatrix} 0 \\ 0 \\ F_m \end{bmatrix} + \begin{bmatrix} 0 \\ 0 \\ -g \end{bmatrix} + \xi_v, \tag{8.3}$$

$$\dot{\Omega}_b = R_0(\Omega_b)\omega, \tag{8.4}$$

$$\dot{\omega} = \begin{bmatrix} \frac{J_y - J_z}{J_x} qr \\ \frac{J_z - J_x}{J_y} pr \\ \frac{J_x - J_y}{J_z} pq \end{bmatrix} + \begin{bmatrix} \frac{M_x}{J_x} \\ \frac{M_y}{J_y} \\ \frac{M_z}{J_z} \end{bmatrix} + \xi_\omega, \tag{8.5}$$

where the specific forms of the matrix $R_0(\Omega_b)$ can be expressed as

$$R_0(\Omega_b) = \begin{bmatrix} 1 & \tan\theta \sin\phi & \tan\theta \cos\phi \\ 0 & \cos\phi & -\sin\phi \\ 0 & \sin\phi/\cos\theta & \cos\phi/\cos\theta \end{bmatrix} \tag{8.6}$$

and the parameters are listed in Table 8.1.

8.2.2 The actuator fault model

In this chapter, the quadrotor UAV inputs $u = \begin{bmatrix} F_m & M_x & M_y & M_z \end{bmatrix}^T$ can be simplified and thereby described as

$$\begin{bmatrix} F_m \\ M_x \\ M_y \\ M_z \end{bmatrix} = R_u f_s = \begin{bmatrix} 1 & 1 & 1 & 1 \\ -d_\phi & -d_\phi & d_\phi & d_\phi \\ d_\theta & -d_\theta & d_\theta & -d_\theta \\ c_{\tau f} & -c_{\tau f} & -c_{\tau f} & c_{\tau f} \end{bmatrix} \begin{bmatrix} f_1 \\ f_2 \\ f_3 \\ f_4 \end{bmatrix}, \tag{8.7}$$

where $f_s(s = 1, \ldots, 4)$ is the force produced by each rotor acting on the quadrotor body. d_φ is the half of roll motor-to-motor distance, d_θ is the half of pitch motor-to-motor distance, and $c_{\tau f}$ is a fixed constant reflecting the relationship between the thrust force f_s and its corresponding torque.

The slow speed of the rotor due to phase break is an important reason for quadrotor UAV tasking failure during missions. The main reasons for phase break include excessive temperature, excessive load, and aging of the coil. For example, because of the aging of the coil insulation layer, some coils are short circuited during flight, which eventually induces phase break and slows down the rotors' speed. In fact, this speed change is often time varying.

According to propeller dynamics, the actuator faults caused by slower rotor speed can be modeled as a time-varying partial loss of effectiveness (LOE) in the rotors and thereby represented as

$$f_s^* = \Gamma_u f_s, \tag{8.8}$$

where $f_s = \begin{bmatrix} f_1 & f_2 & f_3 & f_4 \end{bmatrix}^T$ denotes the commanded thrust force generated by the sth rotor, and $f_s^* = \begin{bmatrix} f_1^* & f_2^* & f_3^* & f_4^* \end{bmatrix}^T$ stands for the actual thrust force generated by the sth rotor. $\Gamma = diag(\alpha_s)(s = 1, \ldots, 4)$ and $\alpha_s \in (0, 1]$ is an unknown parameter representing the occurrence of a partial LOE fault in the sth rotor. The case of $\alpha_s = 1$ represents a healthy rotor, whereas $\alpha_s < 1$ represents a faulty rotor with LOE.

By using Eq. (8.8), the actual system inputs can be expressed as

$$u^* = R_u f_s^* = R_u \Gamma_u f_s. \tag{8.9}$$

Assumption 1. The unstructured modeling uncertainties and environment disturbance ξ_v and ξ_ω in Eq. (8.3) and Eq. (8.5), respectively, are unknown but assumed to be bounded by some known functions. For $t \geq 0$, $|\xi_v| \leq \bar{\xi}_v$, $|\xi_\omega| \leq \bar{\xi}_\omega$ and the bounding functions $\bar{\xi}_v$ and $\bar{\xi}_\omega$ are known, continuous, and bounded.

Assumption 1 represents the class of modeling uncertainty considered. To generate the adaptive threshold to distinguish the influence of fault and the modeling uncertainty during the FDD process, the boundary of the uncertainty of unstructured modeling should be known *a priori*.

8.3 Naso-based FTC

The proposed FTC scheme is shown in Fig. 8.2. It can be seen that the FDD module includes two main components: a nonlinear fault detection module to determine fault occurrence, and a set of nonlinear adaptive fault diagnosis estimators to identify fault rotor and estimate unknown fault parameters. Once the fault detection estimator detects an actuator fault, four estimators are activated to locate the failed rotor. After fault diagnosis, fault parameters are employed to adjust controller output signals.

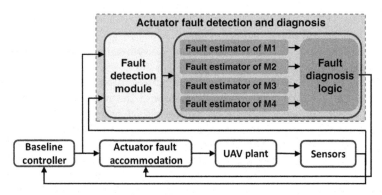

FIGURE 8.2   Structure of the FTC.

8.3.1   The fault detection module

In this section, a nonlinear state observer is designed to generate the state residual between the estimated and the actual states. According to Assumption 1, an adaptive threshold is proposed to improve the robustness of the scheme. Thus, the fault detection module can determine the fault occurrence accurately by monitoring the state residual.

First, consider the state vector $\zeta = \begin{bmatrix} v_z & p & q & r \end{bmatrix}^T$, where v_z represents the quadrotor velocity in the vertical direction of IF. Therefore, regardless of the effect of the failure, by substituting Eq. (8.7) and Eq. (8.8) into Eq. (8.3) and Eq. (8.5), one can obtain the following:

$$\dot{\zeta} = f(\bar{x}, t) + BR_u f_s + \xi(\bar{x}, t), \qquad (8.10)$$

where f_s is the commanded forces of each rotor, $B = diag\left\{ \frac{\cos\phi\cos\theta}{m}, \frac{1}{J_x}, \frac{1}{J_y}, \frac{1}{J_z} \right\}$, and the known nonlinearity $f(\bar{x}, t)$ is defined as

$$f(\bar{x}, t) = \begin{bmatrix} -g \\ \frac{J_y - J_z}{J_x} qr \\ \frac{J_z - J_x}{J_y} pr \\ \frac{J_x - J_y}{J_z} pq \end{bmatrix}. \qquad (8.11)$$

Subsequently, by using Eq. (8.10), the nonlinear state observer is designed as

$$\begin{bmatrix} \dot{\hat{v}}_z \\ \dot{\hat{p}} \\ \dot{\hat{q}} \\ \dot{\hat{r}} \end{bmatrix} = -\Delta \begin{bmatrix} \hat{v}_z - v_z \\ \hat{p} - p \\ \hat{q} - q \\ \hat{r} - r \end{bmatrix} + f(\bar{x}, t) + BR_u f_s, \qquad (8.12)$$

where $\begin{bmatrix} \hat{v}_z & \hat{p} & \hat{q} & \hat{r} \end{bmatrix}^T$ represents estimated velocity in IF and angular rates in BF, $\Delta = diag(\Delta_i)$, with $\Delta_i > 0$ for $i = 1, ..., 4$. Based on Eq. (8.10) and Eq. (8.12), the state estimation error dynamics can be described as

$$\dot{\varepsilon}(t) = -\Delta\varepsilon(t) + \xi(\bar{x}, t), \qquad (8.13)$$

where $\varepsilon(t) = \begin{bmatrix} v_z - \hat{v}_z & p - \hat{p} & q - \hat{q} & r - \hat{r} \end{bmatrix}^T$ denotes the residual of nonlinear state estimator.

According to Assumption 1, $\xi(\bar{x}, t)$ given by Eq. (8.10) is bounded. Meanwhile, the matrix Δ is also stable. Thus, the error dynamics given by Eq. (8.13) are stable. By using Eq. (8.13), the residual of nonlinear state estimator $\varepsilon(t)$ satisfies

$$|\varepsilon_i(t)| \leq e^{-\Delta_i(t-t_0)}|\varepsilon_i(t_0)| + \int_{t_0}^{t} \left| e^{-\Delta_i(t-\tau)}\xi_i(\bar{x}, \tau) \right| d\tau, \qquad (8.14)$$

where $\varepsilon_i(t)$ represents the ith component of $\varepsilon(t)$. According to Assumption 1 and Eq. (8.14), an adaptive threshold $\bar{\varepsilon}_i(t)$ can be defined as

$$\bar{\varepsilon}_i(t) = e^{-\Delta_i(t-t_0)}|\varepsilon_i(t_0)| + \int_{t_0}^{t} \left| e^{-\Delta_i(t-\tau)}\bar{\xi}_i(\bar{x}, \tau) \right| d\tau, \qquad (8.15)$$

where the first term of the equation is only related to the state residual, and the upper bounds from Assumption 1 determine the second term of the equation. Thus, the proposed threshold can be adaptively adjusted according to the state residual at the last moment and Assumption 1. If any one of the residuals $\varepsilon_i(t)$ exceeds the adaptive threshold $\bar{\varepsilon}_i(t)$, the fault detection module will conclude that a fault has occurred and fault diagnosis estimators will be activated.

8.3.2 The fault diagnosis module

In this section, a NASO is adopted to estimate the unknown fault parameters of each rotor. It is worth mentioning that there is no independent design of reconfigurable control—only the control signal of the original basic controller is adjusted. Hence, in this section, the basic controller is stable to the system by default.

After the fault is detected, the equation of state can be written as:

$$\dot{\zeta} = f(\bar{x}, t) + BR_u(I - \Gamma_u)f_s + \xi(\bar{x}, t). \qquad (8.16)$$

With respect to the ith rotor, the equation of state can be written as

$$\dot{\zeta}^i = f(\bar{x}, t) + BR_u f_s - \alpha^i BR_u \Lambda^i f_s + \xi(\bar{x}, t), \qquad (8.17)$$

where α^i represents the fault parameter of the ith rotor and Λ^i denotes the transformation matrix of the ith rotor. For example, if $i = 1$, $\Lambda^i = diag(1, 0, 0, 0)$.

As shown in Fig. 8.2, once the fault is detected, a set of four nonlinear adaptive estimators are activated to identify the fault rotor and estimate unknown fault parameters. Based on Eq. (8.17), each estimator is designed for the corresponding actuator failure.

Theorem 1. Consider the faulty system described by Eq. (8.16). Define z_1 and z_2 as the states of the designed NASO. If the observer of the ith rotor is formed as

$$\begin{cases} \dot{z}_1^i = -\Psi e^i + f(\bar{x}, t) + BR_u f_s - z_2^i BR_u \Lambda^i f_s \\ \dot{z}_2^i = \left(BR_u \Lambda^i f_s \right)^T e^i, \end{cases} \tag{8.18}$$

where $\Psi = diag(\eta_s^i)(s = 1, \ldots, 4)$ is a positive coefficient matrix of the ith rotor, and $e^i = z_1^i - \zeta$. The terms ζ^i and α^i will be estimated within finite time through z_1^i and z_2^i, respectively.

Proof. Consider the Lyapunov function as

$$V_1 = \frac{1}{2} \left(e^i \right)^T e^i + \frac{1}{2} z_2^2, \tag{8.19}$$

together with Eq. (8.17). Thus, the derivative of V_1 can be obtained as

$$\begin{aligned} \dot{V}_1 &= \left(e^i \right)^T \dot{e}^i + z_2^i \dot{z}_2^i \\ &= \left(e^i \right)^T \left(-\Psi e^i + f(\bar{x}, t) + BR_u f_s - z_2^i BR_u \Lambda^i f_s - \dot{\zeta} \right) + z_2^i \left(BR_u \Lambda^i f_s \right)^T e^i \\ &= -\Psi \left(e^i \right)^T e^i + \left(e^i \right)^T \left(f(\bar{x}, t) + BR_u f_s - \dot{\zeta} \right) \\ &\quad + \left(e^i \right)^T \left(-z_2^i BR_u \Lambda^i f_s \right) + z_2^i \left(BR_u \Lambda^i f_s \right)^T e^i \\ &= -\Psi \left(e^i \right)^T e^i + \left(e^i \right)^T \left(f(\bar{x}, t) + BR_u f_s - \dot{\zeta} \right), \end{aligned} \tag{8.20}$$

where Ψ is a positive coefficient matrix, ensuring that the first term of the equation is less than 0. The baseline controller is stable, guaranteeing that the second term of the equation is less than 0. Hence, the stability of the NASO can be guaranteed.

Moreover, it is proven that z_1^i and z_2^i can approach to ζ^i and α^i within finite time, respectively.

Remark 1. The fault diagnosis scheme presented in this chapter is applicable not only for the constant fault but also for the time-varying fault. In comparison with the work of Avram et al. [12], this study has a deeper analysis of quadrotor UAV actuator faults, whereas the proposed NASO has better tracking performance in the case of time-varying faults.

8.3.3   The fault accommodation module

As depicted in Fig. 8.3, the trajectory control of quadrotor is implemented by using a dual-loop architecture. More specifically, the outer loop controls the X

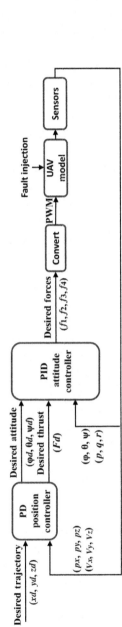

FIGURE 8.3 Structure of the baseline controller.

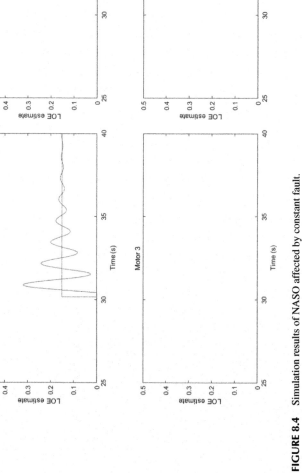

FIGURE 8.4 Simulation results of NASO affected by constant fault.

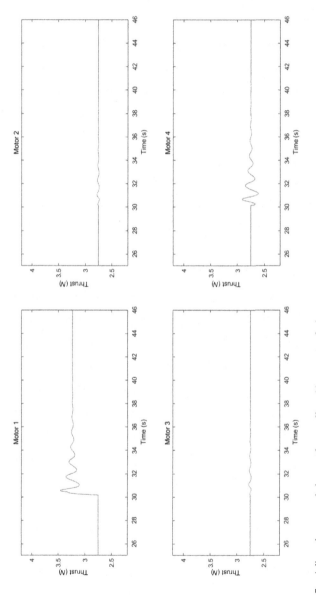

FIGURE 8.5 Adjusted commanded motor forces affected by constant fault.

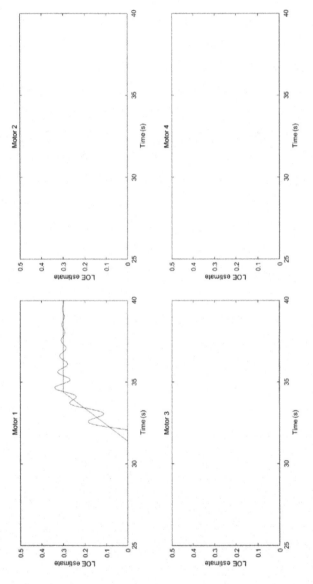

FIGURE 8.6   Simulation results of NASO affected by time-varying fault.

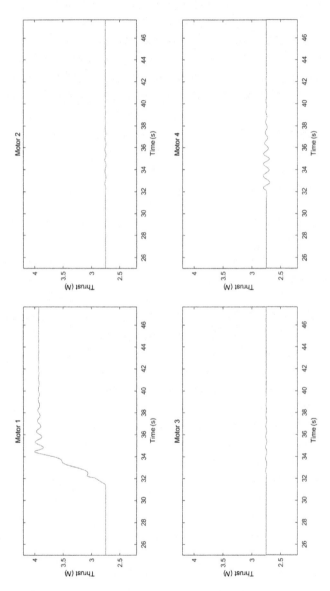

FIGURE 8.7   Simulation results of NASO affected by time-varying fault.

FIGURE 8.8 Flight test environment.

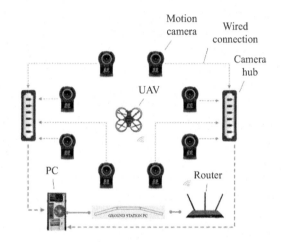

FIGURE 8.9 Layout of the experimental test environment.

and Y positions by generating the desired roll and pitch angles. The altitude and attitude controller can generate the required rotor speed for the quadrotor UAV to track the required attitude and altitude.

To sum up, based on the nominal quadrotor model given by Eqs. (8.1) through (8.4) under healthy conditions, the PID baseline controller is designed for inner and outer loops providing satisfactory tracking performance. The PID controller calculates the required rotor speed, which is used by the motor servo control system to generate the force and moment acting on the quadrotor to track a set of reference trajectories.

As shown in Fig. 8.2, after the fault actuator is isolated by the fault diagnosis component, the matching adaptive estimators can provide an estimate of the unknown fault amplitude. In consequence, the estimated values can be used by the fault accommodation module to adjust the baseline control signals. On

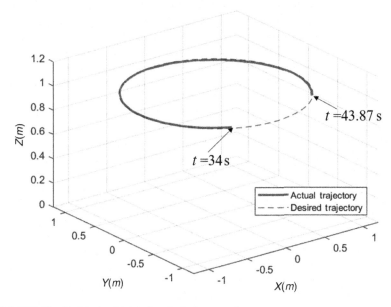

FIGURE 8.10   Position tracking before the fault occurred.

the premise of generality, it is assumed that the partial LOE occurs in the first actuator and the fault is detected at time t_d. Hence, for $t \geq t_d$, the output signal of baseline controller is modified as

$$\tilde{f}_s = \left(I_4 - z_2^i \Lambda^i\right)^{-1} f_s,   \tag{8.21}$$

where f_s is the commanded thrust force generated by the baseline controller, $\tilde{f}_s$ is the adjusted commanded thrust force sent to the actuator control system, and z_2^i is the fault parameter estimate provided by the NASO corresponding to the ith rotor.

Remark 2. According to Eq. (8.21), it is worth noting that the proposed fault accommodation scheme cannot deal with the situation where the actuator fails completely due to the requirement of inversion. In addition, the inaccuracy of the estimator has not been taken into consideration in the current stage. The significance of this method lies in that time-varying faults can be handled. Based on the estimated parameters, adjustment of actuator allocation and mission replanning under saturation is one of our future works.

8.4   Validation

To validate the effectiveness of the developed active FTC scheme, both numerical simulation and real-world flight test are conducted.

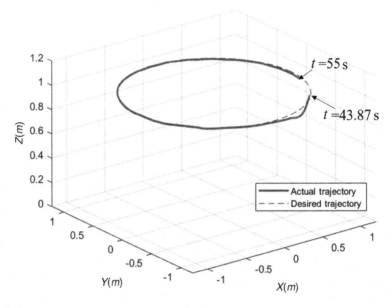

FIGURE 8.11   Position tracking after the fault occurred.

8.4.1   Numerical simulation results

(1) *Constant fault.* The results of estimated fault parameters and the corresponding adjusted control signals under the constant faults are shown in Fig. 8.4 and Fig. 8.5, respectively. As can be observed from Fig. 8.4, the estimated fault parameter can be eventually converged to the actual one successfully. As illustrated in Fig. 8.5, after 15% constant LOE fault occurs on motor 1, the adjustment of control signal increases by 17.65%. Hence, from Figs. 8.4 and 8.5, the constant LOE fault is successfully compensated by the proposed FTC scheme.

(2) *Time-varying fault.* The result of estimated fault parameter and the adjust control signal under the time-varying fault are shown in Fig. 8.6 and Fig. 8.7, respectively. As is visible in Fig. 8.6, the fault detection time is 0.66 s after the fault occurrence, whereas the estimated result is always kept in a reasonable range that is close to the real one. From Fig. 8.7, it is illustrated that the adjust control signals become larger as the faults increase. Thus, from Figs. 8.6 and 8.7, the actuators governed by the proposed FTC can satisfactorily handle the time-varying faults.

8.4.2   Flight test

The developed algorithms herein are implemented on the quadrotor UAV platform, whereas the experimental environment is illustrated in Fig. 8.8.

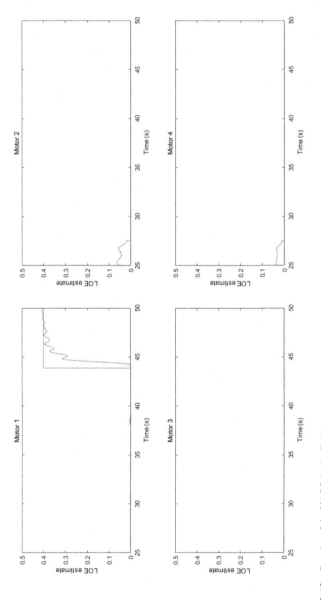

FIGURE 8.12   Results of the NASO in the flight test.

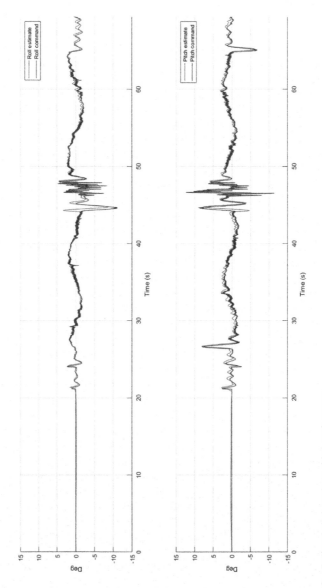

FIGURE 8.13 Roll and pitch angle tracking performance.

Fig. 8.9 shows the layout of the experimental environment. The experiments are conducted in an indoor environment without GPS. Hence, a network of eight motion cameras, which can locate the mark balls on the quadrotor UAV, is exploited for position capture. Furthermore, a ground station is adopted for command set and real-time status monitoring. The connection between the quadrotor UAV and the station is achieved by a router.

Fig. 8.10 and Fig. 8.11 present the tracking performance of the quadrotor before and after a 40% constant fault occurred, respectively. As checked in the figures, it can be concluded that the UAV can guarantee a good trajectory tracking performance before and after the fault occurrence. The mean absolute tracking error is 0.0372 m and the maximum shifting distance is 0.253 m during the flying test.

Fig. 8.12 shows the results of estimated fault parameters in the flight test. As can be seen from this figure, the fault detection time is 0.36 s after the fault takes place. It is worth mentioning that the fault detection module has misdiagnosed for motors 2 and 3 during 25 s to 27 s. This situation is caused by the ground effect of the quadrotor UAV. The ground effect affects the flow characteristics of air around the UAV, which eventually causes the rotor to not produce the desired thrust.

Fig. 8.13 shows the satisfactory performance of roll and pitch angle tracking. The residual increases sharply at 43.87 s because of the fault. It is also interesting to find that the residual is quickly reduced with the aid of fault diagnosis and the accommodation module.

8.5   Conclusion

A NASO-based active FTC system is presented for quadrotor UAV. The benefits of the proposed FTC scheme include the following: (1) the adaptive fault detection threshold enhances the robustness of the FDD module to external disturbance, (2) the NASO is able to guarantee the tracking performance for time-varying fault, and (3) the proposed fault accommodation scheme can complete the compensation of the fault without changing the baseline controller. These improvements offer the potential to enhance the safety of quadrotor UAV. The numerical simulation and flight test demonstrate that the proposed FTC scheme can effectively deal with actuator LOE faults.

References

[1]   G. Vachtsevanos, L. Tang, G. Drozeski, L. Gutierrez. From mission planning to flight control of unmanned aerial vehicles: Strategies and implementation tools. Annual Reviews in Control 2005;29(1):101--115.

[2]   Y.M. Zhang, A. Chamseddine, C.A. Rabbath, B. W. Gordon, C.-Y. Su, S. Rakheja, C. Fulford, J. Apkarian, P. Gosselin. Development of advanced FDD and FTC techniques with

application to an unmanned quadrotor helicopter testbed. Journal of the Franklin Institute 2013;350(9):2396--2422.

[3] X. Yu, Y. M. Zhang. Sense and avoid technologies with applications to unmanned aircraft systems: Review and prospects. Progress in Aerospace Sciences 2015;74:152--166.

[4] S. Gupte, P. I. T. Mohandas, J. M. Conrad. A survey of quadrotor unmanned aerial vehicles. In *2012 Proceedings of IEEE Southeastcon*. 1–6.

[5] A. Jaimes, S. Kota, J. Gomez. An approach to surveillance an area using swarm of fixed wing and quad-rotor unmanned aerial vehicles UAV(s). In *Proceedings of the2008 IEEE International Conference on System of Systems Engineering*. 1–6.

[6] Y. M. Zhang, J. Jiang. Bibliographical review on reconfigurable fault-tolerant control systems. Annual Reviews in Control 2008;32(2):229--252.

[7] Z. T. Dydek, A. M. Annaswamy, E. Lavretsky. Adaptive control of quadrotor UAVs: A design trade study with flight evaluations. IEEE Transactions on Control Systems Technology 2012;21(4):1400--1406.

[8] X. Yu, J. Jiang. A survey of fault-tolerant controllers based on safety-related issues. Annual Reviews in Control 2015;39:46--57.

[9] J. Jiang, X. Yu. Fault-tolerant control systems: A comparative study between active and passive approaches. Annual Reviews in Control 2012;36(1): 60--72.

[10] S. Li, Y. Wang, J. Tan. Adaptive and robust control of quadrotor aircrafts with input saturation. Nonlinear Dynamics 2017;89(1): 255--265.

[11] F. Chen, R. Jiang, K. Zhang, B. Jiang, G. Tao. Robust backstepping sliding-mode control and observer-based fault estimation for a quadrotor UAV. IEEE Transactions on Industrial Electronics 2016;63(8): 5044--5056.

[12] R. C. Avram, X. Zhang, J. Muse. Nonlinear adaptive fault-tolerant quadrotor altitude and attitude tracking with multiple actuator faults. IEEE Transactions on Control Systems Technology 2017;26(2): 701--707.

[13] D. Lee, H. J. Kim, S. Sastry. Feedback linearization vs. adaptive sliding mode control for a quadrotor helicopter. International Journal of Control Automation & Systems 2009;7(3): 419–428.

[14] H. K. Khalil. Adaptive output feedback control of nonlinear systems represented by input–output models. IEEE Transactions on Automatic Control 1996;41(2): 177–188.

[15] M. S. Mahmoud, H. K. Khalil. Robustness of high-gain observer-based nonlinear controllers to unmodeled actuators and sensors. Automatica 2002;38: 361–369.

[16] P. Freeman, R. Pandita, N. Srivastava, G. J. Balas. Model-based and data-driven fault detection performance for a small UAV. IEEE/ASME Transactions on Mechatronics 2013;18(4):1300–1309.

[17] F. Chen, W. Lei, G. Tao, B. Jiang. Actuator fault estimation and reconfiguration control for the quad-rotor helicopter. International Journal of Advanced Robotic Systems 2016;13(13): 1–12.

[18] H. A. Izadi, Y. Zhang, B. W. Gordon. Fault tolerant model predictive control of quad-rotor helicopters with actuator fault estimation. IFAC Proceedings Volumes 2011;44(1): 6343--6348.

[19] H. Aguilar-Sierra, G. Flores, S. Salazar, R. Lozano. Fault estimation for a quad-rotor MAV using a polynomial observer. Journal of Intelligent & Robotic Systems 2014;73(1-4): 455--468.

[20] X. Yu, J. Jiang. Hybrid fault-tolerant flight control system design against partial actuator failures. IEEE Transactions on Control Systems Technology 2012;20(4): 871--886.

[21] A. Abbaspour, K.K. Yen, P. Forouzannezhad, A. Sargolzaei, A neural adaptive approach for active fault-tolerant control design in UAV, IEEE Transactions on Systems, Man & Cybernetics: Systems (2018), doi:10.1109/TSMC.2018.2850701.

[22] Y. Song, L. He, D. Zhang, J. Qian, J. Fu. Neuroadaptive fault-tolerant control of quadrotor UAVs: A more affordable solution. IEEE Transactions on Neural Networks & Learning Systems 2018;30(7): 1975--1983.

Chapter 9

Defect detection and classification in welding using deep learning and digital radiography

M-Mahdi Naddaf-Sh[a], Sadra Naddaf-Sh[a], Hassan Zargarzadeh[a], Sayyed M. Zahiri[b], Maxim Dalton[b], Gabriel Elpers[b] and Amir R. Kashani[b,1]

[a]Electrical Engineering Department, Lamar University, United States. [b]Artificial Intelligence Lab, Stanley Oil and Gas, Stanley Black and Decker, United States

9.1 Introduction

One of the foremost concerns of the modern-day industrial and infrastructure world is safety [74]. Whether it be in the context of towering skyscrapers, bridges, or pipelines, the consequences of overlooked or undetectable mistakes can be incredibly catastrophic [38]. To ensure safety, you not only have to devise and implement reliable systems, equipment, and methods, but you must also have an accurate means of recognizing the presence of unexpected and potentially fatal flaws. This is because the practice of automating machinery for complex systems can be incredibly difficult and error prone [4]. Perhaps even more important, no matter how precise and consistent equipment otherwise may be, if human beings are allowed to contribute directly in the process, there is always room for mistakes. As technology advances and gives rise to new ways to minimize risk for human operators and maximize efficiency or capacity, it also becomes increasingly difficult to identify the fault modes of supremely precise instrumentation [3].

Reliability of large-scale infrastructure is heavily dependent on reliability of welding massive metal structures together to function as an individual unit [23].

[1]Primarily with help from the Jason Miller, William Aston, Manny Glover, and Shengnan Wang from Stanley Oil and Gas.

Fault Diagnosis and Prognosis Techniques for Complex Engineering Systems. DOI: 10.1016/B978-0-12-822473-1.00007-0
327

Naively, the process is simple—apply sufficient heat to melt two pieces of metal that are fused together after they cool and harden [39]. In general, a separate piece of metal acts as a filler material that is deposited and cooled between the two pieces in question. In practice, the process requires careful adjustment of finely tuned parameters (automatic or manually enforced) that govern an energy source powerful enough to first liquefy the metal and then deposit the filler material without compromising the integrity of either the base metal or the finished product [7]. From the standpoint of a welding operator, the slightest accidental or misjudged movement can be the difference between a successful weld and a very costly mistake.

Quality of welded joints and their inspection assessment are critical elements of industries (e.g., marine, chemical, and aeronautical industries) [77]. Digitizing and automation of monitoring the quality of product is one of the main pillars of Industry 4.0. To achieve this goal, various robotic platforms to increase the consistency, quality, and automation of the process have been developed, including robotic platforms to perform digital radiography.

It is important to note that not all construction projects involve the same fundamental structures or conditions. Of unique interest is the process of welding in pipeline construction [69]. Rather than being concerned with welding two large metal beams at varying degrees, as would be necessary in the construction of a building, welding in pipelines is principally focused with a particular type of weld—that of girth welding. Girth welding involves welding the circumference of two cylindrical pipe joints together to extend the pipeline into a single contiguous unit [16].

In practice, pipelines are constructed in one of two primary environments: on land and underwater. In both cases, multiple teams work in tandem to ensure that the pipeline is welded, properly inspected, and potentially coated with special material to protect the pipeline from corrosion [82]. When pipelines are constructed on land, the pipeline is often buried under ground after each joint is welded. Pipelines that are constructed underwater or offshore are produced on large vessels from massive segmented spools and laid underwater [48]. In the latter case, it becomes especially important to protect against the effects of saltwater or electrochemical corrosion with an extra layer of coating. At the end of the day, welding in both environments requires very accurate testing to ensure long life of reliable operation because the cost of repairs skyrocket by the time leaks or major damage occur [57].

Thanks to advances in robotics engineering, there are actually two primary means to perform welds in the modern day. Although there are still many occasions for manual welding, automatic welding is proven to be more cost efficient, more consistent, and faster in many cases, especially in pipeline construction [6]. In this case, the ability to perform automated welds inside the pipe produces a greater quality joint that is fused both internally and externally and is often necessary because it is impossible to fit a person within small-diameter pipelines.

In addition, at the very least, it is safer to eliminate the need for a person to travel within a pipe, let alone operate welding equipment inside of it [80].

9.1.1 Welding Process

In general, manual welding is noticeably less precise and more prone to error. Hand movements are random and often unpredictable. By contrast, automated welding is designed to be performed on the basis of pre-configured parameters of operation. Complex factors due to how the weld is deposited around the pipe and effects due to gravity require careful and reproducible patterns that are nearly impossible for a person to re-create every time and almost guaranteed for high-quality equipment [51]. No two welding technicians will perform a weld under the same conditions in the same way, and neither will a technician reproduce exactly what was done yesterday [41]. As much as an improvement as automated welding is over manual welding, just in the space of automated welding, there is already a surprisingly large set of potential flaws that are classifiable and well documented [54].

It is clear that a set of methodologies ought to be established for testing the quality or properties of a weld without destroying the end product. Otherwise, a major purpose for testing in the first place would be defeated. In fact, there is a field dedicated precisely to this endeavor, namely the field of non-destructive testing (NDT) [11]. NDT techniques are designed to probe material with some sort of stimulus and interpret the response, either manually or automatically through software [58]. The idea is that anomalies in the welding process are impossible to detect with the naked eye, especially when they are located internally away from the visible surface. Using techniques and the research from applied physics, we have a plethora of ways at our disposal to observe how a material behaves or how something is affected indirectly through the material in question [33]. Most importantly, we can do so without damaging something very costly to repair later down the line [21].

NDT not only provides a means to test material properties but is also intentionally non-destructive. This means that we can actually save on the cost of more expensive repairs as we have the means to preemptively repair a given weld before it becomes too costly and ensure the quality of welds without any harm or foul should they be in flawless condition [67]. Furthermore, testing in general is necessary because it would be fundamentally impossible to easily detect flaws that are known to be problematic in practice. This is because they are embedded deep in the material, hidden from sight [66]. Certain techniques also afford a level of precision that allows for measured judgment calls based on acceptability standards [70].

Although it is clear that a certain weld may contain a defect, it is also possible through NDT techniques such as automated ultrasonic technology (AUT) and real-time radiography (RTR) to actually measure the geometric properties of the defect [20]. In this way, one can assess whether the defect is really worth a full weld reattempt or repair or if it is absolutely necessary [1]. Is it substantially long or deep? Is it likely to form the basis of a future crack that will eventually and completely undermine the integrity of the weld? These questions are now possible to answer because the entire material and the conditions and characteristics that gave rise to the defects are fully intact and explorable. Alternative methods might either weaken the material, destroy the possibility of assessing defect characteristics, or disallow the ability to test a specific weld altogether [64].

In welding, NDT ranges across a wide spectrum of techniques built on a potentially very different physical phenomenon [32]. For instance, the same fundamental ultrasound technology that enables us to observe an unborn fetus in its mother's womb can be applied fundamentally to weld inspection [37]. In NDT, this is called *AUT*. In pipeline girth (circumferential) welding, AUT can be used to identify defects based on irregularities in the patterns of waves that reflect from surfaces in the pipe [61]. AUT technology itself has many forms and variations that adjust characteristics such as the patterns of the wave pulses and the orientation or number of transducers [2].

Another perhaps even more familiar medical technique that has a similar counterpart in weld NDT is that of X-ray scanning. Here, high-energy radiation is deposited directly onto material in front of a screen that shows visual contrast between regions where the material is dense and very little of the radiation is able to pass through and regions where the material is relatively less dense where the screen absorbs the radiation. This technology corresponds to a technique called *radiographic testing* (RT) [40]. In RTR, the same fundamental RT technology is employed, but the radiation absorption pattern is captured electronically rather than on film [46]. Many other techniques exist, such as electromagnetic or eddy current testing (ET), magnetic particle testing (MT), and acoustic emission testing (AE), and the list goes on [9].

9.1.2   Digital Radiography

Compared to other NDT techniques, RT/RTR remains a popular mainstay of inspection in practice [60]. Other techniques might be too new to have a real presence and trustworthiness among both practitioners and project stakeholders, although this is changing quickly as other techniques are gaining widespread acceptance in the inspection community [13]. In addition, they might be too costly to maintain or support in the field [22]. In some cases, they might even be applicable only in limited circumstances or too uninformative to form the basis of support for or against critical judgment calls [28]. Similar to most other NDT techniques, the discipline of RTR is divided into two primary responsibilities: using equipment to scan the weld and properly interpreting the scan. Although there is always technological advancement in the efficient and high-quality

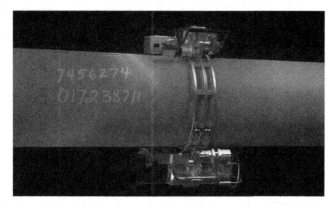

FIGURE 9.1 Digital X-ray detector and source on a robotic platform (Stanley Oil & Gas).

capture of radiographic weld scans, the most important task remains the latter, as this is what determines the perceived weld quality [35].

In practice, RT is a very well established technique. Entire codes or specifications for the practice of and techniques of inspection have been developed as an attempt to standardize the practice of reliable interpretation and operation [73]. One of the most well defined standards relate to the interpretation of radiographic weld scans. In these images, experienced operators are trained to visually detect a whole host of defects that are caused by different factors related to the material and welding process. To the untrained eye, defects are difficult to classify, even imperceptible, unless told exactly what to look for. Through years of experience, individual operators learn how to identify, classify, and prioritize these defects, in accordance with one of the major inspection codes that provide visual examples, descriptions, and guidelines [18].

Even today, almost the entire process, with the exception of image capture itself, is managed and performed by people. Fig. 9.1 is showing an example of X-ray image capturing devices. People are both susceptible to visual fatigue and uniquely trained over long periods of time with an exclusive skillset that takes years to transfer to the next generation of qualified technicians [36]. A fundamental question that we can ask regarding RTR inspection is what role, if any, could the interpretation of radiographic scans be delegated to a machine? The idea of allowing a machine to assist a human operator in the process of interpreting responses from the application of NDT inspection techniques for the purpose of identifying problematic flaws is known as assisted defect recognition (ADR) [83]. The key is that although we still ensure that human beings are fully and directly in control of the final assessment, people are still willing and able to use whatever tools in their disposal they have to aid them in a task so difficult to consistently reproduce or perform perfectly.

Human operators, depending on their role and level of skill, are expected to visually parse up to hundreds of weld scan images daily. This is a lot to ask from the eyes and the brain and is bound to at least introduce the possibility of reasonable errors that are unfortunately essentially unacceptable from a standard of quality [52]. As the process of radiographic scanning is nearly fully automated in practice, people, although specially trained, are expected to keep up with the same pace and operate with consistent and high accuracy [15]. This is at least one reason it seems reasonable that people should be assisted in the process of interpretation and analysis of radiographic weld scans.

Given that a single individual is incapable of being at thousands of places in the world at once, we must employ the talents of many individuals to process weld images from multiple projects. Moreover, each individual has a unique project background, level of experience, predisposition, preference, and opinion [65]. Despite the presence of NDT inspection codes and standard, it is almost totally unexpected to expect all of these individuals to agree upon the same internal standard formed over time and through experiences [78]. Even if this were not the case, it guaranteed that individuals trained on the basis of separate codes, which are not entirely consistent with one another, will differ with one another in the interpretation of certain key conditions [49]. It takes substantial time for someone to internalize an external standard and develop the ability to apply it as second nature professionally. If operators are trained under the API 1104 inspection standard for pipeline welding, they do not automatically acquire the knowledge embodied in the ASME standard [50]. Even more pertinent, they must contradict themselves relative to a previous standard on occasion to adhere to a new one. People are not often capable of maintaining multiple potentially incompatible standards mentally.

The fundamental approach is a data-driven machine learning model. The advantage of this is that the experience of technicians on former projects across a wide variety of conditions are now directly available, more than any one individual could provide, even with sufficient time. On some level, there is no real need to acquire the understanding that an experienced inspection operator possesses to develop a system that is capable of learning from that operator [81]. There are varying ways and degrees that a machine might assist a human by focusing on tasks that machines are proven to be successful with.

Specially designed software might simply identify the presence of anomalies or more usefully classify them into distinct categories. Operators would be able to see the most likely classification candidates. Going one step further, this classification could be based on specific classes that are defined in common inspection codes. As an operator is advised through ADR, the operator may also be quickly directed to key references or guidelines in a given code or explicit examples encountered in previous projects. When it becomes clear that different codes deviate from one another, such a tool could be trained much like a person is trained on competing standards, far more than any one individual could master in a lifetime. In essence ADR offers the opportunity to enlist the help of all professionals everywhere in every decision, even when they are completely unavailable [53].

As the inspection community gains increased faith in the accuracy and efficacy of such machine learning models and associated software tooling, the possibilities for safe and valuable technological aids in the process of NDT weld inspection and analysis grow accordingly.

The focus of this research is at the intersection of most of the key concepts discussed previously. In particular, the goal was the development of a practically valuable and useful model for defect recognition in radiographic image scans on pipeline girth welds to assist inspection technicians and operators in visual analysis. To accomplish this, deep learning techniques were employed for automatically classifying defects built from a library of a large number of labeled annotations. The value of a data-driven model is rooted in the ability to learn from professional experts directly and efficiently [29]. Additionally, a deep learning model—provided enough training data—is capable of extracting highly complex features that intuitively correspond to the same features for which professional inspection technicians develop a reliable intuition [71]. This intuition, by its very nature, is unfortunately at the same time difficult to explain to another person. Even without years of professional experience, it becomes possible to translate the collective experience of the inspection expert community into a language that can be understood.

NDT using radiography (RT) is one of the oldest techniques to evaluate welding quality, which is used predominantly in inspection of welded joints. Utilization of RT techniques is inevitable to certify the safety and reliability of manual welds in these structures. However, the process of assessment of X-ray images is both time consuming and at times can be subjective for the expert operators due to various reasons [76]. The advent of novel image processing techniques and pattern recognition has been applied to accelerate the process of improving the image quality to help in weld defect diagnosis and increase the accuracy of defect detection. Nonetheless, due to the complexity of the task, many efforts have been made to develop such an automated intelligent system to recognize defective welded joints.

9.2 Literature Review

Previous research on weld defect detection was mostly performed using techniques such as segmentation and texture features extraction and classification, one of which has reached the highest accuracy of 90.91% by Mery et al. [55]. Artificial neural networks (ANNs) are used predominantly with enhancements on classifier description among various research. Kumar et al. [44] used gray-level co-occurrence (GLCM) texture features as ANN input and reached an accuracy of 86.1% for defect classification.

In another work, Kumar et al. [43] added geometrical features and enhanced the accuracy to 87.34%. Zapata et al. [85] utilize two neural classifiers: an adaptive network-based fuzzy inference system (ANFIS) and ANN. In their work, ANN achieved accuracy of 78.9%, and using the 12 chosen geometrical features extracted from X-ray images as input, ANFIS reached an accuracy of 82.6%. To integrate both texture and geometrical features and to select only useful features, preventing an increase in computational complexity, Valavanis et al. [75] employed the sequential backward selection (SBS) technique to design the classifier of the ANN, which resulted in 85.4% accuracy.

Wang and Liao [79] designed a weld flaw detection system, in which they extracted 12 different features from radiography images such as size, intensity, direction, location, and shape for the input of networks. They applied two well-known networks for detection of weld flaws: fuzzy k-nearest neighbor (fuzzy K-NN) and multi-layer perceptron (MLP) neural networks, which attained accuracy of 91.57% and 92.39%, respectively. However, traditional feature extraction methods used in these systems are not able to extract discriminant features.

The solution to improve pattern recognition in weld flaw detection is to take advantage of deep neural networks that can extract higher-level and more abstract features. Hou et al. [30] employed deep convolutional neural networks (DCNNs) on three resampled datasets based on the GDXray [56] weld dataset, and the highest achieved accuracy was 97.2% on the dataset using the synthetic minority over-sampling technique (SMOTE). For the remaining datasets, two techniques of random over-sampling (ROS) and random under-sampling (RUS) for modifying samples were applied, and best-case performances were 96.3% and 79.9% for each method, respectively.

Integration of computer vision methods using DCNNs shows exceptional promise for use in several applications like object detection, crack detection and NDT but requires many images for the training process [87–89]. Although utilizing CNNs improves the defect detection accuracy, there are known drawbacks in the conventional approaches for studying the defects. For instance, Lin et al. [47] studied application of a DCNN for detection of casting defects. The authors reported 96% accuracy achieved through using an eight-layer CNN for detecting defects in X-ray images of a casted object. They could improve the detection rate up to 8% by applying their CNN-based method over an ANN-based method. Utilizing the proposed method not only makes it possible to increase the quality but also reduces the scrap rate during the casting process.

As reviewed earlier, methods based on deep learning for NDT improve false detection accuracy within X-ray images, whether using different CNN architectures or pre-trained models [31]. Zhang et al. [86] deployed a CNN classification model to detect weld defects. The goal was to detect defects for aluminum alloy in robotic arc welding using an 11-layer CNN. In this method, the authors reported 99.38% accuracy for a single type of welded material. For the database, the authors used a CCD camera instead of radiography images. In industry, using images of a CCD camera for various types of materials in weld is not common, because it is not possible to detect in-depth defects like lack of infusion. As another example in the work of Zhang et al. [84], the imaging sensor is improved by utilizing a UVV band visual sensor system, but this method is

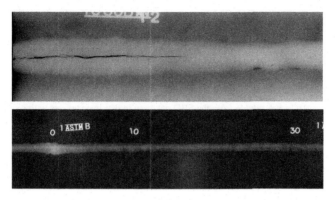

FIGURE 9.2   Two samples of images in the GDXray database (top) and the SBD database (bottom). The bottom image is cropped to be able to compare with the top image. Defects are more visible in the GDXray database.

not capable of penetrating through material deep enough to detect defects like an internal cavity.

Zhang et al. [34] studied generative adversarial nets (GANs) in weld defect detection. Based on the definition of defect, they categorized defects into three groups: cracks, porosity, and burn through. In their definition, series of single dots considered as porosity, linear shadow that can be found vertical or parallel to the edge of welds defined as crack, and burn through defined as shadows of anomalous shape on X-ray images. The definition is not applicable in the real-world X-ray images, and many defect classes will be neglected due to the limited bounded definition. Although the proposed method could reach above 94%, the trained CNN results on other sets of X-ray weld datasets are not reported to check the generalization of the network.

However, most reported works used the GDXray database, which is public [56], and the WDXI dataset [25], which is not available publicly. As depicted in Fig. 9.2 (top image), defects in images of the GDXray database are handpicked and defects are easily visible due to their extreme nature. In addition, weld images in the GDXray database are limited to only 88 samples. Although GDXray has become a standard benchmark for testing the performance of different algorithms [30], defects in real-world X-ray images are not as obvious as those in the GDXray dataset. As is shown in Fig. 9.2 (bottom image), it is more challenging to detect defects in real-world X-ray images that are not handpicked and do not contain oversized defects. Moreover, in previous studies, whether traditional image processing algorithms or methods based on deep learning were used, differences in wall thickness of welded objects and scattering or waving in the weld root due to geometry of pipes and variation of welding pattern of manual welder are not present and neglected as a feature. In such cases, due to the change of thickness, intensity of the image varies significantly in each segment of the weld, as is shown in Fig. 9.3.

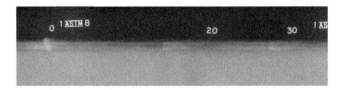

FIGURE 9.3 An image sample from the SBD database in that two surfaces with different thicknesses are welded.

In this chapter, the main goal is set to locate and identify discontinuities and defects (detection) in a realistic X-ray image dataset (the SBD dataset) and to determine the defect type (DT). Due to the aforementioned differences with the GDXray database, previous methods either failed to detect the defects present in the SBD dataset or the overall accuracy of the methods was dramatically lowered. To address this problem, a DCNN is designed and trained to be able to detect defects in the SBD database.

9.3 Database Preparation

To develop and evaluate the proposed network architecture, 5000 full-sized $15,360 \times 1024$ (*Width $\times$ Height*) pixel images of a welded pipeline containing weld defects are gathered. All full-sized images were cropped into 224×224 pixel patches, each patch including the weld's center. A total of 100,000 patches were categorized as non-defected and defected by an expert with API and ASME welding certificates. In the first phase, we selected a smaller set of non-defective and defective patches in a balanced number. This database was named *SBD-1*. In addition, to evaluate the network's performance against different types of defects, another database was developed (SBD-2), containing more than 20,000 discontinuity patches, including 11 different types of major defects recognized by ASME and API. DTs and quantity are shown in Table 9.1. Fig. 9.4 depicts 10 samples of non-defective patches, and Fig. 9.5 shows 11 patches along with the DT.

9.4 Experimental Study

AlexNet was first proposed in 2012 by Krizhevsky et al. [42] and was the winner of the ImageNet [14] LSVRC-2010 contest. In this work, for the first time, rectified linear units (ReLUs) were used as the activation function. The network was trained on 1.2 million high-resolution images with the goal of classifying into 1000 categories. The eight-layer architecture of the network, which is an extension of Lenet-5 [45], consists of 5 convolutional layers followed by three fully connected (FC) ones, and the network has 60 million parameters. As input, it takes 224 RGB images, and the final FC layer is a 1000-way Softmax. After

TABLE 9.1 Defect types

Defect Type	Total Number of Patches
Elongated Slag Inclusions (ESIs)	5884
Hollow Bead (HB)	5204
Isolated Slag Inclusions (ISIs)	3776
Gas Porosity (GP)	2528
Inadequate Penetration (IP)	1216
External Undercut (EU)	1692
Porosity (P)	785
Scattered Porosity (SP)	632
Inadequate Penetration Due to High-Low (IPD)	938
Internal Undercut (IU)	902
Internal Concavity (IC)	814

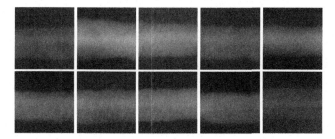

FIGURE 9.4 Ten samples of non-defect images.

that, using Deeper CNN has become prevalent among researchers. As a proof that deepening the network will improve model accuracy, VGG-16 and VGG-19 were proposed by Simonyan and Zisserman [72] on the same ImageNet dataset. The network has approximately 138 million parameters. In comparison to AlexNet, they have added 11 convolutional layers to VGG-16 and 14 to VGG-19 to improve overall accuracy of the network. In 2015, a deeper model with 152 layers, which is known as ResNet-50 [27], was presented by Microsoft Research. In this work, they leveraged skip connections and batch normalization to address the problem of saturation and prevent compromising of model generalization in deep networks.

Several CNN architectures like VGG-16, VGG-19, AlexNet, and ResNet were applied to SBD-1 and SBD-2. The maximum accuracy obtained using these well-known architectures was 86% for SBD-1 and 75% for SBD-2. Because the

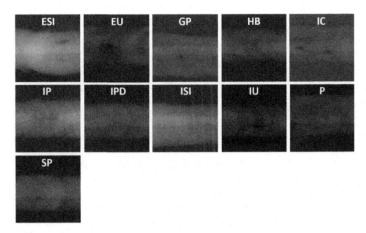

FIGURE 9.5 Classes of welding defects based on the ASME standard.

object of interest in this problem is not a common object (i.e., weld defects), methods based on transfer learning, such as in the work of Ferguson et al. [17], would not be an optimal solution. To address this problem, a network architecture is proposed in the following section. To obtain optimized values for the number of layers, filter size, and other hyperparameters (HPs), a Bayesian optimization algorithm was used [63]. In the following sections, design steps for the SBD-2 database are presented. The steps will be the same for the SBD-1 database with exception of the number of classes.

9.4.1 Deep Learning Architecture

Fig. 9.6 illustrates the overall CNN architecture including input and multiple convolutional layers, followed by a batch normalization layer, ReLU layer, max-pooling layer, FC layer, SoftMax layer, and output layer for the classification task.

The first layer is the input layer that receives the input of 224×224 patches for classification. In SBD-1, this classification would be between two classes, and in SBD-2, it would be a multi-class classification problem.The defect features in each data batch were extracted using multiple convolutional layers. This layer consists of various sets of neurons whose weights and biases will be updated relative to the defect features. In the convolutional layer, the neuron input consists of small sectors from the previous layer called the *filter* (kernel). The size of the filter, s_f, can be tuned from 1×1 pixels up to the size of the input image. In the convolutional layer, the filter moves along the input and builds a convoluted feature map. To increase the number of feature maps, multiple filters should be used, and each filter has different weights and biases to be able to extract various features of the image. The stride (amount of horizontal

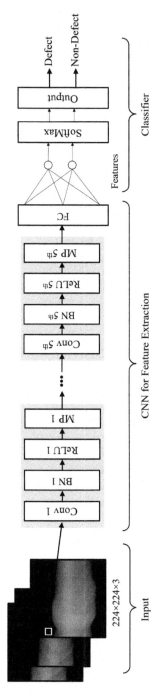

FIGURE 9.6 Network architecture used for weld defect detection. Five convolutional layers are used in this architecture. Conv: convolutional layer; BN: batch normalization layer; MP: max-pooling layer; FC: fully connected layer.

and vertical movement of the filter on the input per convolution) is set to 3 pixels.

After the convolutional layer, a batch normalization layer is used for reducing the CNN sensitivity to initial HP values and decreasing training processing time. Following the batch normalization layer, a ReLU activation layer is added to apply a zero threshold to all negative values in the batch normalization layer, which means that the inputs from the previous layer b go through $max(0, b)$. The max-pooling layer downsamples the input by dividing it into rectangular pooling regions to compute the maximum of each region of gathered feature matrices. After designing the feature extractor, the FC layer is used to map the features matrix in the last layer in the form of a $1 \times c$ vector, where $c = 11$ is the number of DTs in SBD-2. For representing the probability distribution over multiple classes in the output of a classifier, a generalized model of binary logistic regression classifier (Softmax function) is utilized after the FC [8, 21]. Considering the input of the Softmax function as a sample defect patch (DP) that belongs to one of the 11 DTs, $DP{\in}DT_j$, where $j{\in}\{1, ..., 11\}$, then the DP prior probability [24] is defined as $P(DT_j)$, which shows the probability of $DP{\in}DT_j$ and conditional probability as $P(DP_j, \bar{\phi}|DT_j)$, where $\bar{\phi} = [\bar{\omega}, \bar{b}]$ is the parameter vector that consists of weights $\bar{\omega}$ and biases $\bar{b}$. The Softmax function is described as follows:

$$S_j(DP, \bar{\phi}) = P\left(DT_j \mid DP, \bar{\phi}\right)$$
$$= \frac{P\left(DP, \bar{\phi} \mid DT_j\right)P\left(DT_j\right)}{\sum_{m=1}^{j} P\left(DP, \bar{\phi} \mid DT_m\right)P(DT_m)} = \frac{\exp\left(r_j(DP, \bar{\phi})\right)}{\sum_{m=1}^{L} \exp\left(r_m(DP, \bar{\phi})\right)}, \quad (9.1)$$

where $r_j(DP, \bar{\phi}) = \ln(P(DP, \bar{\phi}|DT_j)P(DT_j))$ and S_j is a probability distribution as the Softmax function output, where $0 \leq S_j \leq 1$ and $\sum_{j=1}^{13} S_j(DP, \bar{\phi}) = 1$.

Following the Softmax function, the classification output layer (cross entropy function) is used to assign each input to one of the $n = 11$ mutually exclusive DTs using the loss function shown in the following:

$$l(\bar{\phi}) = - \sum_{i=1}^{p} \sum_{j=1}^{n} d_{ij} \ln S_j\left(DP_i, \bar{\phi}\right), \quad (9.2)$$

where p is the number of samples and d_{ij} is a matrix that shows with what probability the ith sample of DP belongs to the jth DT.

9.4.2 Training

Stochastic gradient descent with momentum (SGDM) was used to train the CNN for classification. This method updates the CNN's weights and biases to minimize the loss function that measures the difference between true-classified and false-classified DPs. The SGDM uses a subset of training data (mini-batch). The gradient derived from the data within the mini-batch is used for updating the weights and biases. Each update to the weights and biases is defined as one iteration. The gradient descent update law is described as

$$\bar{\phi}_{k+1} = \bar{\phi}_k - \lambda l(\bar{\phi}_k) + \eta(\bar{\phi}_k - \bar{\phi}_{k-1}), \tag{9.3}$$

where subscript k represents the iteration number, the initial learning rate is $0 < \lambda < 1$, $\bar{\phi}$ is a vector that contains the weights and biases, $l(\bar{\phi})$ is the loss function, and $0 \leq \eta \leq 1$ is the momentum, which defines the level of contribution from the previous step. For λ values close to 0, the learning processed is slowed. and values close to 1 lead to either diverging or suboptimal weights. Moreover, to prevent over-fitting of the CNN during the training process, L2 regularization [8, 62] is utilized as follows:

$$l(\bar{\phi}_{k+1}) = l(\bar{\phi}_k) + \frac{\tau(\bar{w}_k^T \bar{w}_k)}{2}, \tag{9.4}$$

where τ is the regularization factor. To address over-fitting and feature memorization, and improve the generalization of the Softmax classifier during the training process, a modified data augmentation procedure is used during each iteration [21], where the DPs were translated randomly in the horizontal and vertical directions by a maximum of by ± 10 pixels.

9.4.3 Network HP Optimization

HPs in the proposed CNN architecture and SGDM are the filter size s_f, number of filters N_f, and number of CNN layers ND, η, τ, and λ. The search range for HPs was defined as $ND \in \{1, 2, \ldots, 20\}, S_f \in \{1, 2, \ldots, 15\}$, $N_f \in \{1, 2, \ldots, 100\}, 0 \leq \eta \leq 1, 0 \leq \tau \leq 1$, and $0 < \lambda < 1$. The possible values for ND, s_f, and N_f are integers, and for η, τ, and λ are logarithmically spaced values between 0 and 1. The classification error is the number of misclassified DPs by the classifier (Softmax).

The objective of optimization is to find optimal values for the HPs such that the classification error is minimized. Thus, the objective function can be considered a function with HPs as the input and the classification error as the output. Modeling of this objective function is algebraically complicated and computationally intensive. The BOA is capable of performing optimizing the HPs to minimize the classification error, whereas the objective function is considered as a black box [5]. To perform the BOA, a validation set was defined that consists of 15% randomly selected DPs from the training set. The inputs of the objective function are the training set and the validation set.

As shown in Fig. 9.7 the objective function trains the CNN and returns the classification error on the validation set. By modeling the calculated error using a Gaussian process (GP) as mentioned in the work of Gelbart et al. [19] and in multiple iterations Z, where $z = \{1, 2, \ldots, 83\}$, the BOA finds the optimal values

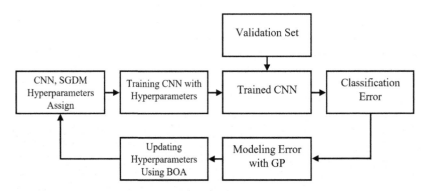

FIGURE 9.7 Block diagram of optimizing hyperparameters using BOA.

for HPs that minimize the classification error. The kernel function that was used for the GP is the automatic relevance determination (ARD) Matéra 5/2 in the work of Rasmussen [68]. In addition, the acquisition function $(q_z(HP))$ that is used for the GP is the expected improvement function E(.) [59], as follows:

$$q_z(HP) = \arg\max_{HP} E\big(\max\big\{0, f_{z+1}(HP) - f_z^{\max}(HP)\big\}\big), \qquad (9.5)$$

where $f_z^{\max}(HP)$ is the current maximum observed value for the objective function. The next estimation for maximizing the objective function is obtained by using the acquisition function. The GP posterior is updated in each iteration using Eq. (9.5):

$$P(f_z \mid u) = \frac{P(u \mid f_z)P(f_z)}{P(u)}, \qquad (9.6)$$

where $u = \{(HP_z, f_z), z = 1 : 100\}$.

The extrema of $f_z(HP)$ was obtained numerically at sampled values of the function. A closed-form expression of the objective function is not required within the BOA mathematical structure [10]. The objective function and acquisition function for two of the SGDM HPs (i.e., η and λ) during the optimization process are shown in Fig. 9.8. As depicted in Fig. 9.8(A) the observed points are demarcated by blue dots $(f_z(HP))$, the model mean that is obtained from the observations is depicted as the red surface, and subsequent evaluation point addition is demarcated with a black dot. Moreover, Fig. 9.8(B) illustrates the acquisition function. The objective function is shown to reach a minimum at the 53rd iteration; this point is demarcated with a black star. Fig. 9.8(B) shows the maximum feasible value that is generated upon minimizing the classification error. Fig. 9.9 visualize how well False Positive rate decreased after HP optimization. Most enhancement is performed on SP and ISI, and class IU did not show any improvement.

The total number of iterations was set to 80. Each iteration calculates the classification error among 700 randomly selected DPs from 11 defect batches.

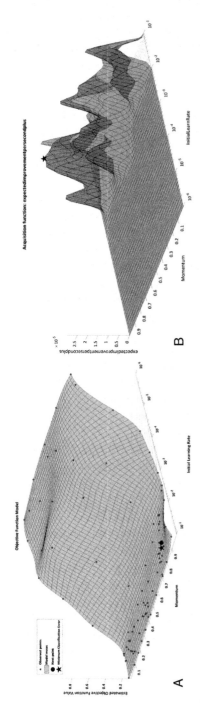

FIGURE 9.8 (A) The observation function model. (B) Acquisition function for two parameters of the SGDM. The starred point is the calculated optimal values at the 53rd iteration for η and λ.

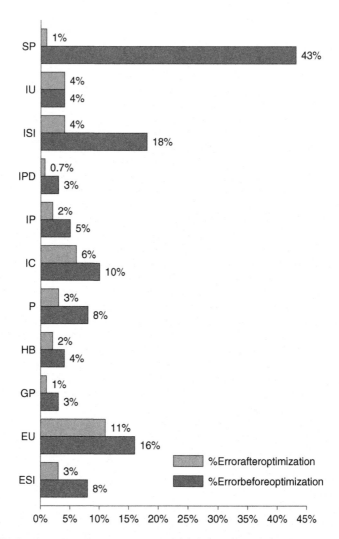

FIGURE 9.9 Comparison between error among 11 defect categories before and after optimization.

The BOA was evaluated statistically using the Wald method [26] by representing the images in the test set as independent events with a known probability of success. The number of misclassified images were represented with a binomial distribution. By applying the trained CNN with optimized HPs on the test set and computing the number of correctly classified DPs, the test error E_r is defined as follows:

$$E_t = 1 - \frac{1}{b}\sum_{i=1}^{b} Ds_i, \tag{9.7}$$

where Ds and b are the number of correctly classified DPs and the total number of DPs in the test set, respectively. Note that to evaluate the trained CNN performance on the test set without exposing the CNN to the optimization process, E_s is used to obtain the standard error. This approach helps to increase the optimization speed. The standard error is represented as follows:

$$E_s = \sqrt{\frac{E_t}{Ds}(1 - E_t)}. \tag{9.8}$$

Moreover, as the target of this research, to obtain a ± 8 error margin, a confidence interval of 92% is defined to calculate the generalization error E_G defined as

$$E_G = E_t \pm 0.92 E_s. \tag{9.9}$$

The final HP values for the CNN were $S_{f1} = 7$, $S_{f2} = 7$, $S_{f3} = 9$, $N_{f1} = 235$, $N_{f2} = 105$, and $N_{f3} = 98$. In addition, the optimized values for SGDM were $\lambda = 0.0016737$, $\eta = 0.040151$, and $\tau = 0.003134$. Applying the optimal HP values to the CNN and SGDM yields a CNN with 20 layers and 91.20% accuracy in eight epochs. Moreover, the minimized value for the loss function was 0.021 in eight epochs. As well, the generalized error E_G interval for the test set was [0.517 0.0115].

The same steps followed for SBD-1 and the resulted HP values for the CNN were $S_{f1} = 3$, $S_{f2} = 5$, $S_{f3} = 9$, $N_{f1} = 107$, $N_{f2} = 87$, and $N_{f3} = 159$. The optimized values for SGDM were $\lambda = 0.002128$, $\eta = 0.025287$, and $\tau = 0.003876$. Applying the optimal HP values to the CNN and SGDM yields a CNN with 20 layers and 95.86% accuracy in eight epochs. Moreover, the minimized value for the loss function was 0.037 in eight epochs.

9.5 Experimental Implementation

There is an increasingly valuable resource for managing both training and inference pipelines based on sophisticated models, namely large-scale cloud computing platforms. Through services that are specifically designed to offload high capacity data storage and intensive computations to scaleable remote specially designed servers, it becomes possible to develop and serve complex models such as the aforementioned and chosen deep learning CNN. In the course of the model's development, multiple AWS were employed to handle all stages of production, including the means to easily supply the model through the input of dedicated experts. For such advanced models, it simply is not possible anymore to train efficiently on local generic machines with the amount of data necessary so that the model can learn or extract sophisticated features and maximize accuracy in its predictions.

In particular, the storage service known as S3 allowed for systematic massive-scale large file data hosting. In practice, the electronically captured radiographic DICOM (Digital Imaging and Communications in Medicine) images assessed and analyzed by inspection operators are quite unwieldy to manage on local machines and are essentially impossible to collect at one time. As technicians capture data through equipment designed for RTR, these images are directed to a database of files ready for processing to either be used as training or testing data for the model or for the opportunity for additional annotations and detailed labeling. As these images are collected from projects that are confidential and display information that should not be seen by third parties, a mandatory redaction step is applied using optical character recognition models to maximize the accuracy of redaction for exclusively confidential information. Once these images have been redacted, they are ready for labeling and annotation or further processing.

Through the power of microservices architecture and services that enable scalable computation effects such as AWS Lambda, each processing step in the development of the model can be produced independently. A separate series of proprietary software tools were responsible for extracting images from large data storage and preparing the input in a preprocessing step for direct application as training data into the model. As these images are essentially long, rectangular unwrapped scans of the weld and pipe circumference, it was necessary to employ a method for essentially extracting the region of the image where the actual weld exists between the the walls of the pipe and producing specifically sized slices that follow the weld precisely.

These slices are then related to labeled defect annotations in one of two ways. First, experts in RTR inspection are asked to annotate and classify suspected defects from the slices directly or from the original image, even outside the context of the project where the weld was originally captured. Second, specific metadata in the image recording the annotations produced from proprietary inspection and annotation tooling software in the project itself is parsed and cross referenced with the exact location of the slice relative to the original image. Finally, after the appropriate labels are applied to the corresponding image slices, each centered on the weld and each containing the visual pattern of the defect class given by the label, the labeled training data is applied to the model.

9.6   Conclusion

In this work, two realistic welding quality datasets for training deep learning models were created based on radiography images collected from various projects and NDT expert-annotated datasets SBD-1 and SBD-2. An optimized CNN was designed to find defects in the weldment and heat-affected zones and was subsequently trained and evaluated based on prepared datasets. An accuracy of 96% was achieved. In addition, the robustness of the proposed method was tested across 11 different types of welding discontinuity. Moreover, the results were compared with the non-optimized CNN architecture. In the case of using an optimized CNN, the classification accuracy was improved by 10%. In addition, this approach was found to be applicable to real-world datasets. This method improved not only the defect detection speed but also the accuracy of defect detection in high-volume projects.

References

[1] A. Aljaroudi, F. Khan, A. Akinturk, M. Haddara, P. Thodi. Risk assessment of offshore crude oil pipeline failure. *Journal of Loss Prevention in the Process Industries* 37 (2015) 101–109.

[2] W.M. Alobaidi, E.A. Aluam, H.M. Al-Rizzo, E. Sandgren. Applications of ultrasonic techniques in oil and gas pipeline industries: A review. *American Journal of Operations Research* 5 (4) (2015) 274.

[3] M. Arntz, T. Gregory, U. Zierahn. Revisiting the risk of automation. *Economics Letters* 159 (2017) 157–160.

[4] T. Backström, M. Döös. A comparative study of occupational accidents in industries with advanced manufacturing technology. *International Journal of Human Factors in Manufacturing* 5 (3) (1995) 267–282.

[5] R. Baptista, M. Poloczek. Bayesian optimization of combinatorial structures. arXiv:1806.08838, 2018.

[6] R. Beeson, Miller Electric Manufacturing Co., Appleton, WI (US). Pipeline welding goes mechanized. Welding Journal (Miami) 78 (11) (1999) 47–50.

[7] K. Benyounis, A. Olabi, M. Hashmi. Effect of laser welding parameters on the heat input and weld-bead profile. *Journal of Materials Processing Technology* 164 (2005) 978–985.

[8] Don E. Bray, Don McBride, Nondestructive testing techniques, NASA Scientific and Technical Information/Recon Technical Report A93 (1992) 17573.

[9] E. Brochu, V.M. Cora, N. De Freitas. A tutorial on Bayesian optimization of expensive cost functions, with application to active user modeling and hierarchical reinforcement learning. arXiv:1012.2599, 2010.

[10] L. Cartz. *Nondestructive Testing.* ASM International, Novelty, OH, 1995.

[11] J. De Raad, F. Dijkstra. Mechanized ultrasonic testing on girth welds during pipeline construction. *Materials Evaluation* 55 (8) (1997) 890–895.

[12] J. Deng, W. Dong, R. Socher, L.-J. Li. K. Li, L. Fei-Fei. ImageNet: A large-scale hierarchical image database. In *Proceedings of the IEEE Conference on Computer Vision and Pattern Recognition*, 2009. IEEE, Los Alamitos, CA. 248–255.

[13] M.C. Domke, J.H. Messinger, S. Soorianarayan, T.E. Lambdin, S. L. Sbihli. Systems and methods for analyzing data in a non-destructive testing system. US Patent 9,217,999, 2015.

[14] D. Fairchild, M. Macia, N. Bangaru, J. Koo. Girth welding development for x120 linepipe. *International Journal of Offshore and Polar Engineering* 14 (1) (2004) ISOPE-04-14-018.

[15] M.K. Ferguson, A. Ronay, Y.-T.T. Lee, K.H. Law. Detection and segmentation of manufacturing defects with convolutional neural networks and transfer learning. *Smart and Sustainable Manufacturing Systems* 2 (2018) 10.

[16] D.S. Forsyth, H.T. Yolken, G.A. Matzkanin. A brief introduction to nondestructive testing. *AMMTIAC Quarterly* 1 (2) (2006) 7–10.

[17] M.A. Gelbart, J. Snoek, R.P. Adams. Bayesian optimization with unknown constraints. arXiv:1403.5607, 2014.

[18] O. Gericke. Determination of the geometry of hidden defects by ultrasonic pulse analysis testing. *Journal of the Acoustical Society of America* 35 (3) (1963) 364–368.

[19] I. Goodfellow, Y. Bengio, A. Courville. *Deep Learning.* MIT Press, Cambridge, MA, 2016.

[20] R. Gordon, R. Holdren, M. Johnson, M. Lozev. Reducing pipeline construction costs: New technologies. *Welding in the World* 47 (5-6) (2003) 7–14.

[21] L. Gourd. *Principles of Welding Technology.* Edward Arnold, London, UK, 1986.

[22] A. Graves, S. Fernández, F. Gomez, J. Schmidhuber. Connectionist temporal classification: Labelling unsegmented sequence data with recurrent neural networks. In *Proceedings of the 23rd International Conference on Machine Learning,* 2006, 369–376.

[23] W. Guo, H. Qu, L. Liang. WDXI: The dataset of X-ray image for weld defects. In *Proceedings of the 14th International Conference on Natural Computation, Fuzzy Systems, and Knowledge Discovery (ICNC-FSKD),* 2018. IEEE, Los Alamitos, CA. 1051–1055.

[24] F.E. Harrell Jr. Regression Modeling Strategies: With Applications to Linear Models, Logistic and Ordinal Regression, and Survival Analysis. Springer Nature, Switzerland AG, 2015.

[25] K. He, X. Zhang, S. Ren, J. Sun. Deep residual learning for image recognition. In *Proceedings of the IEEE Conference on Computer Vision and Pattern Recognition,* 2016, 770–778.

[26] Y. He. M. Pan, F. Luo, G. Tian. Pulsed eddy current imaging and frequency spectrum analysis for hidden defect nondestructive testing and evaluation. *NDT & E International* 44 (4) (2011) 344–352.

[27] M.A. Henn, H. Zhou, B.M. Barnes. Data-driven approaches to optical patterned defect detection. *OSA Continuum* 2 (9) (2019) 2683–2693.

[28] W. Hou, Y. Wei, Y. Jin, C. Zhu. Deep features based on a DCNN model for classifying imbalanced weld flaw types. *Measurement* 131 (2019) 482–489.

[29] W. Hou, D. Zhang, Y. Wei, J. Guo, X. Zhang. Review on computer aided weld defect detection from radiography images. *Applied Sciences* 10 (5) (2020) 1878.

[30] D.C. Howard. Non-destructive testing of pipeline. US Patent 4,098,126, 1978.

[31] O. Hunaidi, M. Bracken, A. Wang. Non-destructive testing of pipes. US Patent 7,328,618, 2008.

[32] H. Zhang, Z. Chen, C. Zhang, J. Xi, X. Le. Weld defect detection based on deep learning method. In *Proceedings of the IEEE 15th International Conference on Automation Science and Engineering (CASE),* 2019. IEEE, Los Alamitos, CA. 1574–1579.

[33] J. Jarmulak, E.J.H. Kerckhoffs, P.P. van't Veen. Case-based reasoning for interpretation of data from non-destructive testing. *Engineering Applications of Artificial Intelligence* 14 (4) (2001) 401–417.

[34] J.F. Jarvis. Visual inspection automation. In *Proceedings of the IEEE Computer Society's 3rd International Computer Software and Applications Conference,* 1979. IEEE, Los Alamitos, CA. 251–255.

[35] J.A. Jensen. Medical ultrasound imaging. *Progress in Biophysics and Molecular Biology* 93 (1-3) (2007) 153–165.

[36] B.Y. Jeong. Occupational deaths and injuries in the construction industry. *Applied Ergonomics* 29 (5) (1998) 355–360.

[37] M.I. Khan. *Welding Science and Technology.* New Age International, Delhi, India, 2007.

[38] S. Knight, S.G. Drake. X-ray inspection apparatus for pipeline girth weld inspection. US Patent 8,923,478, 2014.

[39] K. Kobayashi, S. Ishigame, H. Kato. Skill Training System of Manual Arc Welding. Entertainment Computing: Technologies and Application. Springer US, Boston, MA, 2003, pp. 389–396.

[40] A. Krizhevsky, I. Sutskever, G.E. Hinton. ImageNet classification with deep convolutional neural networks. In *Advances in Neural Information Processing Systems.* 1097–1105.

[41] J. Kumar, R. Anand, S. Srivastava. Flaws classification using ANN for radiographic weld images. In *Proceedings of the International Conference on Signal Processing and Integrated Networks (SPIN)*, 2014. IEEE, Los Alamitos, CA. 145–150.

[42] J. Kumar, R. Anand, S. Srivastava. Multi-class welding flaws classification using texture feature for radiographic images. In *Proceedings of the International Conference on Advances in Electrical Engineering (ICAEE)*, 2014. IEEE, Los Alamitos, CA. 1–4.

[43] Y. Lecun, L. Bottou, Y. Bengio, P. Haffner. Gradient-based learning applied to document recognition. *Proceedings of the IEEE* 86 (11) (1998) 2278–2324.

[44] G. Light. *Demonstration of Realtime Radiography on Pipeline Girth Welds*. Technical Report. Southwest Research Institute, San Antonio, TX.

[45] J. Lin. Y. Yao, L. Ma, Y. Wang. Detection of a casting defect tracked by deep convolution neural network. *International Journal of Advanced Manufacturing Technology* 97 (1-4) (2018) 573–581.

[46] T.R. Lin, B. Guo, S. Song. A.J. Chacko. Ghalambor. Offshore Pipelines. Gulf Professional Publishing. Burlington, MA, 2005.

[47] R. Lumb. Non-destructive testing of high-pressure gas pipelines. *Non-Destructive Testing* 2 (4) (1969) 259–268.

[48] B. Ma, J. Shuai, J. Wang, K. Han. Analysis on the latest assessment criteria of ASME B31G-2009 for the remaining strength of corroded pipelines. *Journal of Failure Analysis and Prevention* 11 (6) (2011) 666–671.

[49] T. Matsutani, F. Miyasaka, T. Oji, Y. Hirati. Mathematical modelling of gta girth welding of pipes. *Welding International* 11 (8) (1997) 615–620.

[50] E. Megaw. Factors affecting visual inspection accuracy. *Applied Ergonomics* 10 (1) (1979) 27–32.

[51] J. Meier, I. Tsalicoglou, R. Mennicke. The future of NDT with wireless sensors, AI and IoT. In *Proceedings of the 15th Asia Pacific Conference for Non-Destructive Testing*, 2017.

[52] P.F. Mendez, T.W. Eagar. Penetration and defect formation in high-current arc welding. *Welding Journal* 82 (10) (2003) 296.

[53] D. Mery, M.A. Berti. Automatic detection of welding defects using texture features. *Insight: Non-Destructive Testing and Condition Monitoring* 45(10) (2003) 676–681.

[54] D. Mery, V. Riffo, U. Zscherpel, G. Mondragón, I. Lillo, I. Zuccar, H. Lobel, M. Carrasco. GDXray: The database of X-ray images for nondestructive testing. *Journal of Nondestructive Evaluation* 34 (4) (2015) 42.

[55] C. Mgonja. The consequences of cracks formed on the oil and gas pipelines weld joints. *International Journal of Engineering Trends and Technology* 54 (2017) 223–232.

[56] Introduction to Nondestructive Testing: A Training Guide, John Wiley & Sons, Hoboken, New Jersey, 2005.

[57] J. Mockus. Bayesian Approach to Global Optimization: Theory and Applications, 37, Springer, Netherlands, 2012.

[58] L. Morgan. Testing defects in automated ultrasonic testing and radiographic testing. *Insight: Non-Destructive Testing and Condition Monitoring* 60 (11) (2018) 606–612.

[59] L. Morgan, P. Nolan, A. Kirkham, R. Wilkerson. The use of automated ultrasonic testing (AUT) in pipeline construction. *Insight: Non-Destructive Testing and Condition Monitoring* 45 (11) (2003) 746–763.

[60] M. Naddaf-Sh, S. Hosseini, J. Zhang, N.A. Brake, H. Zargarzadeh. Real-time road crack mapping using an optimized convolutional neural network. *Complexity* 2019 (2019) 2470735.

[61] J. Nestleroth. Pipeline in-line inspection challenges to NDT. *Insight: Non-Destructive Testing and Condition Monitoring* 48 (9) (2006) 524.

[62] L. Norros. Human and organisational factors in the reliability of non-destructive testing (NDT). In *RATU2: The Finnish Research Programme on the Structural Integrity of Nuclear Power Plants*. VTT Technical Research Centre of Finland, Espoo, Finland, 271.

[63] H.G. Pisarski, C.M. Wignall. Fracture toughness estimation for pipeline girth welds. In *Proceedings of the International Pipeline Conference*, Vol. 36207 (2002) 1607–1611.

[64] J. Quirk. Achieving greater efficiency in NDT inspections. *Sensor Review* 19 (4) (1999) 268–272.

[65] C.E. Rasmussen. Gaussian processes in machine learning. Summer School on Machine Learning, Springer, Berlin, Heidelberg, 2003, pp. 63–71.

[66] W.G. Roe. Welding system. US Patent 3,278,721, 1966.

[67] A. Seto, T. Masuda, S. Machida, C. Miki. Very low cycle fatigue properties of butt welded joints containing weld defects: Study of acceptable size of defects in girth welds of gas pipelines. *Welding International* 14 (1) (2000) 26–34.

[68] F. Shaheen, B. Verma, M. Asafuddoula. Impact of automatic feature extraction in deep learning architecture. In *Proceedings of the International Conference on Digital Image Computing: Techniques and Applications (DICTA)*, 2016. IEEE, Los Alamitos, CA. 1–8.

[69] K. Simonyan, A. Zisserman. Very deep convolutional networks for large-scale image recognition. arXiv:1409.1556, 2014.

[70] S.T. Snyder. Alternative acceptance criteria for pipeline girth welds. *Inspection Trends* 2006 (2006) 20–22.

[71] M.Törner, A. Pousette. Safety in construction—A comprehensive description of the characteristics of high safety standards in construction work, from the combined perspective of supervisors and experienced workers. *Journal of Safety Research* 40 (6) (2009) 399–409.

[72] I. Valavanis, D. Kosmopoulos. Multiclass defect detection and classification in weld radiographic images using geometric and texture features. *Expert Systems with Applications* 37 (12) (2010) 7606–7614.

[73] P. Valentin. *Weld Classification Based on Grey Level Co-occurrence and Local Binary Patterns*. Master's Thesis, Aalborg University, Aalborg, Denmark, 2017.

[74] R. Vilar, J. Zapata, R. Ruiz. An automatic system of classification of weld defects in radiographic images. *NDT & E International* 42 (5) (2009) 467–476.

[75] M. Wall. Human factors guidance to improve reliability of non-destructive testing in the offshore oil and gas industry. In *Proceedings of the 7th European-American Workshop on Reliability of NDE*.

[76] G. Wang, T.W. Liao. Automatic identification of different types of welding defects in radiographic images. *NDT & E International* 35 (8) (2002) 519–528.

[77] R. Wang, R.-J. Guo. Developments of automatic girth welding technology in pipelines. *Dianhanji/Electric Welding Machine* 41 (9) (2011) 53–55.

[78] D. R. Williams Jr., L. R. Gutzwiller, M. U. Hazen, B. S. Anderson, A. McIntyre, T. Abeles. Classifying data with deep learning neural records incrementally refined through expert input. US Patent 9,324,022, 2016.

[79] D. Yapp, S. Blackman. Recent developments in high productivity pipeline welding. *Journal of the Brazilian Society of Mechanical Sciences and Engineering* 26 (1) (2004) 89–97.

[80] S. Yella, M. Dougherty, N. Gupta. Artificial intelligence techniques for the automatic interpretation of data from non-destructive testing. *Insight: Non-Destructive Testing and Condition Monitoring* 48 (1) (2006) 10–20.

[81] Y. Zhang, D. You, X. Gao, N. Zhang, P.P. Gao. Welding defects detection based on deep learning with multiple optical sensors during disk laser welding of thick plates. *Journal of Manufacturing Systems* 51 (2019) 87–94.

[82] J. Zapata, R. Vilar, R. Ruiz. Performance evaluation of an automatic inspection system of weld defects in radiographic images based on neuro-classifiers. *Expert Systems with Applications* 38 (11) (2011) 8812–8824.

[83] S. Naddaf-Sh, M-M. Naddaf-Sh, A.R. Kashani, H. Zargarzadeh, An Efficient and Scalable Deep Learning Approach for Road Damage Detection, 2020 IEEE International Conference on Big Data (Big Data) (2020) 5602–5608, doi:10.1109/BigData50022.2020.9377751.

[84] M-M. Naddaf-Sh, H. Myler, H. Zargarzadeh , Design and Implementation of an Assistive Real-Time Red Lionfish Detection System for AUV/ROVs, Complexity (2018) doi:https://doi.org/10.1155/2018/5298294.

[85] M. M. Dargahi, A. Khaloo, D. Lattanzi, Color-space analytics for damage detection in 3D point clouds, Structure and Infrastructure Engineering (2021) 1–14 doi:https://doi.org/10.1080/15732479.2021.1875488.

Chapter 10

Real-time fault diagnosis using deep fusion of features extracted by PeLSTM and CNN

Funa Zhou[a], Zhiqiang Zhang[b] and Danmin Chen[c]

[a] School of Logistic Engineering, Shanghai Maritime University, China. [b] School of Computer and Information Engineering, Henan University, China. [c] School of Software, Henan University, China

10.1 Introduction

Fault diagnosis is one of the critical means to secure the safety and efficient operation of large-scale automation systems. Therefore, fault diagnosis has received much attention from experts in both academic and engineering fields [1–5,13,33,36]. The data-driven fault diagnosis method has become a promising tool in engineering applications because no accurate physical model and information of correct expert knowledge are required [8,19,26,29].

Deep learning is an efficient data feature representation tool. It can be applied to data-driven fault diagnosis. There are currently four kinds of fault diagnosis methods that use deep learning: deep neural network (DNN)-based methods, deep belief network (DBN)-based methods, convolutional neural network (CNN)-based methods, and long short-term memory (LSTM) neural network–based methods [1,2,12,15,17,27,40]. Different from DNN and DBN, CNN can well extract local feature successively by using a convolution layer and pooling layer. LSTM does well in sequence feature extraction by designing a forget gate.

Jiang et al. [14] reshaped the vibration signal of bearing into 2-D matrix data and fed them into CNN for fault diagnosis. Zhang et al. [38] first converted bearing vibration signal into a 2-D matrix and then fed it into CNN for fault diagnosis. However, an accurate fault diagnosis result can be achieved only by reshaping the vibration signal into large 2-D matrix data, which will seriously affect the computational complexity of the diagnosis algorithm and does not take real-time performance into account [38]. Eren et al. [6] used the original time series data as the input of 1-D CNN to achieve real-time fault diagnosis without manually extracting features in advance. Peng et al. [22] proposed a new

Fault Diagnosis and Prognosis Techniques for Complex Engineering Systems. DOI: 10.1016/B978-0-12-822473-1.00003-3

multiscale CNN for feature extraction of a strong coupling vibration signal with a low signal-to-noise ratio. Han et al. [9] used a wavelet transform spectrum of the vibration signal as the input of CNN for fault diagnosis of rolling bearing. Kou et al. [16] developed a CNN-based fault diagnosis method for bearing working under different working conditions without manual intervention. Hsueh et al. [11] first converted the original current signal into a 2-D grayscale image, then applied the deep CNN model to automatically extract robust features from the grayscale image to diagnose fault of the induction motor. However, the preceding methods cannot extract the autocorrelation feature involved in 1-D sequence data, which may yield an inaccurate fault diagnosis result.

Wen et al. [30] converted 1-D sequence data into 2-D images and then used CNN to extract features. However, the training samples are required to be reshaped into a 2-D image in advance, which cannot secure a real-time fault diagnosis [30]. Jiang et al. [14] used CNN directly to extract features involved in original bearing data, but it cannot reach a real-time fault diagnosis either.

It is worth noting that the preceding methods utilize CNN as a unique feature extraction tool that is incapable of comprehensive feature extraction. Li et al. [18] integrated DNN and CNN to design a fault diagnosis algorithm using deep learning that can improve the diagnosis accuracy. The feature extracted by CNN and DNN can be fused to get more accurate feature representation and fault diagnosis. However, this method can only be applied to offline fault classification rather than online real-time fault diagnosis.

LSTM is an important branch of the recurrent neural network (RNN). It is an efficient tool for extracting long-term dependency trend features involved in 1-D sequence data. Yang et al. [32] designed a rotary mechanical fault diagnosis method using LSTM to extract long-term time dependency from all available data sampled by multiple sensors to detect and classify faults. Wang et al. [28] designed an LSTM network to extract features involved in gear fault data, which is used for gear fault diagnosis. By using LSTM's forgetting mechanism, Yu et al. [35] designed a layered LSTM algorithm for overcoming the shortcomings of feature extraction in shallow networks. Thus, a fault diagnosis algorithm of rolling bearings is constructed [27]. Using LSTM to extract the feature of long-term autocorrelation for original nonlinear vibration data, Fu et al. [7] designed a fault diagnosis algorithm for a high-speed train bogie. Luo and Hu [20] proposed a rolling bearing fault diagnosis method based on LSTM to simplify the fault diagnosis process. The well-trained network is used to discriminate multiple categories of original data. Thus, it can effectively improve the accuracy of fault diagnosis [29]. Xiao et al. [31] designed a fault diagnosis algorithm of the three-phase asynchronous motor by using LSTM to extract features involved in the motor acceleration signal. Thus, the relationship between the original vibration signal and the health state can be established [30]. Yin et al. [34] proposed a fault diagnosis method for a wind turbine gearbox by designing an optimized LSTM neural network with cosine loss. The loss is converted from Euclidean space to angular space through cosine loss, thereby eliminating effects and improving the accuracy of fault diagnosis for the gearbox [34]. Polat [23] used deep-level LSTM to diagnose fault by extracting features involved in the vibration signals of CNC machine tools. Yu et al. [35] proposed a layered LSTM algorithm for overcoming the shortcomings of the shallow structure without any preprocessing operations or manual feature extraction. Thus, an end-to-end fault diagnosis system framework for rolling bearing is proposed. Zhao et al. [40] developed a fault diagnosis method based on LSTM to directly classify the original process data without specific feature extraction and classifier design. It can also adaptively learn the dynamic information involved in the original data [40]. Bruin et al. [4] constructed LSTM for signals from multiple tracking circuits in a geographical area; faults can be diagnosed by exploring the spatial and temporal dependencies of the faults, and these dependencies can be directly learned from the data to perform fault diagnosis for railway track circuits. The preceding works use LSTM as a feature extraction tool to extract features of the raw data sequence. However, the local feature involved in data cannot be well extracted by using LSTM as a unique feature extraction tool. The integration of LSTM and CNN can obtain more accurate feature representation.

Although LSTM has achieved great success in fault diagnosis by extracting features involved in 1-D sequence data, its inherent structure makes it fail to deal with long sequence data [21]. The reason is that the previous information contained at the front of the sequence cannot be transferred to the end of the sequence when traditional LSTM is adopted. This memory bottleneck will limit the feature extraction capabilities of LSTM [37]. Zhang et al. [39] proposed a parallel LSTM (PLSTM) by processing all observations in the sequence at the same time to overcome the memory bottleneck problem. However, since feature extraction is performed on the complete sequence at the same time, difficulties corresponding to insufficient data utilization and information loss will be encountered.

For the purpose of safety monitoring, data collected by various types of sensors for health monitoring of rotary machines are stored in the database, such as 1-D vibration signals collected by various accelerometers. However, the waveform of some critical variables monitored are usually displayed in real time. According to the refresh rate of the display, a screenshot of the display at the monitoring center can illustrate the waveform of the vibration signal. The screenshot image contains some trend information for the vibration signal, which can be accurately extracted by CNN.

It can be concluded from the preceding analysis that it is important to design an efficient feature extraction network for 1-D sequence data. Developing a deep fusion mechanism to incorporate the 2-D screenshot image and the original 1-D sequence may be helpful to extract more comprehensive features required by a more accurate fault diagnosis result.

The main contributions of this chapter are as follows:

(1) We aim to develop an efficient feature extraction technique for health monitoring of rolling bearing utilizing a 1-D vibration signal. The proposed PLSTM with Peephole (PeLSTM) can prevent useless information transfer. It not only can solve the memory bottleneck problem of traditional LSTM for long sequences but also can make full use of all possible information helpful for feature extraction.

(2) A fusion network with a new training mechanism is designed to fuse features extracted from PeLSTM and CNN respectively to further explore the potential feature related to both autocorrelation and local cross-correlation information. By designing a new loss function and global optimization mechanism for the training process, the fusion network can incorporate a 2-D screenshot image into comprehensive feature extraction. It can provide a more accurate fault diagnosis result since the 2-D screenshot image is another expression form of the 1-D vibration sequence involving additional trend and locality information.

(3) A real-time screenshot image is fed into the input of CNN to secure a real-time online fault diagnosis, which is the primary requirement in the engineering field of health monitoring.

The rest of this chapter is organized as follows. Section 10.2 introduces some basic theory of CNN and LSTM. Section 10.3 presents a real-time fault diagnosis method based on the deep feature fusion of PeLSTM and CNN. Section 10.4 presents our experimental verification. Section 10.5 provides our conclusion and plans for future work.

10.2 Basic theory

This section briefly introduces some basics of CNN and LSTM.

10.2.1 Convolutional neural network

As is known to all, CNN is a special feed-forward neural network for 2-D image feature extraction, especially for large-sized images. Fig. 10.1 shows the schematic chart of CNN.

Compared with traditional DNN, there are no specific hidden layers in the structure of CNN. Multiple layers of convolution operation and pooling operation are connected in sequence. CNN can extract local features involved in the input image with a reduced computational complexity since weight sharing of convolution operation and downsampling of the pooling operation is used. The resulting feature map is expanded into a 1-D vector, which can be fed as the input of the fully connected layers and the following classifier. The parameters of the network can be updated by minimizing the loss function through a back-propagation algorithm.

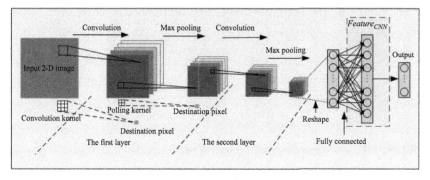

FIGURE 10.1   Schematic diagram of CNN.

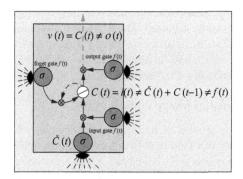

FIGURE 10.2   Schematic diagram of an LSTM cell.

10.2.2   Long short-term memory

LSTM is a variation of RNN. It can sequentially extract autocorrelated features involved in the 1-D sequence data [10]. Each LSTM cell is composed of three gates defined by activation functions, which are used for transferring information for current input, for the previous input, and for the output of the cell, respectively. The schematic chart is shown in Fig. 10.2.

The function of the forget gate is to determine what kind of previous information can be transferred through the cell. The function of the input gate is to determine what kind of input information needs to be transferred through the cell. The parameters of a well-trained LSTM include weights and bias of each gate.

10.3   Deep fusion of feature extracted by PeLSTM and CNN

Real-time accurate diagnosis is the primary performance index required for fault diagnosis algorithms. How to make full use of all available multimodal data to accurately extract the feature involved in them is one of the critical means to get

0.0085	0.4236	0.0130	-0.2652	0.2372	0.5909	-0.0930	-0.4069	0.2794	0.4370	-0.3529	-0.6124
0.1539	0.1425	0.1182	0.0597	-0.1397	0.0524	0.2867	-0.0471	-0.1206	0.0597	0.2367	0.0394
-0.2940	0.1145	0.3594	-0.2424	-0.4772	0.1048	0.3825	-0.2485	-0.3980	0.1771	0.4581	0.0301
0.0747	-0.1693	-0.1174	0.0650	-0.1271	0.0244	0.6120	-0.4378	-1.3324	1.1468	1.6211	-1.8449
0.9567	-0.2834	-0.6989	1.2382	0.0199	-1.5967	0.4045	1.5456	-0.7866	-1.3332	1.0684	1.0968
-0.1389	0.4231	-0.0032	-0.3545	0.3151	0.2989	-0.3610	-0.0244	0.5332	-0.0459	-0.3155	0.1624
0.2140	-0.2079	-0.3094	-0.1689	0.3350	-0.1125	0.0179	0.2144	0.0686	0.0694	0.0670	-0.0512
0.1259	0.1263	-0.0508	-0.0435	0.0451	-0.0182	0.1580	0.1559	0.0219	0.1750	0.2469	-0.0057
-0.0585	0.3066	0.1653	0.1105	0.0451	0.2778	0.1170	-0.0244	0.0743	0.0560	0.0678	-0.0621
0.3854	1.5951	-0.9896	-1.2962	1.7048	0.8910	-1.1789	-0.8788	1.6865	8.1218e-04	-0.8727	0.5596
-0.9417	0.1872	1.1898	-0.7135	-0.8187	0.5389	0.5413	-0.3200	-0.3890	0.6294	0.0731	-0.1559

$$X_{1D}(t-1) \quad X_{1D}(t) \quad X_{1D}(t+1) \qquad \bullet \quad \bullet \quad \bullet \qquad X_{1D}(N)$$

FIGURE 10.3 1-D sequence data stored in the database.

more accurate fault diagnosis results. For this goal, by incorporating the fault diagnosis error, the output error of CNN, and the output error of PeLSTM, a global optimization mechanism is developed to train the deep fusion network such that a more comprehensive feature involved in 1-D sequence data and the 2-D screenshot image can be achieved. The designing mechanism of an improved LSTM, called *PeLSTM*, is illustrated in detail for accurate feature extraction of 1-D sequence data, which can further secure the efficiency of deep fusion for 1-D sequence data and the 2-D screenshot image.

10.3.1 2D screenshot image construction

During the operation process of rotating machinery, the vibration signals collected for monitoring is a kind of 1-D sequence signal, and the existing CNN-based fault diagnosis method is obliged to use 1-D sequence data by reshaping it into 2-D matrix data row by row or transforming it into the time-frequency image via fast Fourier transform. This method relies on the collected sequence in a long period time window, so it can only be used for offline fault classification rather than online fault diagnosis. However, there are many kinds of monitoring sensors collecting 1-D sequence data, but only a few 1-D signals of critical monitored variables can be shown on the display of the monitoring center. Since the 2-D screenshot image is captured in real time, feeding it into CNN may achieve real-time fault diagnosis. Moreover, some real-time dynamic trend information of the 1-D signal can also be involved in the 2-D screenshot image rather than the 1-D signal itself.

Fig. 10.3 shows the 1-D sequence stored in the database of the monitoring center, and it can be seen that each sample can be expressed as: $X_{1D}(t) \in R^{1 \times 1}$.

For CNN of the 1-D sequence signal, the common means is to reshape the 1-D sequence into 2-D matrix data by stacking in row as shown in Eq. (10.1):

$$X_{2D,Matrix}(t)$$

$$= \begin{bmatrix} X_{1D}(t+1) & X_{1D}(t+2) & \cdots & X_{1D}(t+l) \\ X_{1D}(t+l+1) & X_{1D}(t+l+2) & \cdots & X_{1D}(t+l+3) \\ \cdots & \cdots & \cdots & \cdots \\ X_{1D}(t+(l-1)*l+1) & X_{1D}(t+(l-1)*l+2) & & X_{1D}(t+l*l) \end{bmatrix} \in R^{l \times l},$$

$$(10.1)$$

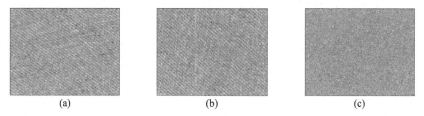

(a)   (b)   (c)

FIGURE 10.4   2-D matrix data by different reshaping means.

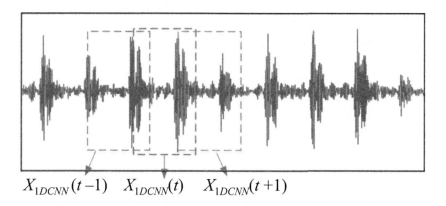

$$X_{1DCNN}(t-1) \quad X_{1DCNN}(t) \quad X_{1DCNN}(t+1)$$

FIGURE 10.5   Samples fed into 1-D CNN.

where t is the start time of the sample and l is the number of sample points collected in each row. It is invalid for real-time fault diagnosis at time t.

It can be seen from Eq. (10.1) that the samples fed in CNN are stacked by reshaping the 1-D sequence signal using a rather large window size, so it is not real-time observation related to the vibration sensor. The 2-D matrix reshaped by different means can involve different information, as shown in Fig. 10.4. Fig. 10.4(a) through (c) correspond to reshaping in row, reshaping in column, and reshaping in random, respectively. This figure indicates that for a given window size, the 2-D matrix data using different reshaping means are different. So reshaping is not a good choice for extracting local features of 1-D sequence data by using CNN. In addition, the dynamic sequential feature may be completely lost.

As shown in Eq. (10.2) and Fig. 10.5, the sample fed in 1-D CNN is not a real-time observation at time t either.

$$X_{1DCNN}(t) = [X_{1D}(t*1+1)X_{1D}(t*1+2)\cdots X_{1D}((t+1)*l)] \in R^{l \times l} \tag{10.2}$$

It can be obtained from Eqs. (10.1) and (10.2) that 1-D CNN does better than traditional CNN since the sample fed into 1-D CNN ranges over a relatively

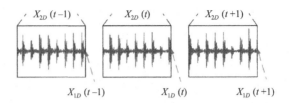

FIGURE 10.6 2-D screenshot image in the monitoring center.

short period of time. However, the existing method for the fusion of CNN and LSTM uses a separately training mechanism. Since 1-D CNN and LSTM are connected in series, inaccurate feature extraction from 1-D CNN will deteriorate the following feature extraction from LSTM. Therefore, it is not a preferred choice.

Fig. 10.6 shows the 2-D screenshot image stored in the database of the monitor center. This figure indicates that each sample $X_{2D}(t) \in R^{l \times l}$ is directly related to the real-time observation at time t. Feeding the 2-D screenshot image $X_{2D}(t) \in R^{l \times l}$ into CNN can achieve real-time feature extraction. Furthermore, the more accurate features can be extracted since dynamic trend information at time t can also be illustrated in the screenshot $X_{2D}(t) \in R^{l \times l}$.

10.3.2 The feature fusion algorithm based on CNN and PeLSTM

CNN and LSTM can extract different kinds of features involved in data described in different forms. They both can be applied for fault diagnosis, but there are still some difficulties encountered in the application. How to overcome these difficulties is a challenging problem for further study.

CNN pays more attention to local neighborhood feature extraction by using the pooling operation, whereas the autocorrelated dynamic feature is not considered. Due to its inherent structure, LSTM does well in autocorrelated feature extraction without considering too much local feature representation. Fuse the feature extracted by CNN and LSTM respectively to extract more accurate features, which is the basis requirement of accurate fault diagnosis.

Fig. 10.7(a) shows a schematic chart for existing fault diagnosis based on 1-D CNN and LSTM. However, LSTM feature extraction has a strong dependence on the output of 1-D CNN. Inaccurate features extracted by 1-D CNN will deteriorate the feature extraction of the following LSTM. Separately training 1-D CNN and LSTM cannot secure a comprehensive feature involved in different modes of the signal. Yet not making full use of the 2-D screenshot image may result in an inaccurate feature. To overcome these shortcomings, an improved PLSTM, called *PeLSTM*, is designed and the deep fusion mechanism of the features extracted by PeLSTM and CNN is developed for more comprehensive feature extraction from multimodal data stored in the monitoring center. Fig. 10.7(b) shows a schematic diagram of existing fault diagnosis based on

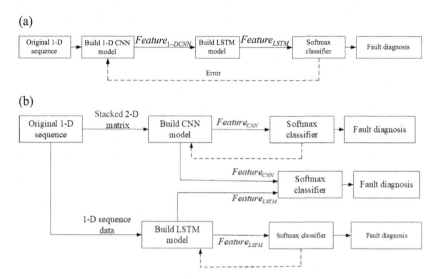

FIGURE 10.7 Block diagram comparison of different fusion mechanisms. (a) Existing fault diagnosis method combining 1DCNN and LSTM. (b) Existing fault diagnosis method for combining CNN and LSTM. (c) Fault diagnosis using deep feature fusion of PeLSTM and CNN.

CNN and LSTM. It cannot come to a real-time fault diagnosis result since the sample fed into CNN is a 2-D matrix reshaped from 1-D sequence data. Using LSTM to extract features from 1-D sequence data cannot avoid the memory bottleneck problems. However, training CNN and LSTM separately cannot guarantee comprehensive features from multimodal data. To solve the preceding problems, a global optimization mechanism is developed to train all four related networks simultaneously. Fig. 10.7(c) shows the schematic of the deep fusion method developed in this work.

The detail algorithms are as follows.

Step 1: Designing a parallel PeLSTM. Since the 1-D sequence signal is one of the most commonly used data for mechanical system monitoring, extracting a satisfying dynamic sequential feature from 1-D sequence data is critical for accurate fault diagnosis. However, there is a memory bottleneck problem when LSTM is used to process long sequences: the useful information at the front of the sequence cannot be transferred to the back end of the sequence. PLSTMs can partially solve the preceding problems by processing the sequence at the same time. However, the state of the memory unit of traditional PLSTM may be affected by useless information. To overcome this problem, PeLSTM is designed. The peepholes will prevent the memory unit from transferring useless information.

Taking the LSTM cell in Fig. 10.8 as an example, design a peephole denoted with a red line to prevent useless information from transferring. During the process of updating for $C(t)$, the information flow through the forget gate $f(t)$

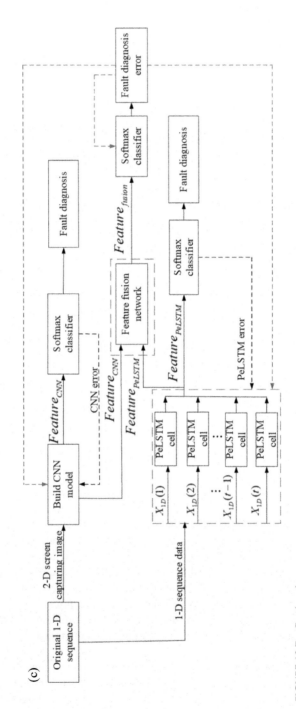

FIGURE 10.7 Continued.

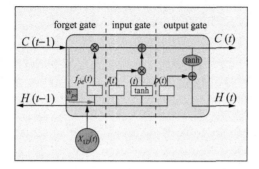

FIGURE 10.8 PeLSTM cell structure.

without peephole is shown in Eq. (10.3):

$$f(t) = \sigma(u_f H(t-1) + w_f X_{1D}(t) + b_f), \tag{10.3}$$

where the forget gate is defined by the sigmoid function σ, u_f, w_f represent the weight parameters, b_f represents the bias, and $H(t-1)$ represents the output at the previous sampling time. It can be seen that $C(t-1)$ is not transferring through the forget gate, which will cause some useless information in $C(t-1)$ to occupy memory and it may be transferred to $C(t)$. To overcome this shortcoming, a peephole connection is designed to filter $uC(t-1)$. After adding the peephole, information flow through the forget gate $f_{pe}(t)$ can be described as follows:

$$H_{pe}(t-1) = w_{pe} C(t-1) \tag{10.4}$$

$$f_{pe}(t) = \sigma(H_{pe}(t-1) + w_f X_{1D}(t) + u_f H(t-1) + b_f) \tag{10.5}$$

It can be seen from Fig. 10.8 that the updating process from $C(t-1)$ to $C(t)$ can be described as follows:

$$C(t) = i(t) * \tilde{C}(t) + C(t-1) * f_{pe}(t) \tag{10.6}$$

Therefore, in the process of memory unit updating, once any useless information leaks from $C(t-1)$, it will be introduced into the forget gate by the peephole and then cut off by the forget gate. For 1-D sequence data sampled in a given time window $X_{1D} = (X_{1D}(1), X_{1D}(2), ..., X_{1D}(t), ..., X_{1D}(m))$, m samples can be fed into the m PeLSTM cell simultaneously by exchanging information with adjacent front and back samples. Repeat this step until the information transferring within the sequence is completed. Fig. 10.9 shows a schematic of the designed PLSTM with peephole. This figure shows the p-th round of the sequence information exchanging process. The red line shows the peephole connection, which is designed in front of each forget gate.

Take the p-th information exchange process at time t as an example to illustrate its internal information exchange process. Each cell of PeLSTM has

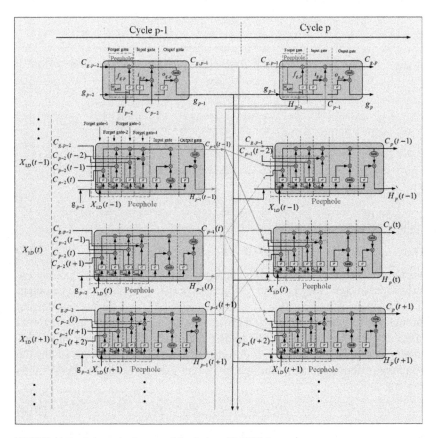

FIGURE 10.9 Schematic diagram of the designed PeLSTM.

four forget gates. The first forget gate can be expressed by Eq. (10.7) and Eq. (10.8):

$$H_{1,pe}(t) = w_{1,pe} C_{p-1}(t+1), \qquad (10.7)$$

$$f_{1,p}(t) = \sigma(H_{1,pe}(t) + w_{f_{1,p}} \xi_p(t) + u_{f_{1,p}} X(t) + v_{f_{1,p}} g_{p-1} + b_{f_{1,p}}), \qquad (10.8)$$

where $H_{1,pe}(t)$ represents the peephole connection of the first forgetting gate $f_{1,p}(t)$, which corresponds to the information filter process of $C_{p-1}(t+1)$. It not only can prevent useless information from transferring but also can make fuller use of information in feature extraction by reusing the information. $\xi_{p-1}(t) = [H_{p-1}(t-1), H_{p-1}(t), H_{p-1}(t+1)]$ represents the information of three adjacent samples in step $p-1$. g_{p-1} represents information for all sequences in step p-1. In Eq. (10.8),$w_{f_{1,p}}$, $u_{f_{1,p}}$, $v_{f_{1,p}}$ represent weights and $b_{f_{1,p}}$ represents bias. Eq. (10.9) and Eq. (10.10) can be used to describe the information transferring process of the second forget gate:

$$H_{2,pe}(t) = w_{2,pe}C_{p-1}(t), \tag{10.9}$$

$$f_{2,p}(t) = \sigma(H_{2,pe}(t) + w_{f_{2,p}}\xi_p(t) + u_{f_{2,p}}X(t) + v_{f_{2,p}}g_{p-1} + b_{f_{2,p}}), \tag{10.10}$$

where $H_{2,p\,e}(t)$ represents the peephole connection of the second forgetting gate $f_{2,p}(t)$, which corresponds to the information filter process in $C_{p-1}(t)$. It not only can prevent the transferring of useless information but also can make fuller use of information during the recurrent process of information. Similarly, the third forget gate and the fourth forget gate can be designed to prevent useless information transferring in $C_{p-1}(t-1)$ and $C_{g,p-1}$. In such a way, the peephole of the third forget gate and the fourth forget gate can be designed by Eqs. (10.11) through (10.14):

$$H_{3,pe}(t) = w_{3,pe}cC_{p-1}(t-1), \tag{10.11}$$

$$f_{3,p}(t) = \sigma(H_{3,pe}(t) + w_{f_{3,p}}\xi_p(t) + u_{f_{3,p}}X(t) + v_{f_{3,p}}g_{p-1} + b_{f_{3,p}}), \tag{10.12}$$

$$H_{4,pe}(t) = w_{4,pe}C_{g,p-1}, \tag{10.13}$$

$$f_{4,p}(t) = \sigma(H_{4,pe}(t) + w_{f_{4,p}}\xi_p(t) + u_{f_{4,p}}X(t) + v_{f_{4,p}}g_{p-1}(t) + b_{f_{4,p}}). \tag{10.14}$$

The forget gate is designed to filter the useless information in the corresponding memory unit, and the peephole connection can transfer useful information to the corresponding forget gate when useless information leaks out of the memory unit and cut off its transferring, as shown in Eq. (10.15):

$$P\,e_p(t) = f_{1,p}(t) * C_{p-1}(t+1) + f_{2,p}(t) * C_{p-1}(t) \\ + f_{3,p}(t) * C_{p-1}(t-1) + f_{4,p}(t) * C_{g,p-1}. \tag{10.15}$$

The feature $Pe_p(t)$ transferred in the network after filtering the useless information is combined with the features of the same $i_p(t)$ and $o_p(t)$ to complete the transferring process, as shown in Eq. (10.16):

$$H_p(t) = o_p(t) * \tanh(Pe_p(t) + i_p(t) * \tilde{C}_p(t)). \tag{10.16}$$

In PLSTM, g_p represents the overall characteristics of the sequence in the p-th cycle, as shown in Fig. 10.9. The peephole connections are designed during the propagation of g_p, denoted by the solid red line in the figure. The peephole connection is designed before the forget gate as follows:

$$H_{g,pe}(t) = w_{g,pe}C_{g,p-1}(t), \tag{10.17}$$

$$f_{g,p}(t) = \sigma(H_{g,pe}(t) + w_{fg}g_{p-1} + u_{fg}H_{p-1}(t) + b_{fg}), \tag{10.18}$$

where $H_{g,pe}(t)$ is the peephole connection of the forget gate $f_{g,p}(t)$, which corresponds to the recurrent information in $C_{g,p-1}(t)$. It not only can prevent useless information from transferring but also can make the feature extraction more sufficient by designing the peephole connection. Since the information transferring of $g_p(t)$ is similar to that of the LSTM cell, it has the same structure of forget gate $i_{g,p}$ and output gate $o_{g,p}$. Therefore, the information transferring of $g_p(t)$ can be described as follows:

$$g_p(t) = o_{g,p} * \tanh(f_{g,p} * C_{g,p-1}(t) + \sum_m i_{g,p} * C_{g,p-1}(t)). \quad (10.19)$$

By designing the peephole connection, the information in the corresponding memory unit can be reused to avoid waste of useful information in the recurrent process and make the information more fully utilized. At the same time, useless information can be prevented from occupying weights in the information flow of the memory unit, which can effectively overcome the memory bottleneck problem.

Step 2: Build a PeLSTM model to extract the autocorrelated dynamic feature involved in 1-D sequence data. First, we build a PeLSTM neural network Net_{PeLSTM} with peephole connections as follows:

$$Net_{PeLSTM} = Gen_{PeLSTM}(\theta_1, \theta_2, \cdots, \theta_m; n_1, n_2, \cdots, n_m; Cycles), \quad (10.20)$$

where Gen_{PeLSTM} is a function for generating a neural network and m is the number of cells. $n_1, n_2, \ldots, n_m$ are the number of hidden neurons in the gate structure in each cell. $\theta_1 = \{w_{1,gate}, b_{1,gate}\}$, $\theta_2 = \{w_{2,gate}, b_{2,gate}\}, \ldots, \theta_m = \{w_{m,gate}, b_{m,gate}\}$ are the weight and bias of each gate structure in each cell. $Cycles$ is the number of cycles. Take the output of the last cycle step g_{cycle} as the feature extracted from PeLSTM:

$$F_{PeLSTM} = G_{PeLSTM}(Net_{PeLSTM}, \theta_{PeLSTM}, X_{1D}), \quad (10.21)$$

where G_{PeLSTM} is the nonlinear function to describe the relation of input and output of the PeLSTM network, $\theta_{PeLSTM} = \{\theta_1, \theta_2, \ldots, \theta_m\}$.

Step 3: Use CNN to extract the local feature of the 2-D screenshot image. Build Net_{CNN}, as shown in Eq. (10.22):

$$Net_{CNN} = Gen_{CNN}(K, b_{CNN}; K_c, K_{size}, P_{size}; K_{step}, P_{step}), \quad (10.22)$$

where K_c, K_{size}, and K_{step} represent the parameters of the convolution kernel: the number of channels, the size, and the step size during the convolution operation. P_{size} and P_{step} represent the parameters of the pooling kernel: the size and the step size when completing the pooling operation. The constructed CNN can be trained by samples of the 2-D screen capturing image. Back Propagation is used for parameter adjustment of Net_{CNN} to obtain the convolution kernel K and the bias b_{CNN}. Once Net_{CNN} is well trained, the feature F_{CNN} of the 2-D screen capturing image can be extracted by Eq. (10.23):

$$F_{CNN} = G_{CNN}(Net_{CNN}, K, b_{CNN}, X_{2D}), \quad (10.23)$$

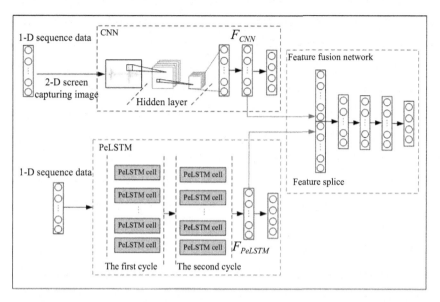

FIGURE 10.10   Schematic diagram of fault diagnosis based on deep feature fusion of PeLSTM and CNN.

where G_{CNN} is the nonlinear output function to describe the relation of CNN's input and output.

Step 4: Design the feature fusion network for CNN and PeLSTM. To fuse the feature extracted by PeLSTM and CNN, a fusion network can be designed via Eq. (10.24):

$$Net_{fusion} = Feedforward(\theta_{fusion}; H_{fusion}, L_{fusion}),\qquad(10.24)$$

where $\theta_{fusion} = \{W_{fusion}, b_{fusion}\}$ are the parameters of the fusion network. The number of hidden neurons is represented by H_{fusion}. L_{fusion} is the number of layers of the fusion network. As shown in Fig. 10.10, the rough feature F_{CNN}, F_{PeLSTM} respectively extracted by PeLSTM and CNN can be fed into Net_{fusion}.

Step 5: The global optimization mechanism to tune the parameters of the four networks. To achieve a deep fusion rather than a simple combining of F_{CNN} and F_{PeLSTM}, a global optimization mechanism to further tune the parameters of Net_{fusion}, Net_{PeLSTM}, Net_{CNN} and the classifier for fault diagnosis is developed as shown in Fig. 10.11.

Use the fused feature F_{fusion} as the input of the Softmax classifier as follows:

$$Net_{softmax} = Feedforward(\theta_s; H_s),\qquad(10.25)$$

where $\theta_s = \{W_s, b_s\}$ is the parameter of the classifier network for fault diagnosis; W_s and b_s are the corresponding weight and bias, respectively; and H_s is the number of hidden layers of Softmax. The global error is generated by comparing

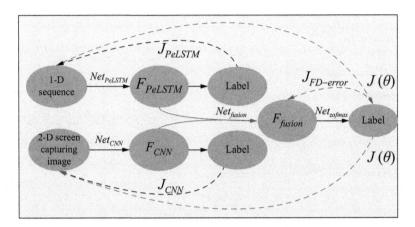

FIGURE 10.11   Global optimization parameter architecture diagram.

the output label $label_{output}$ with the real label $label_{real}$, and the error is back-propagated to PeLSTM, CNN, and the fusion network at the same time. Thus all parameters related to the fault classification network, fusion network, PeLSTM, and CNN can be globally adjusted by minimizing the global loss function defined in Eq. (10.26):

$$J(\theta) = J_{FD-error} + J_{fusion} + J_{PeLSTM} + J_{CNN}, \qquad (10.26)$$

where $J_{FD-error}$, J_{fusion}, J_{PeLSTM}, J_{CNN} are the loss functions of the corresponding four networks defined in the form of cross entropy, as shown in Eq. (10.27):

$$J_{FD-error} = -\frac{1}{K} \sum label_{real} \ln label_{output}$$
$$+ (1 - label_{real})ln(1 - label_{output}). \qquad (10.27)$$

J_{fusion}, J_{PeLSTM}, J_{CNN} are defined similarly to that found in Eq. (10.27). It is worth noting that during the back-propagation process, E_{fusion} used to define J_{fusion} can be divided into two parts as follows:

$$E_{fusion} = E_{PeLSTM} + E_{CNN}. \qquad (10.28)$$

Once the four networks Net_{CNN}, Net_{PeLSTM}, Net_{fusion}, and $Net_{softmax}$ are globally optimized, the well-trained network parameters can be described as

$$Tr_{global} = Train(Net_{PeLSTM}, Net_{CNN}, Net_{fusion}, Net_{softmax};$$
$$J(\theta); X_{1D}, X_{2D}), \qquad (10.29)$$

where $Tr_{global} = \{\theta_{PeLSTM}; K, b_{CNN}; \theta_{fusion}; \theta_{Softmax}\}$ are the well-trained parameters of the four networks. The fusion feature F_{fusion} is generated on the last layer of the fusion network as follows:

$$F_{fusion} = G_{fusion}(Net_{fusion}, \theta_{fusion}, X_{1D}, X_{2D}). \qquad (10.30)$$

Step 6: Online fault diagnosis based on the feature fusion network. Once online samples $X_{online,1D}(t)$ and $X_{online,2D}(t)$ at time t are collected, use the well-trained PeLSTM network Net_{PeLSTM} to extract features involved in online 1-D data as follows:

$$F_{PeLSTM}(t) = G_{PeLSTM}(Net_{PeLSTM}, \theta_{PeLSTM}; Cycles, X_{online,1D}(t)). \qquad (10.31)$$

Use the well-trained CNN network Net_{CNN} to extract features involved in the online 2-D screenshot image as follows:

$$F_{CNN}(t) = G_{CNN}(Net_{CNN}; K, b_{CNN}; X_{online, 2D}(t)). \qquad (10.32)$$

The trained fusion network Net_{fusion} is then used to fuse the feature of the online 1-D sequence data and 2-D screenshot image as follows:

$$F_{fusion}(t) = G_{fusion}(Net_{fusion}, \theta_{fusion}, F_{PeLSTM}(t), F_{CNN}(t)). \qquad (10.33)$$

Finally, the fused feature $F_{fusion}(t)$ obtained by deep feature fusion is used as the input of the fault classifier, and the output of the classifier can be obtained via Eqs. (10.34) and (10.35):

$$h_{\theta, fusion}(t) = \begin{bmatrix} p(label(t) = 1)|F_{fusion}(t); \theta_S) \\ p(label(t) = 2)|F_{fusion}(t); \theta_S) \\ \vdots \\ p(label(t) = L)|F_{fusion}(t); \theta_S) \end{bmatrix}$$

$$= \frac{1}{\sum\limits_{l=1}^{L} \theta_{S_l}^T F_{fusion}} \begin{bmatrix} e^{\theta_{S_1}^T F_{fusion}(t)} \\ e^{\theta_{S_2}^T F_{fusion}(t)} \\ \vdots \\ e^{\theta_{S_L}^T F_{fusion}(t)} \end{bmatrix}, \qquad (10.34)$$

$$lable_X(t) = \underset{k=1,2,\cdots,K}{argmax} \{h_{\theta_s, fusion}(t)|X_{1D}(t), X_{2D}(t); \theta_s)\}, \qquad (10.35)$$

where θ_s is the parameter of the Softmax classifier and $lable_X(t)$ is the online diagnostic result at time t.

The flowchart of fault diagnosis using deep fusion of the feature extracted by PeLSTM and CNN is shown in Fig. 10.12.

Remark 1. The proposed algorithm can improve the accuracy of fault diagnosis results in the following aspects:

(1) An improved PLSTM called *PeLSTM* is designed for more accurate feature extraction by designing a peephole connection before each forget gate to prevent useless information transferring in the cell.

(2) A deep fusion mechanism using a new loss function and global training scheme is designed to incorporate the feature involved in 1-D sequence data

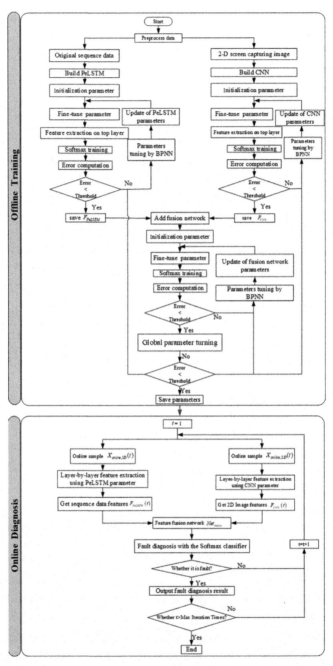

FIGURE 10.12   Flowchart of fault diagnosis using the deep fusion of features extracted by PeL-STM and CNN.

FIGURE 10.13 Experimental platform for the rolling bearing to obtain a vibration signal. From the Bearing Data Center of Case Western Reserve University [3].

and the 2-D screenshot image such that a more comprehensive feature can be extracted.

(3) Since the global training mechanism is used for PeLSTM, CNN, the fusion network, and the classifier, the accuracy of fault classification can be improved once the feature involved in data is accurately extracted.

10.4 Experimental testing

In this section, bearing data and gearbox data are used for experimental research in Sections 10.4.1 and 10.4.2, respectively.

10.4.1 Rolling bearing test and analysis

The rolling bearing data used in this section are downloaded from the Bearing Data Center of Case Western Reserve University [3]. The experimental platform is shown in Fig. 10.13. The fault size ranges from 0.007 to 0.014 to 0.021 to 0.028 in. The motor load varies from 0 to 3 hp. The sample frequency of the raw vibration signal is 12 kHz.

10.4.1.1 Data preprocessing and experimental design

The moving window technique is used to generate the samples required by the related algorithms. To test the algorithm in different experimental scenarios, three kinds of sliding window size 100/400/900 are selected, and the sliding step is set to 20. The original 1-D sequence data is stacked into 2-D matrix data with window size 10*10/20*20/30*30, and the size of the 2D screenshot is 28*28. Table 10.1 lists the model and the data used in the related algorithms.

TABLE 10.1 Model and data used in the related algorithms.

Experimental serial no.	Fault type	Fault size (in.)	Training sample	Test sample
1	Normal Bearing/Ball Fault/Inner Ring Fault/ Outer Ring Fault	0/0.007/0.007/0.007	8000	4000
2	Normal Bearing/Ball Fault/Inner Ring Fault/ Outer Ring Fault	0/0.014/0.014/0.014	8000	4000
3	Normal Bearing/Ball Fault/Inner Ring Fault/ Outer Ring Fault	0/0.21/0.021/0.021	8000	4000
4	Normal Bearing and different fault size of ball Fault	0/0.007/0.014/0.021	8000	4000
5	Normal Bearing and different fault size of Inner Ring Fault	0/0.007/0.014/0.021	8000	4000
6	Normal Bearing and different fault size of Outer Ring Fault	0/0.007/0.014/0.021	8000	4000
7	Multiple Faults of various Fault size	0/0.007/0.014/0.021	20000	10000

The experimental design is shown in Table 10.2. Experiments 1 through 6 are for 10 failure categories. Experiments 1 through 3 are for different sequence length; each class has 2000 training samples. A total of 8000 training samples are available for Experiments 4 through 6. Experiments 7 through 9 are set for different fault sizes, and Experiments 9 through 12 are set for the different types of faults. These 12 experiments are designed for the case of different fault types and different fault sizes with the same fault type, respectively. The network parameters of the related 12 experiments are listed in Table 10.3. Diagnostic accuracy comparison of the related algorithms is presented later in Table 10.6.

10.4.1.2 Analysis of experimental results

Experiments 1 through 3 are the scenarios with different window size, and 2000 samples are used for training. However, in the field of engineering, the samples used to train the network are not easy to obtain, so we designed Experiments 4 through 6, and the training set has 800 training samples. Experiments 7 through 9 are the scenarios for different sizes of fault settings, and Experiments 9 through 12 are for different types of fault settings. Through the experimental results,

TABLE 10.2 Experimental design.

Grouping	No. of layers	No. of neurons	Iteration times	Learning rate
1	5	100/200/100	2000	0.01
2	5	100/200/100	2000	0.01
3	5	100/200/100	2000	0.01
4	5	100/300/100	3000	0.005
5	5	100/300/100	3000	0.005
6	5	100/300/100	3000	0.005
7	5	200/500/200	5000	0.01

TABLE 10.3 Network parameters.

Experiment no.	Diagnostic accuracy using only 1-D data	Diagnostic accuracy using only 2-D data	Diagnostic accuracy using feature fusion by simply splicing	Diagnostic accuracy using extracted common features	Diagnostic accuracy extracted by fusion of common features and two other features
1	80.02%	82.45%	90.02%	97.14%	98.7%
2	81.32%	83.35%	91.87%	97.50%	98.87%
3	83.89%	86.44%	94.45%	98.92%	99.17%
4	73.22%	75.07%	87.45%	93.27%	95.95%
5	74.50%	76.59%	89.39%	94.60%	97.75%
6	75.52%	75.57%	87.62%	94.12%	97.27%
7	70.27%	73.38%	84.28%	90.13%	94.38%

the advantages of the proposed PeLSTM and the fusion algorithm can be clearly seen. The accuracy of the proposed fault diagnosis algorithm using 1-D sequence data is compared with other existed methods. The fault diagnosis results are shown in Table 10.4.

Comparing column 6 and column 2 in Table 10.4, it can be concluded from row 3 that when the window size of the 1-D data sequence is small, the diagnostic accuracy of PeLSTM is 93.63%, whereas the diagnostic accuracy of the traditional DNN with the same training data is only 82.28%. The diagnostic accuracy is improved by 11.35%, indicating that the autocorrelation involved in 1-D sequence data has a great influence on the diagnosis results, and PeLSTM can make full use of this information. For all 12 experiments, the accuracy of PeLSTM is always more than 5% higher than DNN.

TABLE 10.4 Fault diagnosis results.

Algorithm no.	Abbreviation	Algorithm description
1	DDN	DNN using only 1-D sequence data
2	I-D CNN	1DCNN using only 1-D sequence data
3	LSTM	LSTM using only 1-D sequence data
4	PLSTM	Parallel LSTM using 1-D sequence data
5	PeLSTM	Pe LSTM using only 1-D sequence data
6	CNN(2DM)	CNN using 2-D matrix data stacked in a row
7	CNN(2DS)	CNN for fault diagnosis using a 2-D screenshot image
8	LSTM-IDCNN	The method of using the output of 1-D CNN as the input of LSTM in the sequence feature fusion method
9	LSTM-SF-CNN(2DM)	Splicing fusion of LSTM using 1-D sequence data and CNN using 2-D matrix data
10	LSTM-DF-IDCNN	Deep feature fusion of LSTM using 1-D sequence data and I-D CNN
11	PeLSTM-DF-IDCNN	Deep feature fusion of PeLSTM using 1-D sequence data and I-D CNN
12	LSTM-FF-CNN (2DM)	Deep fusion of LSTM and CNN using 1-D sequence data and 2-D matrix data stacked in a row
13	PeLSTM-DF-CNN(2DS)	Deep feature fusion of PeLSTM and CNN using a 2-D screenshot image

Column 3 in Table 10.4 shows the fault diagnosis accuracy using 1-D CNN. Comparing column 6 with column 3 indicates that when the number of training samples is reduced, 1DCNN's ability to extract features is obviously affected, as only 87.63% accuracy can be achieved, whereas the accuracy of PeLSTM in the same experimental scenario can reach 94.93%. For the same experimental scenario, the accuracy of PeLSTM is 9.02% higher than that of DNN and LSTM.

Comparing column 6 and column 4 in Table 10.4, it can be concluded that in the scenario of Experiment 8, when fault size is 0.014, although diagnosis accuracy using LSTM can reach 97.65%, there is still some useless information spread in the LSTM network, affecting the diagnosis effect. Using PeLSTM can further improve the diagnosis accuracy by 1.15% to 98.71%. The reason is that the traditional LSTM cannot prevent the propagation of useless information, which occupies network memory and affects accuracy of fault diagnosis.

Using features extracted from heterogeneous data can effectively improve accuracy of the fault diagnosis result. Table 10.5 compares the fault diagnosis

TABLE 10.5 Comparison of fault diagnosis accuracy of the proposed deep fusion algorithm with other existing fusion methods.

Experiment no.	Fault type	Window size	Fault size	Training sample size	Test sample size
1	Inner race: Roller and outer race of three sizes; normal condition	100	0.007/0.014/0.021/0	20,000	20,000
2	Inner race: Roller and outer race of three sizes; normal condition	400	0.007/0.014/0.021/0	20,000	20,000
3	Inner race: Roller and outer race of three sizes; normal condition	900	0.007/0.014/0.021/0	20,000	20,000
4	Inner race: Roller and outer race of three sizes; normal condition	100	0.007/0.014/0.021/0	8000	4000
5	Inner race: Roller and outer race of three sizes; normal condition	400	0.007/0.014/0.021/0	8000	4000
6	Inner race: Roller and outer race of three sizes; normal condition	900	0.007/0.014/0.021/0	8000	4000
7	Inner race: Roller and outer race; normal condition	400	0.007/0.014/0.021/0	8000	8000
8	Inner race: Roller and outer race; normal condition	400	0.007/0.014/0.021/0	8000	8000
9	Inner race: Roller and outer race; normal condition	400	0.007/0.014/0.021/0	8000	8000
10	Roller faults of three sizes; normal condition	400	0.007/0.014/0.021/0	8000	8000
11	Inner faults of three sizes; normal condition	400	0.007/0.014/0.021/0	8000	8000
12	Outer faults of three sizes; normal condition	400	0.007/0.014/0.021/0	8000	8000

accuracy of the proposed deep fusion algorithm with other existing fusion methods.

In Table 10.5, column 2 illustrates the fault diagnosis accuracy of existing sequential fusion when the output of 1DCNN is fed into LSTM. Column 4 corresponds to deep fusion of 1-D CNN and LSTM. Comparing column 2 and column 4 of Table 10.5, it can be seen from row 2 that under the scene setting of Experiment 1, diagnosis accuracy of deep fusion is 90.05%, whereas diagnosis accuracy corresponding to the traditional fusion method is only 88.96%. It shows that for the same set of feature extraction networks, deep fusion methods are superior to traditional fusion methods.

Table 10.6 summarizes the experiment result analysis mentioned earlier. Column 7 of Table 10.6 shows the accuracy for CNN fault diagnosis using 2-D matrix data stacked in a row, and column 8 shows the accuracy for CNN fault diagnosis using a 2-D screenshot image. Comparing with column 7 and column 8, it can be concluded that a more accurate real-time fault diagnosis can be achieved when a 2-D screenshot image of 1-D waveform rather than 2-D matrix data stacked in a row from 1-D sequence data is fed into CNN for feature extraction. Thus, the 2-D screenshot image is a better choice for the data source fed into CNN since dynamic trend information is involved in the 2-D screenshot image. Column 4 of Table 10.6 shows the accuracy for traditional LSTM fault diagnosis using a 1-D sequence.

Comparing column 2 with column 4, it can be seen that diagnosis accuracy of DNN in all experimental scenarios is lower than that of LSTM, and the difference is more than 5%. The reason is that the forgetting mechanism does well in autocorrelation extraction of 1-D sequence data, whereas DNN can only extract features from the overall perspective. Comparing column 4 and column 8 of Table 10.6, the differences in these experimental results indicate that the accuracy of CNN fault diagnosis using a 2-D screenshot image of the screenshot for 1-D sequence data is worse than that of LSTM using 1-D sequence data. The reason is that the original signal used for diagnosis is a sequence signal. Comparing with CNN, LSTM does better in dynamic trend feature extraction.

Column 6 and column 4 of Table 10.6 indicate that PeLSTM can achieve higher diagnostic accuracy than traditional LSTM. PeLSTM does better in sequence feature extraction since the forget gate for PeLSTM can intelligently forget useless information and avoids the problem of the bottleneck, which is the reason for this difference. Column 6 indicates that even when PeLSTM is used, it cannot achieve a satisfying fault diagnosis result required by the engineer since the accuracy for Experiments 1, 4, and 5are all less than 90%. Thus, it is necessary to fuse features extracted by LSTM and CNN.

Column 9 of Table 10.6 shows the fault diagnosis accuracy using existing fusion methods to combine LSTM and 1-D CNN. Comparing column 9 with column 4, it can be found that the fault diagnosis capability can be improved once it incorporates the advantage of LSTM and 1-D CNN. But comparing column 9 and column 6 shows that for some specific experiments, such as

TABLE 10.6 Model parameters table.

Experiment no.	Model	Training parameters
1	DNN	No. of layers: 6 No. of neurons of each layer: (100/400/900)/1200/800/400/100/(10/4) Learning rate: 0.001
2	I-D CNN	Convolutional layer: K_{size}: **1*3** K_e: **16/32**K_{step}: **1** Pooling layer: P_{size}: 1*2 P_{step}: 2 Learning rate: 0.001
3	LSTM	Sequence length: 100/400/900 Cell no.: 10/20/30 No. of hidden neurons in the cell: 138 Learning rate: 0.001
4	PLSTM	Cycle: 7 No. of hidden neurons in the cell: 138 Learning rate: 0.001
5	PeLSTM	Cycle: 7 No. of hidden neurons in the cell: 138 Learning rate: 0.001
6	CNN(2DM) (CNN using 2-D matrix data stacked in a row)	Convolutional Layer : K_{size}: 3 K_c: 16/32 K_{step}: 1 Pooling Layer: P_{size}: 2*2 P_{step}: 2 Fully connected layer: No. of neurons: 138 Learning rate: 0.001
7	CNN(2DS) (CNN for fault diagnosis using a 2-D screenshot imag)	Convolutional Layer : K_{size}: 3*3 K_c: 16/32 K_{step}: 1 Pooling Layer: P_{size}: 2*2 P_{step}: 2 Fully connected layer: No. of neurons: 138 Learning rate: 0.001
8	LSTM-I-D CNN (series feature fusion method where the input of LSTM is the output of IDCNN)	Convolutional Layer: K_{size}: 1*3 K_c: 16/32 K_{step}: 1 Pooling Layer: P_{size}: 1*2 P_{step}: 2 LSTM: No. of hidden neurons in the cell: 138 Learning rate: 0.001
9	LSTM-SF-CNN(2DM) (splicing fusion of LSTM using 1-D sequence data and CNN using 2-D matrix data)	Convolutional Layer: K_{size}: 3*3 K_c: 16/32 K_{step}: 1 Pooling Layer: P_{size}: 2*2 P_{step}: 2 LSTM: No. of hidden neurons in the cell: 138

Experiment 3 and Experiment 11, due to the simple combination instead of using the deep fusion mechanism, the fault diagnosis accuracy is even lower than the method using only PeLSTM. Column 10 of Table 10.6 shows the fault diagnosis accuracy for deep feature fusion of LSTM using 1-D sequence data and CNN using 2-D matrix data. It can be seen from column 8 and column 10 that fault diagnosis accuracy for most experiments is satisfying except in Experiment 3 when the deep feature fusion mechanism is adopted. The reason is that the method in column 10 still uses the original 1-D sequence as the data source since the 2-D matrix is stacked from 1-D sequence data in the means row by row. This indicates that fault diagnosis using a data source with a single modal cannot achieve satisfying diagnosis accuracy even when PeLSTM and the deep fusion mechanism are used for advanced feature extraction means. Column 14 of Table 10.6 tells us that fault diagnosis accuracy based on deep feature fusion of PeLSTM using 1-D sequence data and CNN using a 2-D screenshot image is superior to the other 12 methods mentioned in Table 10.6. Diagnosis for all 12 experiments is higher than 90%, which is satisfying for the engineer.

There are 12 experiments designed in this section to show the influence of window size of the sequence, number of training samples, and fault size. Taking column 14 of Table 10.6 as an example, we tried to analyze these specific influences to fault diagnosis accuracy. Rows 2 through 4 indicate that given the number of training samples, a large window size will result in more accurate fault diagnosis since much information is involved in a relatively long sequence. Comparing rows 5 through 7 with rows 2 through 4, it can be concluded that once the sequence length and the number of training samples are given, fault with large fault size is much easier to detect. The experimental results shown in rows 11 through 13 indicate that in addition to a high diagnosis result for different fault types, the algorithm designed in this chapter can achieve a better distinguished capability of a unique fault with different fault size, which is quite helpful for the prognosis and maintenance of mechanical equipment. In other words, no matter whether the training sample size is small or the sample sequence length is changed, column 14 in Table 10.6 can achieve high diagnostic accuracy, which will help us get the following conclusions:

(1) The 2-D screenshot image involves more useful fault features when CNN is used to extract local features. Thus, it is helpful for real-time and accurate fault diagnosis.
(2) The designed PeLSTM is a good feature extraction tool for 1-D sequence data.
(3) Deep feature fusion using a global optimization mechanism does well in the fusion of heterogeneous data, such as 1-D sequence data and a 2-D screenshot image. Thus, it is helpful for accurate fault diagnosis.

To improve the readability of the experiments listed in Table 10.6, Fig. 10.14 shows the fault diagnosis classification chart taking Experiment 6 as an example.

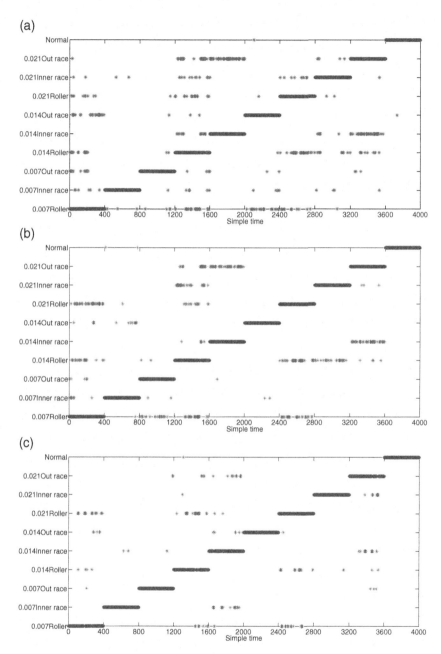

FIGURE 10.14   Fault diagnosis classification chart for Experiment 6 with 10 types of faults, with 800 training samples for each type and a window size of 900.

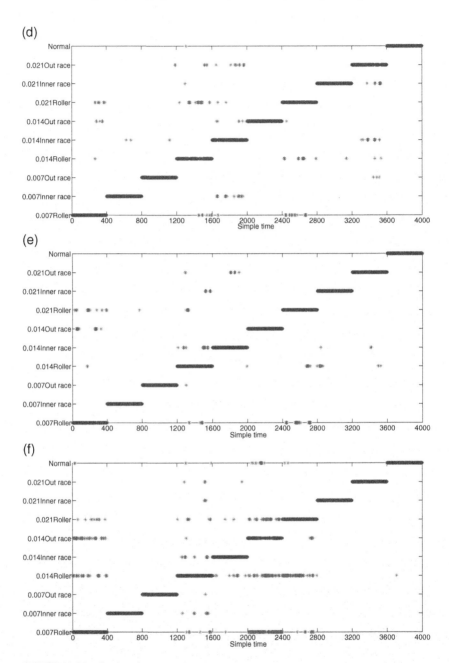

FIGURE 10.14 Continued

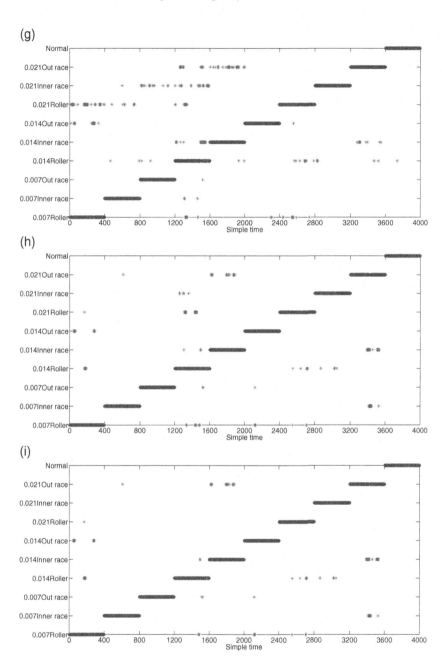

FIGURE 10.14   Continued

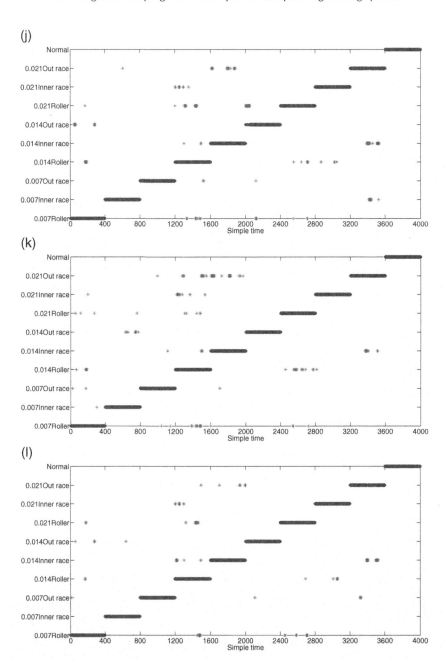

FIGURE 10.14 Continued

(m)

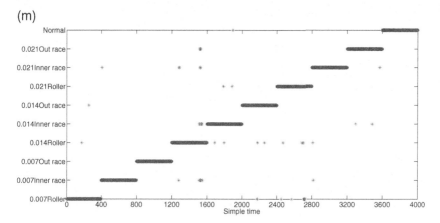

FIGURE 10.14 Continued

The classification results are represented by red stars in the figure. The blue circles represent the true fault categories of the sample. The coincidence of blue circle and red star indicates that the classification is correct. Parts (a) through (m) of Fig. 10.14 correspond to row 7 in Table 10.6. Fig. 10.14(a) is the result of traditional DNN fault diagnosis. Fig. 10.14(b) is the result of 1-D CNN fault diagnosis. Fig. 10.14(c) is the result of LSTM fault diagnosis. Fig. 10.14(d) is the result of PLSTM fault diagnosis, and Fig. 10.14(e) is the diagnosis result of PeLSTM designed in this chapter. Comparing Fig. 10.14(e) with Fig. 10.14(a) through (d), it can be concluded that Fig. 10.14(e) shows more coincidences of stars and circles, indicating that there are fewer samples with misclassification of PeLSTM, which has obvious advantages. Fig. 10.14(f) uses CNN as a feature extraction tool and 2-D matrix data as training data, and Fig. 10.14(g) is the fault diagnosis result using a 2-D screenshot image as the input of CNN. It can be seen from Fig. 10.14(f) that the red stars are dense, which indicates that the misclassification rate is relatively high. Although there are more inconsistent red stars in Fig. 10.14(f) than in Fig. 10.14(c) and (e), they are still significantly less than in Fig. 10.14(f). Comparing Fig. 10.14(f) with Fig. 10.14(g), the fault diagnosis result using two-dimensional screenshots as CNN input is significantly higher than the diagnosis result using two-dimensional matrix as CNN input, indicating that the 2-D screenshot image involves more useful information than that of the 2-D matrix data.

Fig. 10.14(l) is a fault diagnosis result based on feature fusion of CNN(2DS) and PeLSTM. Comparing Fig. 10.14(l) with 10.14(c) and (f) indicates that after merging of features extracted by CNN and LSTM, the diagnosis result is improved. However, it uses CNN as a diagnostic tool and 2-D matrix data stacked from 1-D sequence data as training samples, and this diagnosis algorithm is not in real time. The existing fusion algorithm using 1-D CNN and LSTM is represented in Fig. 10.14(h).

This method uses the output of 1-D CNN as the input of LSTM, so the fused feature extracted by LSTM depends on the output accuracy of 1-D CNN. By comparing with Fig. 10.14(c), it is shown that Fig. 10.14(h) has a better diagnostic result because it merges the local feature extracted using 1-D CNN and the autocorrelated feature extracted using LSTM. Fig. 10.14(m) shows a fault diagnosis result based on deep feature fusion of PeLSTM using 1-D sequence data and CNN using a 2-D screenshot image. It can be seen from Fig. 10.14(m) that the misclassification rate is quite small. The reason is that the networks participating in deep feature fusion can make full feature extraction from heterogeneous data. The fault diagnosis result of this proposed method is the best one in these 13 algorithms since the 1-D dynamic trend feature and 2-D local feature are combined well extracted and fused via a fusion network trained by a mechanism of global optimization. However, the 2-D screenshot image rather than 2-D matrix data stacked from the 1-D sequence is adopted to achieve a real-time diagnosis required by related engineers. Fig. 10.15 shows the comparison bar chart.

Remark 2. This chapter uses common fault diagnosis algorithms such as DNN, CNN, and LSTM for comparison with existing results. The proposed algorithm is compared with the existing feature fusion algorithms such as LSTM-1DCNN and LSTM-DF-1DCNN. After doing multiple sets of experiments under any experimental scene settings, it can be concluded that (1) a 2-D screenshot image involves more useful fault features when CNN is used to extract local features; (2) the designed PeLSTM is a good feature extraction tool for 1-D sequence data, and (3) deep fusion using the global optimization mechanism does well in the fusion of heterogeneous data, such as 1-D sequence data and the 2-D screenshot image.

10.4.2 Gearbox test and analysis

The proposed algorithm can also be applied to diagnose fault of the gearbox. Fault and normal data were collected from the $QPZZ - II$ rotating machinery vibration experimental platform [24]. $QPZZ - II$ can simulate the following faults: pitting, broken tooth, wear, and combining fault of pitting and wear. In this experiment, the data are sampled in the case when speed is 880 r/min and the current is 0.05 A. The vibration signal is sampled at the output shaft motor side. The health state of the gearbox can be divided into six categories: (1) Normal, (2) Pitting, (3) Broken tooth, (4) Wear, (5) Pitting and Wear, and (6) Broken teeth and Wear.

10.4.2.1 Data preprocessing

The original data are the 1-D vibration signal, and it can be a very long sequence, so the moving window technique is used for data preprocessing. The means of data preprocessing is the same as that of bearing data as listed in Section 10.4.1. In other words, the window size of each sample is 100/400/900. The test samples

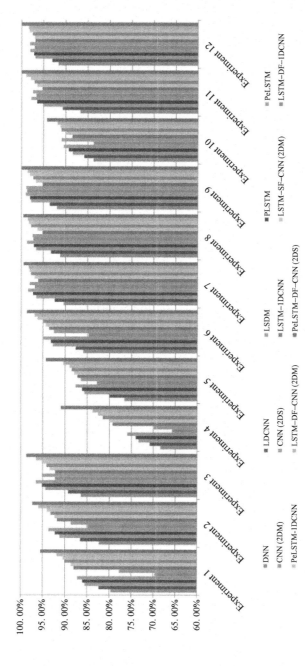

FIGURE 10.15 Comparison of different fault diagnosis methods for rolling bearing.

TABLE 10.7 Comparison of bearing fault diagnosis accuracy using 1-D sequence data.

Experiment no.	DNN	1-D CNN	LSTM	PLSTM	PeLSTM
1	79.68%	82.36%	85.51%	86.03%	87.18%
2	82.28%	86.53%	91.16%	92.23%	93.63%
3	86.41%	89.32%	94.38%	95.14%	96.58%
4	68.39%	70.81%	73.35%	74.02%	75.92%
5	76.56%	79.98%	85.68%	86.17%	87.65%
6	85.91%	87.63%	92.80%	93.21%	94.93%
7	90.39%	92.41%	96.56%	97.32%	98.45%
8	91.21%	93.24%	96.65%	97.13%	98.71%
9	91.96%	93.56%	97.80%	98.04%	99.04%
10	83.73%	85.85%	88.45%	89.37%	90.48%
11	86.74%	90.73%	95.13%	96.46%	97.80%
12	91.68%	93.06%	96.84%	97.17%	98.06%

are reshaped in 2-D matrix data with size 10*10/20*20/30*30, and the screenshot image size is 28*28. The comparison of the corresponding 12 experiments can illustrate the superiority of the proposed algorithm in different cases.

10.4.2.2 Experiment result analysis

Just as the experiment design means in 4.1, there are also 12 experiments designed in this section to verify our method for cases of different fault types and same fault types with different fault sizes, respectively. The specific experimental design is shown later in Table 10.9, and the experimental results are shown later in Table 10.10.

Comparing column 6 and column 2 in Table 10.7, it can be seen from row 4 that when the window size of the 1-D data sequence is short, the diagnostic accuracy of PeLSTM is 89.61%, whereas the diagnostic accuracy of the traditional DNN with the same training data is only 79.31%. The diagnostic accuracy is improved by 10.3%, indicating that the autocorrelation involved in 1-D sequence data has a great influence on the diagnosis results. PeLSTM can make full use of this information. For all 12 experiments, the accuracy of PeLSTM is always 5% higher than DNN.

Column 3 in Table 10.7 shows the fault diagnosis accuracy using 1-D CNN. Comparing column 6 with column 3, it can be concluded that when the number of training samples is reduced, 1DCNN's ability to extract features is obviously affected, as only 81.70% accuracy can be achieved, whereas the accuracy of

TABLE 10.8 Comparison of fault diagnosis accuracy of the proposed deep fusion algorithm with other existing fusion methods.

Experiment no.	LSTM-1DCNN	LSTM-SF-CNN (2DM)	LSTM-DF-IDCNN	PeLSTM-DF-1DCNN	LSTM-DF-CNN (2DM)	PeLSTM-DF-CNN (2DS)
1	87.96%	88.62%	90.05%	90.41%	91.84%	95.49%
2	91.74%	92.31%	93.21%	94.02%	95.94%	97.36%
3	92.24%	93.42%	94.12%	95.21%	96.56%	98.74%
4	79.30%	80.04%	81.47%	82.59%	83.77%	90.95%
5	87.22%	87.93%	88.54%	89.01%	90.38%	94.28%
6	93.65%	94.37%	95.03%	96.25%	96.93%	98.68%
7	96.14%	96.96%	97.81%	98.11%	98.33%	99.45%
8	95.17%	96.27%	97.77%	98.30%	98.51%	99.56%
9	95.18%	96.67%	97.31%	98.04%	98.60%	99.97%
10	88.51%	89.87%	90.99%	91.21%	91.86%	94.21%
11	95.26%	96.42%	97.07%	97.92%	98.98%	99.99%
12	96.86%	97.03%	97.48%	98.20%	98.38%	99.94%

PeLSTM in the same experimental scenario can reach 85.45%. For the same experimental scenario, there is improvement of 11.75% and 2.09% relative to that of DNN and LSTM.

Comparing column 6 and column 4 in Table 10.7 shows that in the scenario of Experiment 8, using LSTM can reach 87.83% accuracy, but there is still some useless information transferred in the LSTM network, which will affect diagnosis efficiency. Using PeLSTM to prevent useless information from transferring can further achieve an improvement of 1.91%. The reason is that traditional LSTM cannot prevent the transferring of useless information, which occupies the network memory and affects the accuracy of fault diagnosis.

Table 10.8 compares the fault diagnosis accuracy of the proposed deep fusion algorithm with other existed fusion methods.

In Table 10.8, column 2 illustrates the fault diagnosis accuracy of sequential fusion when 1-D CNN and LSTM are connected in sequence. Column 4 corresponds to deep fusion of 1DCNN and LSTM. Comparing column 2 with column 4 of Table 10.8, it can be seen from row 2 that under the scenario of Experiment 1, diagnosis accuracy of deep fusion is 89.31%, whereas diagnosis accuracy corresponding to the traditional fusion method is only 87.94%, and the accuracy is increased by 1.37%. It shows that for the same set of feature extraction networks, deep fusion methods are superior to traditional fusion methods.

Fault diagnosis and prognosis techniques for complex engineering systems

TABLE 10.9 Experimental design.

DNN	1DCNN	LSTM	PLSTM	PeLSTM	CNN (2DM)	CNN (2DS)	LSTM-1DCNN	LSTM-S F-CNN (2DM)	LSTM-D F-ID CNN	PeLSTM-DF-1DCNN	LSTM-DF-CNN (2DM)	PeLSTM-DF-CNN (2DS)
79.68%	82.36%	85.51%	86.03%	87.18%	69.63%	77.74%	87.96%	88.62%	**90.05%**	90.41%	91.84%	95.49%
82.28%	**86.53%**	91.16%	92.23%	93.63%	84.83%	88.63%	**91.74%**	92.31%	**93.21%**	94.02%	95.94%	97.36%
86.H%	89.32%	94.38%	95.14%	96.58%	92.25%	95.16%	92.24%	93.42%	94.12%	95.21%	96.56%	98.74%
68.39%	70.81%	**73.35%**	74.02%	75.92%	65.75%	69.90%	79.30%	80.04%	81.47%	82.59%	83.77%	90.95%
76.56%	79.98%	85.68%	**86.17%**	87.65%	**82.83%**	86.43%	87.22%	**87.93%**	88.54%	89.01%	90.38%	94.28%
85.91%	87.63%	92.80%	93.21%	94.93%	54.75	92.55%	93.65%	94.37%	95.03%	96.25%	96.93%	98.68%
90.39%	92.41%	96.56%	97.32%	98.45%	96.88%	98.05%	96.14%	96.96%	97.81%	98.11%	98.33%	99.45%
91.21%	93.2%	96.65%	97.13%	98.71o/o	97.25%	97.46%	95.17%	96.27%	97.77%	98.30%	98.51%	99.56%
91.96%	93.56%	97.80%	98.04%	99.04%	98.71%	98.91%	**95.18%**	96.67%	97.31%	98.04%	98.60%	99.97%
83.73%	85.85%	88.45%	**89.37%**	90.48%	**83.70%**	90.96%	88.51%	89.87%	90.99%	91.21%	91.86%	94.21%
86.74%	**90.73%**	95.13%	96.46%	97.80%	96.65%	97.40%	**95.26%**	96.42%	91.0%	97.92%	98.98%	99.99%
91.68%	93.06%	96.84%	97.17%	98.06%	97.06%	98.11%	96.86%	97.03%	97.48%	98.20%	98.38%	99.94%

TABLE 10.10 Experimental results.

Experiment no.	DNN	I-D CNN	LSTM	PLSTM	PeLSTM
1	72.52%	78.64%	82.75%	83.65%	84.39%
2	75.72%	83.72%	86.86%	87.33%	88.19%
3	79.31%	86.27%	87.42%	88.05%	89.61%
4	63.53%	67.61%	71.16%	72.57/o	73.28%
5	67.10%	69.46%	72.83%	73.37%	74.92%
6	73.70%	81.70%	83.36%	84.24%	85.45%
7	76.74%	82.93%	84.67%	85.43%	86.13%
8	81.27%	85.02%	87.83%	88.59%	89.02%
9	85.63%	88.49%	90.08%	91.29%	92.96%
10	70.24%	73.56%	75.17%	76.65%	77.74%
11	73.25%	76.47%	79.04%	80.41%	81.53%
12	78.67%	82.84%	86.08%	87.10%	89.07%

Comparing with LSTM, the improved PeLSTM does better in feature extraction of 1-D sequence data. Thus, it can be confirmed that after deep fusion, the diagnosis accuracy shown in column 5 is better than that of column 4.

Comparing column 7 with column 5 in Table 10.8, it can be concluded that incorporation of a 2-D image matrix stacked in the deep fusion process can achieve more accurate diagnosis since CNN can extract additional local neighboring features involved in the 2-D image matrix.

Comparing column 7 with column 6 of Table 10.8, it is shown that incorporation of a 2-D screenshot image rather than a 2-D image matrix can achieve a real-time accurate fault diagnosis. It shows that the trend information involved in a 2-D screenshot image can be effectively mined, and the autocorrelation in 1-D sequence data is also effectively extracted. The deep fusion method is more effective in accurate feature extraction when only sensors related to 1-D signals are equipped in the gearbox monitoring system. It can be seen from the column 5 of Table 10.9 that the PeLSTM algorithm proposed in this paper can extract sufficient features from 1-D sequences, and from the column 11, it can be seen that the fusion algorithm proposed in this paper is heterogeneous data fault diagnosis the best solution.

Table 10.10 summarizes the experimental result analysis mentioned earlier. The explanation of Table 10.10 is similar to that of Table 10.6. Comparing Table 10.10 with Table 10.6, it is shown that the fault diagnosis accuracy of gearbox is lower than those of rolling bearing. The reason is that (1) fewer training samples are used to train the related networks, which will inevitably decrease the diagnosis accuracy; (2) wearing fault is usually viewed as a kind of minor fault that is more difficult to detect; and (3) comparing with single fault diagnosis, study for an accurate diagnosis method for combing fault is still a

TABLE 10.11 Comparison of fault diagnosis results based on feature fusion of CNN and LSTM.

Experiment no.	LSTM-1DCNN	LSTM-SF-CNN (2DM)	LSTM-DF-IDC'NN	PeLSTM-DF-IDCNN	LSTM-DF-CNN (2DM)	PeLSTM-DF-CNN (2DS)
1	87.94%	88.02%	89.31%	90.02%	91.08%	93.00%
2	89.27%	90.25%	91.02%	92.51%	93.67%	95.83%
3	91.59%	92.31%	93.62%	94.24%	96.58%	98.78%
4	75.33%	76.43%	77.52%	78.41%	81.33%	82.00%
5	78.97%	79.75%	80.90%	82.64%	84.11%	86.75%
6	88.50%	89.63%	91.06%	92.08%	94.50%	96.33%
7	87.62%	88.06%	89.52%	91.03%	92.96%	94.71%
8	89.07%	90.48%	91.07%	92.65%	94.46%	96.63%
9	92.65%	93.45%	94.38%	96.71%	98.06%	99.83%
10	80.32%	81.39%	82.64%	84.05%	87.38%	89.75%
11	85.51%	86.27%	87.53%	89.57%	91.96%	93.83%
12	89.02%	90.16%	91.01%	92.23%	94.17%	96.25%

challenging problem. Table 10.10 indicates that the proposed method can also do well in minor fault diagnosis and combining fault diagnosis. Table 10.11 is the fault diagnosis accuracy comparison between the fusion algorithm proposed in this paper and the existing fusion algorithm, and Table 10.12 is the experimental design of the gearbox.

To improve the readability of the experiment listed in Table 10.10, a diagnostic classification chart for Experiment 6 is shown in Fig. 10.16. The explanation of Fig. 10.16 is similar to that of Fig. 10.14. Fig. 10.16 indicates that the misclassification rate in Fig. 1.16(m) is the smallest one. The reason is that the deep fusion network does well in comprehensive feature extraction from 1-D sequence data and the 2-D screenshot image by training in a mechanism of global optimization. However, a 2-D screenshot image rather than 2-D matrix data stacked from 1-D sequence is adopted to achieve a real-time diagnosis required by related engineers. Take Fig. 10.16(m) as an example for analysis of detection ability for different kinds of fault. It can been seen from Fig. 10.16(m) that for wear fault, as well as broken and wear fault, the diagnosis accuracy is relatively lower that other five kinds of health state, because wear fault is a minor fault that is difficult to detect. When this minor fault is combined with broken and fault and pitting fault, the difficulties in combining fault detection is significantly increased. Comparing Fig. 10.16(m) with the other parts in Fig. 10.16 shows that even the complex fault is difficult to detect, and the method proposed in this chapter has significant superiority. Fig. 10.17 shows the comparison bar chart.

Experimental results for gearbox fault diagnosis show that the validity of the proposed method can be tested in many cases as long as the monitoring data can

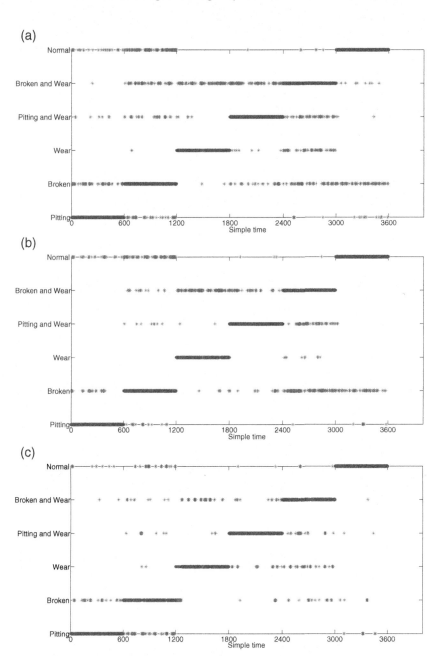

FIGURE 10.16 Gearbox diagnosis accuracy rate graph of Experiment 6 is six faults, with 800 training samples for each fault and a window size of 900.

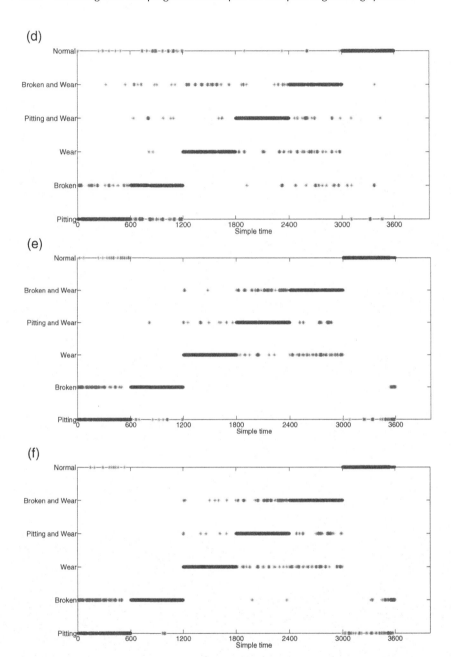

FIGURE 10.16 Continued

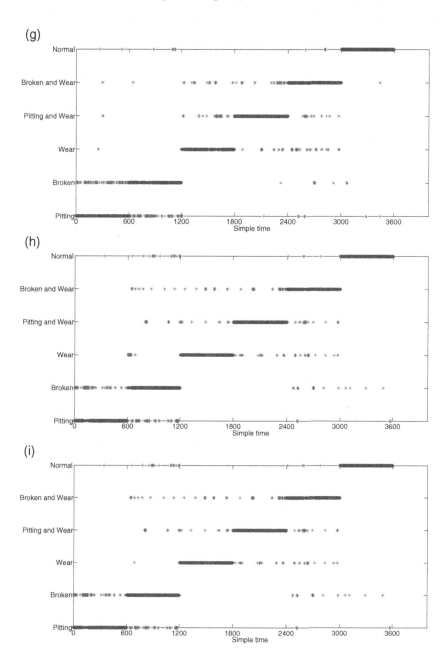

FIGURE 10.16   Continued

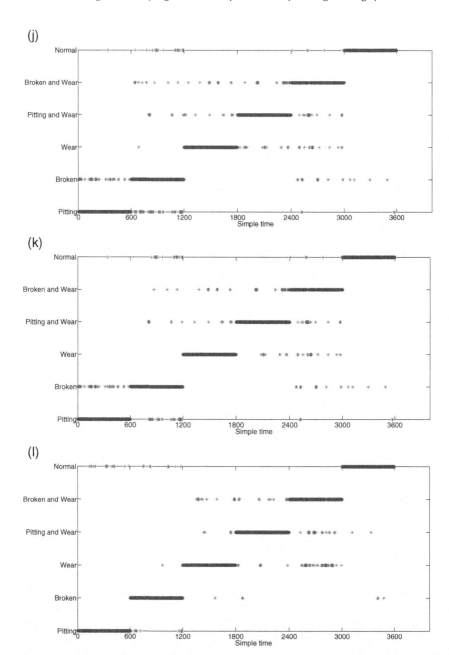

FIGURE 10.16 Continued

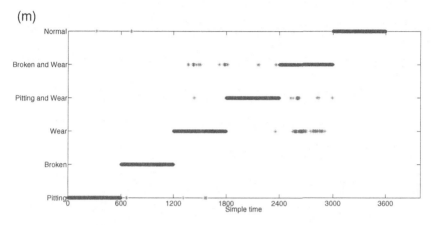

FIGURE 10.16 Continued

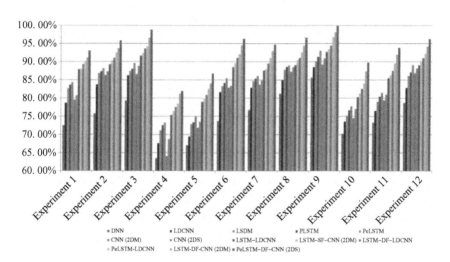

FIGURE 10.17 Comparison of different fault diagnosis methods for the gearbox.

be collected, even when the number of training samples is limited and the faults are complex to detect.

10.5 Conclusion and future work

Deep learning can be applied to fault diagnosis of rotating machinery. The efficiency is up to the training sample size and the means to extract potential features involved in multimodal data available. Since 1-D vibration data collected from the accelerometer are the most common available monitoring signals, it is significant to develop an efficient tool to extract features involved in 1-D sequence data.

TABLE 10.12 Experimental design of gearbox fault diagnosis.

Experiment no.	Window size	Fault type	No. of training samples	No. of test samples
1	100	Normal; pitting; wear; pitting and Wear; broken teeth; broken teeth and wear	12,000	3600
2	400	Normal; pitting; wear; pitting and Wear; broken teeth; broken teeth and wear	12,000	3600
3	900	Normal; pitting; wear; pitting and Wear; broken teeth; broken teeth and wear	12,000	3600
4	100	Normal; pitting; wear; pitting and Wear; broken teeth; broken teeth and wear	4800	3600
5	400	Normal; pitting; wear; pitting and Wear; broken teeth; broken teeth and wear	4800	3600
6	900	Normal; pitting; wear; pitting and Wear; broken teeth; broken teeth and wear	4800	3600
7	100	Normal; pitting; wear; pitting and Wear; broken teeth	8000	2400
8	400	Normal; pitting; wear; pitting and Wear; broken teeth	8000	2400
9	900	Normal; pitting; wear; pitting and Wear; broken teeth	8000	2400
10	100	Normal; pitting; wear; pitting and Wear; broken teeth	3200	2400
11	400	Normal; pitting; wear; pitting and Wear; broken teeth	3200	2400
12	900	Normal; pitting; wear; pitting and Wear; broken teeth	3200	2400

Traditional LSTM may encounter memory bottlenecks that make it an unideal feature extraction tool. This chapter designs a new parallel LSTM with peepholes to overcome the memory bottleneck, as peepholes are designed to prevent useless information transferring. However, using PeLSTM as a unique feature extraction tool will inevitably result in inaccurate diagnosis results because it does not do well in local feature extraction. Existing fusion means of LSTM and CNN cannot achieve a real-time and more accurate diagnosis. Thus, an additional feature fusion network with global training mechanism is designed.

The efficiency of the proposed algorithm is only verified by experimental analysis. Designing an interpretable deep fusion network may guide us to design a better fusion mechanism. However, error control is an important factor in

TABLE 10.13 Comparison of bearing fault diagnosis accuracy using 1D sequence data.

DNN	IDCNN	LSTM	PLSTM	PeLSTM	CNN (2DM)	CNN (2DS)	1DCNN-LSTM-1DCNN	LSTM-SF-CNN (2DM)	LSTM-DF-1DCNN	PeLSTM-DF-1DCNN	LSTM-DF-CNN (2DM)	PeLSTM-DF-CNN (2DS)
72.52%	78.64%	82.75%	83.65%	84.39%	79.53%	80.72%	87.94%	88.02%	89.31%	90.02%	91.08%	93.00%
75.72%	83.72%	86.86%	57.33%	88.19%	86.36%	87.31%	89.27%	90.25%	91.02%	92.51%	93.67%	95.83%
79.31%	86.27%	87.42%	88.05%	89.61%	86.58%	88.81%	91.59%	92.31%	93.62%	94.24%	96.58%	98.78%
63.53%	67.61%	71.16%	72.57%	73.28%	64.08%	68.78%	75.33%	76.43%	77.52%	78.41%	81.33%	82.00%
67.10%	69.46%	72.83%	73.37%	74.92%	71.86%	73.53%	78.97%	79.75%	80.90%	82.64%	84.11%	86.75%
73.70%	81.70%	83.36%	84.24%	85.45%	82.92%	83.39%	88.50%	89.63%	91.06%	92.08%	94.50%	96.33%
76.74%	82.93%	84.67%	85.43%	86.13%	83.75%	84.88%	87.62%	88.06%	89.52%	91.03%	92.96%	94.71%
81.27%	85.02%	87.83%	88.59%	89.01%	87.29%	88.54%	89.07%	90.48%	91.07%	92.65%	94.46%	96.63%
85.63%	88.49%	90.08%	91.29%	92.96%	89.21%	90.87%	92.65%	93.45%	94.38%	96.71%	98.06%	99.83%
70.24%	73.56%	75.17%	76.65%	77.74%	74.50%	76.96%	80.32%	81.39%	82.64%	84.05%	87.38%	89.75%
73.25%	76.47%	79.04%	80.41%	81.53%	79.37%	81.04%	85.51%	86.27%	87.53%	89.57%	91.96%	93.83%
78.67%	82.84%	86.08%	87.10%	89.07%	86.96%	88.18%	89.02%	90.16%	91.01%	92.23%	94.17%	96.25%

engineering [25], and designing a deep fusion network with error constraints is also our future work.

Acknowledgment

This research was partially supported by the NSFC project (grant no. U1604158) and the Shanghai S&T Commission (grant no. 19040501700).

References

[1] W. Abed, S. Sharma, R. Sutton, Neural network fault diagnosis of a trolling motor based on feature reduction techniques for an unmanned surface vehicle, Proceedings of the Institution of Mechanical Engineers, Part I: Journal of Systems & Control Engineering 229 (8) (2015) 738–750.

[2] Z. An, S. Li, J. Wang, Y. Xin, K. Xu, Generalization of deep neural network for bearing fault diagnosis under different working conditions using multiple kernel method, Neurocomputing 352 (2019) 42–53.

[3] Bearing Data Center. Home page. [Online]. 2021. Available at http://csegroups.case.edu/bearingdatacenter/home.

[4] T.D. Bruin, K. Verbert, R. Babuska, Railway track circuit fault diagnosis using recurrent neural networks, IEEE Transactions on Neural Networks 28 (3) (2017) 523–533.

[5] I. Djelloul, Z. Sari, I. Sidibe, Fault diagnosis based on the quality effect of learning algorithm for manufacturing systems, Proceedings of the Institution of Mechanical Engineers, Part I: Journal of Systems & Control Engineering 233 (7) (2019) 801–814.

[6] L. Eren, T. Ince, S. Kiranyaz, A generic intelligent bearing fault diagnosis system using compact adaptive 1D CNN classifier, Signal Processing Systems 91 (2) (2019) 179–189.

[7] Y. Fu, D. Huang, N. Qin, K. Lang, Y. Yang, High-speed railway bogie fault diagnosis using LSTM neural network, Proceedings of the 37th Chinese Control Conference (CCC), (2018) 5848–5852.

[8] R. Guo, K. Guo, J. Dong, Fault diagnosis for the landing phase of the aircraft based on an adaptive kernel principal component analysis algorithm, Proceedings of the Institution of Mechanical Engineers, Part I: Journal of Systems & Control Engineering 229 (10) (2015) 917–926.

[9] J. Han, D. Choi, S. Hong, H. Kim, Motor fault diagnosis using CNN based deep learning algorithm considering motor rotating speed, Proceedings of the IEEE 6th International Conference on Industrial Engineering and Applications (ICIEA) (2019) 440–445.

[10] S. Hochreiter, J. Schmidhuber, Long short-term memory, Neural Computation 9 (8) (1997) 1735–1780.

[11] Y. Hsueh, V.R. Ittangihal, W. Wu, H. Chang, C. Kuo, Fault diagnosis system for induction motors by CNN using empirical wavelet transform, Symmetry 11 (10) (2019) 1212.

[12] R. Huang, Y. Liao, S. Zhang, W. Li, Deep decoupling convolutional neural network for intelligent compound fault diagnosis, IEEE Access 7 (2018) 1848–1858.

[13] H. Jafari, J. Poshtan, H. Sadeghi, Application of fuzzy data fusion theory in fault diagnosis of rotating machinery, Proceedings of the Institution of Mechanical Engineers, Part I: Journal of Systems & Control Engineering 232 (8) (2018) 1015–1024.

[14] H. Jiang, F. Wang, H. Shao, H. Zhang, Rolling bearing fault identification using multilayer deep learning convolutional neural network, Journal of Vibroengineering 19 (1) (2017) 138–149.

[15] S.T. Kandukuri, J.S.L. Senanayaka, V.K. Huynh, H.R. Karimi, K.G. Robbersmyr, Current signature based fault diagnosis of field-oriented and direct torque–controlled induction motor drives, Proceedings of the Institution of Mechanical Engineers–Part I: Journal of Systems & Control Engineering 231 (10) (2017) 849–866.

[16] L. Kou, Y. Qin, X. Zhao, X. Chen, A multi-dimension end-to-end CNN model for rotating devices fault diagnosis on high-speed train bogie, IEEE Transactions on Vehicular Technology 69 (3) (2020) 2513–2524.

[17] J. Lei, C. Liu, D. Jiang, Fault diagnosis of wind turbine based on long short-term memory networks, Renewable Energy 133 (2019) 422–432.

[18] H. Li, J. Huang, S. Ji, Bearing fault diagnosis with a feature fusion method based on an ensemble convolutional neural network and deep neural network, Sensors 19 (9) (2019) 2034.

[19] Y. Li, H.R. Karimi, Q. Zhang, D. Zhao, Y. Li, Fault detection for linear discrete time-varying systems subject to random sensor delay: A Riccati equation approach, IEEE Transactions on Circuits & Systems I: Regular Papers 65 (5) (2017) 1707–1716.

[20] P. Luo, Y. Hu, Research on rolling bearing fault identification method based on LSTM neural network, Materials Science & Engineering 542 (1) (2019;) 012048.

[21] H. Pan, X. He, S. Tang, F. Meng, An improved bearing fault diagnosis method using one-dimensional CNN and LSTM, Journal of Mechanical Engineering 64 (7) (2018) 443–452.

[22] D. Peng, H. Wang, Z. Liu, W. Zhang, M.J. Zuo, J. Chen, Multibranch and multiscale CNN for fault diagnosis of wheelset bearings under strong noise and variable load condition, IEEE Transactions on Industrial Informatics 16 (7) (2020) 4949–4960.

[23] K. Polat, The fault diagnosis based on deep long short-term memory model from the vibration signals in the computer numerical control machines, Journal of the Institute of Electronics & Computer 2 (1) (2020) 72–92.

[24] Pudn.com. QPZZ gearbox data. [Online]. 2021. Available at http://www.pudn.com/Download/item/id/3205015.html.

[25] K. Sun, J. Qiu, H. R. Karimi, H. Gao. A novel finite-time control for nonstrict feedback saturated nonlinear systems with tracking error constraint. IEEE Transactions on Systems, Man & Cybernetics. Early access, December 27, 2019.

[26] K. Tidriri, N. Chatti, S. Verron, T. Tiplica, Model-based fault detection and diagnosis of complex chemical processes: A case study of the Tennessee Eastman process, Proceedings of the Institution of Mechanical Engineers, Part I: Journal of Systems & Control Engineering 232 (6) (2018) 742–760.

[27] H. Wang, S. Li, L. Song, L. Cui, A novel convolutional neural network based fault recognition method via image fusion of multi-vibration-signals, Computers in Industry 105 (2019) 182–190.

[28] W. Wang, X. Qiu, C. Chen, B. Lin, H. Zhang. Application research on long short-term memory network in fault diagnosis. In Proceedings of the 2018 International Conference on Machine Learning and Cybernetics. 360--365.

[29] Y. Wang, M. Liu, Z. Bao, S. Zhang, Stacked sparse autoencoder with PCA and SVM for data-based line trip fault diagnosis in power systems, Neural Computing & Applications 31 (10) (2019) 6719–6731.

[30] L. Wen L, X. Li, L. Gao, Y. Zhang, A new convolutional neural network-based data-driven fault diagnosis method, IEEE Transactions on Industrial Electronics 65 (7) (2017) 5990–5998.

[31] D. Xiao, Y. Huang, X. Zhang, H. Shi, C. Liu, Y. Li. Fault diagnosis of asynchronous motors based on LSTM neural network. In *Proceedings of the 2018 IEEE Prognostics & System Health Management Conference*. IEEE, Los Alamitos, CA, 540--545.

[32] R. Yang, M. Huang, Q. Lu, M. Zhong, Rotating machinery fault diagnosis using long-short-term memory recurrent neural network, IFAC-PapersOnLine 51 (24) (2018;) 228–232.

[33] R. Yang, H. Li, C. He, Z. Zhang, Rolling element bearing weak fault diagnosis based on optimal wavelet scale cyclic frequency extraction, Proceedings of the Institution of Mechanical Engineers, Part I: Journal of Systems & Control Engineering 232 (7) (2018) 895–908.

[34] A. Yin, Y. Yan, Z. Zhang, C. Li, R.-V. Sanchez, Fault diagnosis of wind turbine gearbox based on the optimized LSTM neural network with cosine loss, Sensors 20 (8) (2020) 2339.

[35] L. Yu, J. Qu, F. Gao, Y. Tian, A novel hierarchical algorithm for bearing fault diagnosis based on stacked LSTM, Shock & Vibration 2019 (2019) 1–10.

[36] J. Zarei, M.A. Tajeddini, H.R. Karimi, Vibration analysis for bearing fault detection and classification using an intelligent filter, Mechatronics 24 (2) (2014) 151–157.

[37] B. Zhang, S. Zhang, W. Li, Bearing performance degradation assessment using long short-term memory recurrent network, Computers in Industry 106 (2019) 14–29.

[38] W. Zhang, G. Peng, C. Li, Bearings fault diagnosis based on convolutional neural networks with 2-D representation of vibration signals as input, MATEC web of conferences, EDP Sciences 95 (2017) 13001.

[39] Y. Zhang, Q. Liu, L. Song, Sentence-state LSTM for text representation, Proceedings of the 56th Annual Meeting of the Association for Computational Linguistics (2018) 317–327.

[40] H. Zhao, S. Sun, B. Jin, Sequential fault diagnosis based on LSTM neural network, IEEE Access 6 (2018) 12929–12939.

Index

Page numbers followed by "*f*" and "*t*" indicate, figures and tables respectively.

Printed in the United States
by Baker & Taylor Publisher Services